BIOLOGICAL EXPLORATIONS
A Human Approach
Second Edition

STANLEY E. GUNSTREAM
Pasadena City College

MACMILLAN COLLEGE PUBLISHING COMPANY
New York

MAXWELL MACMILLAN CANADA
Toronto

Editor: Sheri Walvoord
Project Management: J. Carey Publishing Service
Production Manager: Aliza Greenblatt
Cover Designer: Hothouse Designs
Cover Photograph: Globus Brothers/The Stock Market
Original Artwork by Mary Dersch

This book was set in Century Schoolbook by Bi-Comp, Incorporated,
printed and bound by Semline.
The cover was printed by Phoenix Color Corp.
The color insert was printed by Princeton Polychrome.

Macmillan College Publishing Company
866 Third Avenue, New York, New York 10022

Macmillan College Publishing Company is part of
the Maxwell Communication Group of Companies.

Maxwell Macmillan Canada, Inc.
1200 Eglinton Avenue East
Suite 200
Don Mills, Ontario M3C 3N1

Library of Congress Cataloging-in-Publication Data

Gunstream, Stanley E.
    Biological explorations : a human approach / Stanley E. Stanley.—
    2nd ed.
       p.   cm.
    Rev. ed. of: Human biology : laboratory explorations. c1986.
    ISBN 0–02–348525–6
    1. Human biology—Laboratory manuals. I. Gunstream, Stanley E.
    Human biology. II. Title.
    QP44.G86   1993
    612'.0078—dc20   93-25338
                                                              CIP

Printing: 1 2 3 4 5 6 7   Year: 4 5 6 7 8 9 0

Color Insert Credits appear on page 425, which constitutes a continuation of
the copyright page.

PREFACE

Biological Explorations: A Human Approach (formerly *Human Biology: Laboratory Explorations*) is a laboratory manual specifically designed for the laboratory component of courses in: (1) general biology where the human organism is emphasized, and (2) human biology. It is compatible with any modern textbook that emphasizes the human organism. The exercises are appropriate for three-hour laboratory sessions, but they are also adaptable for a two-hour laboratory format.

This laboratory manual is designed not only to enhance learning by students but to simplify the work of instructors. Its design assumes little if any prior experience by students in a biology laboratory.

MAJOR FEATURES

1. The **thirty-one exercises** provide a wide range of options for the instructor, and the range of activities within an exercise further increases the available options.
2. Each exercise is basically **self-directing** which allows students to work independently without direct assistance by the instructor.
3. Each exercise, and its major subunits, are **self-contained** so that the instructor may arrange the sequence of exercises, or the ac-

tivities within an exercise, to suit his or her preferences.
4. Each exercise begins with a list of **objectives** that outlines the minimal learning responsibilities of the student. The objectives also inform the student of the emphasis and scope of the activities to be accomplished and the minimal learning responsibilities.
5. Following the objectives, each exercise starts with a brief discussion of **background information** that is necessary to (a) understand the subject of the exercise, and (b) prepare the student for the activities that follow. The inclusion of the background information minimizes the need for introductory explanations by the instructor and assures that all lab sections receive the same introductory information.
6. Before beginning the laboratory activities, students are directed to demonstrate their understanding of the background information by labeling illustrations and completing the portion of the Laboratory Report that covers this material.
7. Over 200 **illustrations** are provided to enhance students' understanding of both background information and laboratory procedures. Students are asked to color-code significant structures in many illustrations as a means to aid learning. Eight plates of color photos are included in the organismic

diversity section to assist students with the structure of difficult-to-observe organisms.

8. New **key terms** are in bold print for easy recognition by students, and they are defined when first used.

9. Required **materials** are listed for each activity in the exercise. The list is divided into materials needed by each student or student group and materials that are to be available in the laboratory. This list helps the student to obtain the needed materials and guides the laboratory technician in setting up the laboratory. The exercises utilize standard equipment and materials that are available in most biology departments.

10. Activities to be performed by students are identified by an **assignment** heading that clearly distinguishes activities to be performed from the background information. The assignment sections are numbered sequentially within each exercise and on the Laboratory Report to facilitate identification and discussion.

11. Clear **laboratory procedures** guide students through each activity so that minimal assistance is needed from the instructor.

12. A **Laboratory Report** is provided for each exercise to guide and reinforce students' learning. The laboratory reports not only provide a place for students to record observations, collected data, and conclusions, but they also provide a convenient means for the instructor to assess student understanding. The separate laboratory reports may be removed from the manual without removing key information contained in the exercises and needed by students for study.

13. An **Instructor's Guide** accompanies the manual and contains (a) composite lists of equipment and supplies, (b) sources of supplies, (c) special techniques, (d) operational suggestions, and (e) answer keys for the laboratory reports.

ACKNOWLEDGMENTS

A number of users have provided helpful suggestions for improving the second edition. The useful suggestions made by colleagues at Pasadena City College, especially Tom Belzer and Wayne Sawyer, are greatly appreciated. Special recognition is due those whose critical reviews contributed to the present revision:

Paul Bailey, Birmingham-Southern College
William Bessler, Mankato State University
Thomas Fogle, St. Mary's College
Eugene W. Hupp, Texas Woman's University
George Karleskint, St. Louis Community College—Meramec
Keith Klein, Hamline University
Martin Levin, Eastern Connecticut State University
Donald Levine, William Paterson College
James R. Litton, Jr., St. Mary's College
Richard McKeeby, Union County College
John P. Shehan, Iowa Central Community College

The skills and helpfulness of the editorial and production personnel at Macmillan have eased the work of the revision. I especially want to thank Kristin Watts Peri for her suggestions that guided the revision and Sheri Walvoord for shepherding the revision to production. It has been a pleasure to work with Jennifer Carey who skillfully guided the production process and solved numerous problems enroute. Original drawings by Mary Dersch have enhanced the art program in the second edition. The contributions of these talented people are gratefully acknowledged.

Adopters are encouraged to communicate to the author or publisher helpful suggestions that will improve the usefulness of this manual for their courses.

S.E.G.

CONTENTS

PART I

CELL BIOLOGY

1

ORIENTATION

Laboratory study is an important part of a course in biology. It provides opportunities for you to observe and study biological organisms and processes and to correlate your findings with the textbook and lectures. It allows the conduction of experiments, the collection of data, and the analysis of data to form conclusions. In this way, you experience the process of science, and it is this process that distinguishes science from other disciplines.

PROCEDURES TO FOLLOW

Your success in the laboratory depends on how you prepare for and carry out the laboratory activities. The procedures that follow are expected to be used by all students.

Preparation

Before coming to the laboratory, complete the following activities to prepare yourself for the laboratory session.

1. Read the assigned exercise to understand (a) the objectives, (b) the meaning and spelling of new terms, (c) the introductory background information, and (d) the procedures to be followed.
2. Label and color-code illustrations as directed in the manual, and complete the items on the laboratory report related to the background information.
3. Bring your textbook to the laboratory for use as a reference.

Working in the Laboratory

The following guidelines will save time and increase your chances of success in the laboratory.

1. Remove the laboratory report from the manual so that you can complete it without flipping pages. Laboratory reports are three-hole punched so you can keep completed reports in a binder.
2. Follow the directions explicitly and in sequence unless directed otherwise.
3. Work carefully and thoughtfully. You will not have to rush if you are well prepared.
4. Discuss your procedures and observations with other students. If you become confused, ask your instructor for help.

5. Answer the questions on the laboratory report thoughtfully and completely. They are provided to guide the learning process. Just filling in the blanks is not acceptable.

Laboratory Safety and Housekeeping

1. Use equipment with care. Report any problems to your instructor.
2. Inform your instructor of any breakage or spills, and ask for assistance in proper cleanup and disposal procedures.
3. Immediately report any injuries, even minor ones, to your instructor.
4. Tie back long hair and roll up loose sleeves when using open flames.
5. Clean glassware and equipment before and after use. Return each item to its proper location at the end of the session, and clean your workstation.
6. Do not smoke, eat, drink, or apply cosmetics in the laboratory.

Assignment 1

Complete item 1 on Laboratory Report 1 that begins on page 299.

BIOLOGICAL TERMS

One of the major difficulties encountered by beginning students is learning biological terminology. Each exercise has new terms emphasized in bold print so that you do not overlook them. Be sure to know their meanings prior to the laboratory session.

Biological terms are composed of a root word and either a prefix or a suffix, or both. The **root word** provides the main meaning of the term. It may occur at the beginning or end of the term, or it may be sandwiched between a **prefix** and a **suffix.** Both prefix and suffix modify the meaning of the root word. The parts of a term are often joined by adding *combining vowels* that make the term easier to pronounce. The following examples illustrate the structure of biological terms.

1. The term *endocranial* becomes *endo/crani/al* when separated into its components. *Endo* is a prefix meaning within; *crani* is the root word meaning skull; *al* is a suffix meaning pertaining to. Therefore, the literal meaning of endocranial is "pertaining to within the skull."
2. The term *arthropod* becomes *arthr/o/pod* when separated into its components. *Arthr* is a prefix meaning joint; *o* is a combining vowel; *pod* is the root word meaning foot. Therefore, the literal meaning of arthropod is "jointed foot."

Once you understand the structure of biological terms, learning the terminology becomes much easier. *Appendix A contains the meaning of common prefixes, suffixes, and root words.* Use it frequently to help you master new terms.

Assignment 2

Using Appendix A, **complete item 2 on the laboratory report.**

UNITS OF MEASUREMENT

Scientists use the **International System of Units (SI),** commonly called the **metric system,** for making measurements. The basic reference units for these measurements are: **meter** for length, **gram** for mass, **liter** for volume, and **degree Celsius** for temperature. Each basic unit may be preceded by a prefix that modifies the value of the unit by one or more powers of ten as shown in Table 1.1 and summarized below.

$$
\begin{aligned}
\text{kilo-} &= \text{unit} \times 1{,}000 \\
\text{(none)} &= \text{unit} \\
\text{deci-} &= \text{unit} \div 10 \quad 1/10 \\
\text{centi-} &= \text{unit} \div 100 \quad 1/100 \\
\text{milli-} &= \text{unit} \div 1{,}000 \quad 1/1000 \\
\text{micro-} &= \text{unit} \div 1{,}000{,}000 \quad 1/1{,}000{,}000
\end{aligned}
$$

Some English equivalents are included in Table 1.1 for comparison and to allow conversions between the two systems.

The major advantage of the metric system is that it allows easy conversion from one unit to another within a category by multiplying or dividing by the correct power of ten. The following examples show how this is done.

TABLE 1.1
Common Metric System Units

Category	Symbol	Unit	Value	English Equivalent
Length	km *King* ¹⁰⁰⁰	kilometer	1,000 m	0.62 mi
(meter stick)	m *×*	meter*	1 m	39.37 in.
	dm *died*	decimeter	0.1 m	3.94 in.
	cm *carrying*	centimeter ¹⁄₁₀₀ *of meter*	0.01 m	0.39 in.
	mm *millions of*	millimeter	0.001 m	0.04 in.
	μm *microbes*	micrometer *(micron)*	0.000001 m	0.00004 in.
Mass	kg *king*	kilogram *1,000 meters*	1,000 g	2.2 lb
(weight)	g *×*	gram* *(weight of paperclip)*	1 g	0.04 oz
	dg *died*	decigram	0.1 g	0.004 oz
	cg *carrying*	centigram	0.01 g	0.0004 oz
	mg *millions*	milligram	0.001 g	
	μg *microbes*	microgram	0.000001 g	
Volume	l *×*	liter*	1 l	1.06 qt
	ml *millions of*	milliliter	0.001 l	0.03 oz
	μl *microbes*	microliter	0.000001 l	

* Denotes the base unit.

1. To convert 5.75 meters into millimeters, the first step is to determine how many millimeters are in a meter. Table 1.1 shows that 1 mm = 0.001 m or 1/1,000 of a meter. Thus, there are 1,000 mm in 1 m. Since you are converting meters to millimeters, you multiply 5.75 m by a fraction expressing that there are 1,000 mm in 1 m and cancel like units. You use the unit that you want to convert *to* as the numerator of the fraction.

$$\frac{5.75 \text{ m}}{1} \times \frac{1,000 \text{ mm}}{1 \text{ m}} = 5,750 \text{ mm}$$

Note that this equation multiplies 5.75 by 1,000, which moves the decimal 3 places to the right and changes the units from meters to millimeters. Moving the decimal and changing the unit is a quick way to do such problems.

2. To convert 125 centimeters into meters, the first step is to determine how many centimeters are in a meter. Table 1.1 shows that 1 cm = 0.01 m or 1/100 of a meter. Thus, there are 100 cm in 1 m. Since you are converting centimeters into meters, you multiply 125 cm by a fraction expressing that 1 m contains 100 cm and cancel like units. Do you know why 1 m is the numerator of the fraction?

$$\frac{125 \text{ cm}}{1} \times \frac{1 \text{ m}}{100 \text{ cm}} = 1.25 \text{ m}$$

Note that this equation divides 125 cm by 100, which moves the decimal 2 places to the left and changes the units from centimeters to meters.

Length

Length is the measurement of a line, either real or imaginary, extending between two points. Your height, the distance between cities, and the size of a football field involve length. The basic unit of length is the **meter.**

Materials

Per student group
Metric ruler, clear plastic
Penny

Assignment 3

1. ***Complete item 3a on the laboratory report. Perform the following procedures***

and record your data and calculations in item 3 on the laboratory report.

2. Measure the diameter of a penny in millimeters. Convert your answer to centimeters and meters.

3. Measure the length of your little finger as shown in Figure 1.1. Record your answer on the laboratory report and in the class tabulation chart on the chalkboard.

4. Biologists usually analyze data statistically to determine if the findings are significant. We won't do that here, but you will look at certain characteristics of the data that have been collected.

5. **Record the class data in item 3d on the laboratory report.**

6. From the class data, determine the *range* of finger lengths and the *average* finger length. (The average is calculated by dividing the sum of the finger lengths by the number of fingers measured.) **Complete item 3e on the laboratory report.**

7. Plot the little finger lengths against the number of students with each finger length on the graph in item 3f on the laboratory report. This is done for each finger length by placing a dot where an imaginary vertical line extending from the horizontal axis at a particular finger length intersects with an imaginary horizontal line extending from the vertical axis at the number of students having that finger length.

8. **Complete item 3 on the laboratory report.**

Mass

Mass is the characteristic that gives an object inertia, the resistence to a change in motion. It is the quantity of matter in an object. Mass is not the same as weight because weight is dependent upon the force of gravity acting on the object. The mass of a given object is the same on earth as on the moon, but its weight is much less on the moon because the moon's force of gravity is less than the earth's. As a nonscientist, you probably can get by using mass and weight interchangeably, although this is technically incorrect.

In this section, you will use a triple-beam balance to measure the mass of objects. Obtain a balance and locate the parts labeled in Figure 1.2. Note that each beam is marked with graduations. The beam closest to you has 0.1-g and 1.0-g graduations; the middle beam has 100-g graduations; and the farthest beam has 10-g graduations. There is a movable mass attached to each beam. When the pan is empty and clean and the movable masses are moved to zero—as far to the left as possible—the balance mark on the right end of the beam should align with the balance mark on the upright post. If not, rotate the adjustment knob under the pan at the left until it does. Now the balance is ready to use.

Procedure for Measuring Mass

All three beams are used if you are measuring an object with a mass over 100 g; the first and

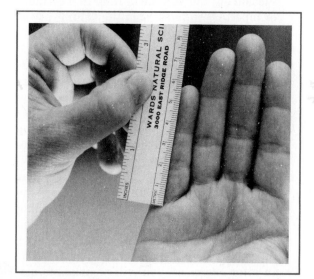

Figure 1.1. Measuring the little finger. Align the edge of the ruler along the midline of the finger.

Figure 1.2. A triple-beam balance. (a) Adjustment knob. (b) Pan. (c) Movable mass.

third beams are used when measuring an object with a mass between 10 and 100 g; and the first beam only is used when measuring an object with a mass of 10 g or less. Here is how to do it.

Place the object to be measured in the center of the pan. Move the movable mass on the middle (100 g) beam to the right, one notch at a time, until the right end of the beam drops below the balance mark. Then move the mass back to the left one notch. Now slide the movable mass on the third (10 g) beam to the right, one notch at a time, until the right end of the beam drops below the balance mark. Then, move the mass back to the left one notch. Finally, slide the movable mass on the first (1 g) beam slowly to the right until the right end of the beam aligns with the balance mark. The mass of the object is then determined as the sum of the masses indicated on the three beams.

Materials

Per student group
Triple-beam balance
Wood blocks with different densities, numbered 1 to 3

Assignment 4

1. Measure the mass of each of the three numbered blocks of wood and record your data.
2. ***Complete item 4 on the laboratory report.***

Volume

Volume is the space occupied by an object or a fluid. The liter is the basic unit of volume, but milliliters are the common units used for small volumes. Graduated cylinders and pipettes are used to measure fluids. Length may also be involved in volume determinations because 1 cubic centimeter (cm^3 or cc) equals 1 ml.

Materials

Per student group
Beaker, 250 ml
Graduated cylinders, 10 and 50 ml
Medicine dropper
Pipettes, 1, 5, and 10 ml
Test tube
Test tube rack

Triple-beam balance
Wood blocks with different densities, numbered 1 to 3

Assignment 5

1. Measure the length, width, and depth of the wood blocks in centimeters. Then determine the volume of each one. Volume = length × width × depth.
2. The density of an object is mass per unit of volume. Using the mass of each block from item 4a on the laboratory report, calculate the density of each wood block.

$$\text{Density} = \frac{\text{Mass}}{\text{Volume}}$$

3. ***Complete items 5a to 5c on the laboratory report.***
4. Examine the pipettes and graduated cylinders to determine the meaning of the graduations of each.
5. Fill the beaker about half full with tap water. Pour a little water into the 10-ml graduated cylinder. Look at the top of the water column from the side and note that it is curved rather than flat. This curvature is known as the **meniscus,** and it results because water molecules tend to "creep up" and stick to the side of the cylinder. When measuring the volume of fluid in a cylinder or pipette, you must read the volume at the *bottom of the meniscus* as shown in Figure 1.3.
6. While observing the meniscus from the side, add water to the 10-ml graduated cylinder and fill it to the 8-ml mark. Repeat until you can do this with ease.
7. Determine the volume of the test tube by filling it with water and then pouring the water into the 50-ml graduated cylinder.
8. Empty the 10-ml cylinder and shake out as much water as possible. Determine the number of drops in 1 ml of water by adding water, drop by drop, to the 10-ml graduated cylinder with the medicine dropper (dropper pipette).
9. In later exercises, you will be asked to dispense a *dropper* of fluid. This means as much fluid as a medicine dropper will hold

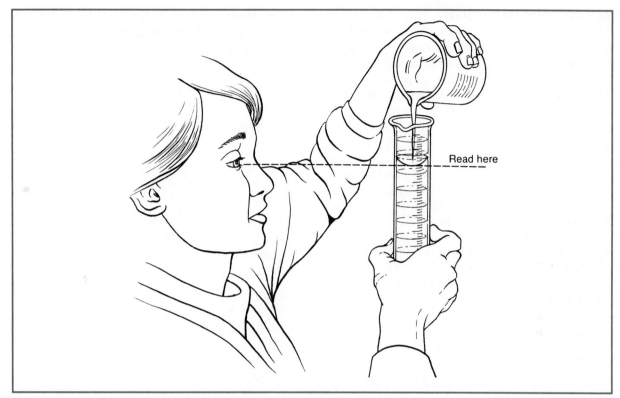

Figure 1.3. Reading the volume of water at the bottom of a meniscus

when the rubber bulb is fully depressed and released while the tip is submerged in the fluid. Determine the volume of a *dropper*.

10. Determine the mass of 30 ml of water using the 50-ml graduated cylinder and a triple-beam balance.

11. **Complete item 5 on the laboratory report.**

Temperature

Scientists use a Celsius thermometer to measure temperature, while nonscientists usually use a Fahrenheit thermometer. The differences in the scales of the two systems of temperature measurement are shown below.

	°C	°F
Boiling point of water	100	212
Freezing point of water	0	32

You can convert from one system to the other by using these formulas.

Celsius to Fahrenheit

$$°F = \frac{9 \times °C}{5} + 32$$

Fahrenheit to Celsius

$$°C = \frac{5 \times (°F - 32)}{9}$$

Materials

Per student group
Beaker, 250 ml
Celsius thermometer

Assignment 6

1. Examine the graduations on the Celsius thermometer.
2. Measure the temperature (°C) of air in the room and cold tap water. Convert the temperatures to °F.
3. **Complete item 6 on the laboratory report.**

SCIENTIFIC METHOD

Many of the laboratory activities in this manual allow you to play the role of a biologist as you perform experiments, collect data, and make conclusions. You will use the **scientific method** of inquiry, a process that tests in a systematic manner possible answers to questions raised about nature. It also allows other scientists to duplicate the testing process.

The scientific method involves several sequential steps that are diagramed in Figure 1.4 and summarized briefly below. *Hypothetical examples* are provided for each step.

1. The process begins with making careful, thoughtful **observations** of nature directly or indirectly.

   Example: Observations suggest that a daily supplement of vitamin C may reduce the risk of cancer.

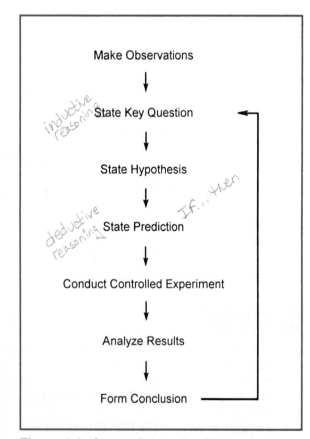

Figure 1.4. Steps of the scientific method

2. Observations raise questions in the biologist's mind that lead to a statement of the **key question** to be answered. This key question is sometimes called "the problem."

   Example: Does a daily supplement of 50 mg of vitamin C reduce the risk of cancer?

3. A **hypothesis** is stated. This is a statement of the anticipated answer to the key question.

   Example: A normal diet plus a daily supplement of 50 mg of vitamin C reduces the risk of cancer.

4. A **prediction** is made based on the hypothesis. It is usually phrased in an "if . . . then" manner. This tells the biologist what to expect in specific situations.

   Example: If daily supplements of 50 mg of vitamin C reduce the risk of cancer, then mice receiving the vitamin C supplement will develop fewer cancers than mice not receiving the supplement when both groups of mice are exposed to a carcinogen (cancer-causing substance).

5. A **controlled experiment** is designed and conducted to test the *experimental null hypothesis* if the hypothesis is testable by experiment. If not, additional observations are made to test the hypothesis. A null hypothesis is a negative restatement of the prediction.

   Example: Daily supplements of 50 mg of vitamin C do not reduce the frequency of cancer in mice when they are exposed to a carcinogen.

   Testing the hypothesis by a controlled experiment is highly preferable. There are three kinds of variables (conditions) in a controlled experiment. The **independent variable** (in our example, vitamin C) is the condition that is being evaluated for its effect on the **dependent variable** (in our example, cancer). **Controlled variables** are all other conditions that could affect the results but do not because they are kept constant. A controlled experiment consists of two parts. The *experimental group* is exposed to the independent variable, but the *control group* is not. All other variables are controlled (kept constant).

Example: From a group of laboratory mice that have been bred to have identical hereditary compositions, 50 randomly selected mice are placed in the experimental group and 50 randomly selected mice are placed in the control group. The experimental group receives a supplement of 50 mg of vitamin C each day, but the control group does not. All other variables are kept identical for each group, i.e., exposure to a carcinogen, normal diet, temperature, humidity, light–dark cycles, water availability, and so forth. After 120 days, the mice are examined for cancers.

6. **Results** are collected and analyzed.

Example: Among the experimental group, 5 mice developed cancers. Among the control group, 20 mice developed cancers.

7. A **conclusion** as to whether the null hypothesis is accepted or rejected is made based on the results. A conclusion often leads to the formation of a new hypothesis and additional experiments.

Example: The null hypothesis is rejected. The frequency of cancer was reduced in mice receiving a daily supplement of 50 mg of vitamin C.

Assignment 7

1. **Complete item 7a on the laboratory report** regarding the above hypothetical example of the scientific method.
2. The results of the hypothetical experiment raise new key questions. What new key questions are raised in your mind?
3. **Complete items 7b to 7e on the laboratory report.**

2

THE MICROSCOPE

A **microscope** is a precision instrument and an essential tool in the study of cells, tissues, and minute organisms. It must be handled and used carefully at all times. Most of the microscopic observations in this course will be made with a **compound microscope,** but a **dissecting microscope** will be used occasionally. A microscope consists of a lens system, a controllable light source, and a mechanism for adjusting the distance between the objective lens and the object to be observed.

To make the observations required in this course, you must know how to use a microscope effectively. This exercise provides an opportunity for you to develop skills in microscopy.

THE COMPOUND MICROSCOPE

The major parts of the compound microscope are shown in Figure 2.1. Refer to this figure as you read this text, and *label the parts* indicated on the figure. Your microscope may be somewhat different from the one illustrated.

The **base** rests on the table and, in most microscopes, contains a built-in **light source** and a **light switch.** Some microscopes have a light intensity (voltage) control knob on the base, usually associated with the light switch. The **arm** rises from the base and supports the stage, lens system, and control mechanisms. The **stage** is the flat surface on which microscope slides are placed for viewing. **Stage clips,** or a **mechanical stage,** hold the slide in place.

Most microscopes have a **condenser** located below the stage. It concentrates the light on the object and may be raised or lowered by the **condenser control knob.** Usually, the condenser should be raised to its highest position. An **iris diaphragm** is built into the base of the condenser. The **iris diaphragm control lever** (a rotatable wheel in some microscopes) varies the amount of light entering the condenser and the lens system.

The **body tube** is supported by the arm and

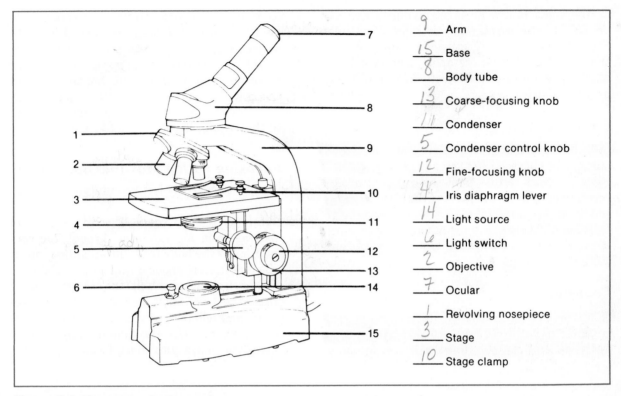

9	Arm
15	Base
8	Body tube
13	Coarse-focusing knob
11	Condenser
5	Condenser control knob
12	Fine-focusing knob
4	Iris diaphragm lever
14	Light source
6	Light switch
2	Objective
7	Ocular
1	Revolving nosepiece
3	Stage
10	Stage clamp

Figure 2.1. Compound microscope

has an **ocular lens** at the upper end and a **revolving nosepiece** with the attached **objective lenses** at the lower end. The nosepiece is rotated to bring different objectives into viewing position. The objectives usually click into viewing position.

Student microscopes usually have three objectives. The shortest is the **scanning objective,** which has a magnification of 4×. The **low-power objective** with a 10× magnification is intermediate in length. The **high-power** (high-dry) **objective** is the longest and usually has a magnification of 40×, but may have a 43× or 45× magnification in some microscopes. In this manual, the high-power objective is often called the 40× objective. Some microscopes have an **oil-immersion objective** (100× magnification) that is a bit longer than the high-power objective.

There are two focusing knobs. The **coarse-focusing knob** has the larger diameter and is used to bring objects into rough focus when using the 4× and 10× objectives. The **fine-focus-**

ing knob has a smaller diameter and is used to bring objects into fine focus. It is the *only* focusing knob used with the high-power and oil-immersion objectives.

Magnification

The magnification of each lens is fixed and inscribed on the lens. The ocular usually has a 10× magnification. The powers of the objectives may vary but usually are 4×, 10×, and 40×. The **total magnification** is calculated by multiplying the power of the ocular by the power of the objective.

Resolving Power

The quality of a microscope depends on its ability to **resolve** (distinguish) objects. Magnification without resolving power is of no value. Modern microscopes increase both magnification and resolution by a careful matching of light source and precision lenses. Most microscopes have a blue light filter located in either the condenser or the light source since resolving power increases as the wavelength of light decreases. Student microscopes usually can resolve objects that are 0.5 µm or more apart. The best light microscopes can resolve objects that are 0.1 µm or more apart.

Contrast

Sufficient **contrast** must be present among the parts of an object for the parts to be distinguishable. Contrast results from the differential absorption of light by the parts of the object. Sometimes, stains must be added to a specimen to increase the contrast. A reduction in the amount of light improves contrast when viewing unstained specimens.

Focusing

A microscope is focused by increasing or decreasing the distance between the specimen on the slide and the objective lens. The focusing procedure used depends on whether your microscope has a **movable stage** or **movable body tube** (see Figure 2.2). Both procedures are described below. Use the one appropriate for your microscope. As a general rule, you should start focusing with the low-power (10×) objective unless the large size of the object requires starting with the 4× objective.

Focusing with a Movable Body Tube

1. Rotate the 10× objective into viewing position.
2. While using the coarse-focusing knob and *looking from the side* (not through the ocular), lower the body tube until it stops or until the objective is about 3 mm from the slide.
3. While looking through the ocular, slowly raise the body tube by turning the coarse-focusing knob toward you until the object becomes visible. Use the fine-focusing knob to bring the object into sharp focus.

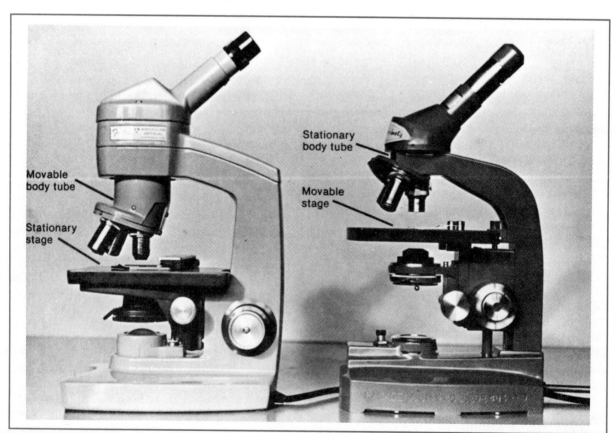

Figure 2.2. Comparison of two microscope designs

Focusing with a Movable Stage

1. Rotate the 10× objective into viewing position.
2. While using the coarse-focusing knob and *looking from the side* (not through the ocular), raise the stage to its highest position or until the slide is about 3 mm from the objective.
3. While looking through the ocular, slowly lower the stage by turning the coarse-focusing knob away from you until the object comes into focus. Use the fine-focusing knob to bring the object into sharp focus.

Switching Objectives

Your microscope is **parcentric** and **parfocal.** This means that if an object is centered and in sharp focus with one objective, it will be centered and in focus when another objective is rotated into the viewing position. However, slight adjustments to recenter and refocus (with the fine-focusing knob) may be necessary. As you switch objectives from 4× to 10× to 40× to increase magnification, the (1) working distance, (2) diameter of the field, and (3) light intensity are *reduced* as magnification increases. Note this relationship in Figure 2.3.

Slide Preparation

Specimens to be viewed with a compound microscope are placed on a **microscope slide** and are usually covered with a **cover glass.** Specimens may be mounted on slides in two different ways. A **prepared slide** (permanent slide) has a permanently attached cover glass, and the specimen is usually stained. A **wet-mount slide** (temporary slide) has the specimen mounted in a liquid, usually water, and covered with a cover glass. In this course, you will observe commercially prepared permanent slides and wet-mount slides that you will make. Wet-mount slides are prepared as shown in Figure 2.4.

Care of the Microscope

You should carry a microscope upright in front of you, not at your side. Use one hand to support the base and the other to grasp the arm. See Figure 2.5. Develop the habit of cleaning the lenses prior to using the microscope. Use only special lint-free lens paper. If the lens paper

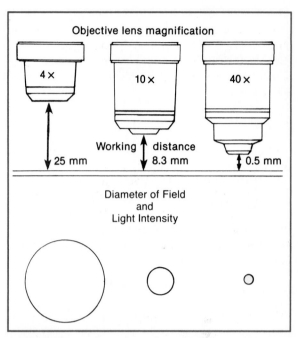

Figure 2.3. Relationship of objective power to (1) working distance, (2) diameter of field, and (3) light intensity

does not clean the lenses, inform you instructor. If any liquid gets on the lenses during use, wipe it off immediately and clean the lenses with lens paper.

When you are finished using the microscope, perform these steps:

1. Remove the slide. Clean and dry the stage.
2. Clean the lenses with lens paper.
3. Rotate the nosepiece so that no objective projects beyond the front of the stage.
4. Raise the stage to its highest position *or* lower the body tube to its lowest position in accordance with the type of microscope you are using.
5. Unplug the light cord and loosely wrap it around the arm below the stage. Add a dust-cover, if present.
6. Return the microscope to the correct cabinet cubicle.

Assignment 1

1. Label Figure 2.1.
2. Obtain the microscope assigned to you. Carry it as described above and place it on the table

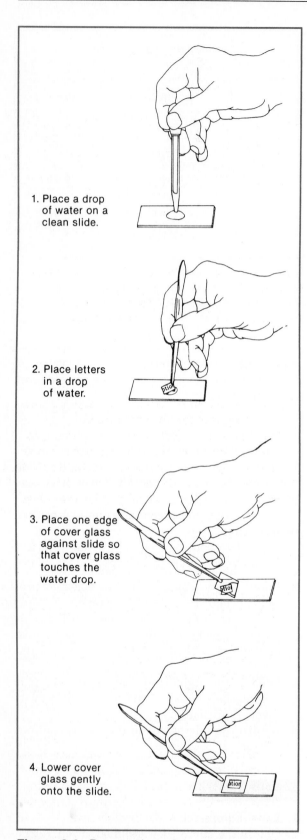

1. Place a drop of water on a clean slide.

2. Place letters in a drop of water.

3. Place one edge of cover glass against slide so that cover glass touches the water drop.

4. Lower cover glass gently onto the slide.

Figure 2.4. Preparation of a wet-mount slide

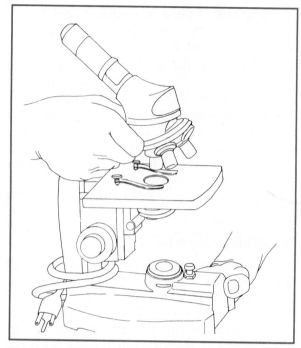

Figure 2.5. How to carry the microscope. Grasp the arm of the microscope with one hand and support the base with the other.

in front of you. Locate the parts shown in Figure 2.1. Clean the lenses with lens paper. Try the knobs and levers to see how they work.

3. Raise the condenser to its highest position and keep it there. Plug in the light cord and turn on the light. If your microscope has a voltage control knob, adjust it to an intermediate position to prolong the life of the bulb.

4. Rotate the 4× objective into viewing position and look through the ocular. The circle of light that you see is called the **field of view** or simply the **field.**

5. While looking through the ocular, open and close the iris diaphragm and note the change in light intensity. Repeat for each objective and note that light intensity decreases as the power of the objective increases. Therefore, you will need to adjust the light intensity when you switch objectives. *Remember to use reduced light intensity when you are viewing unstained and rather transparent specimens.*

6. If your microscope has a voltage control knob, repeat item 5 while leaving the iris dia-

phragm open but changing the light intensity by altering the voltage.

7. ***Complete item 1 on Laboratory Report 2 that begins on page 303.***

Developing Microscopy Skills

The following microscopic observations are designed to help you develop skill in using a compound microscope.

Materials

Per student
Compound microscope
Dissecting instruments
Medicine dropper
Metric ruler, clear plastic
Water in dropping bottle

Per lab
Kimwipes
Lens paper
Microscope slides and cover glasses
Newspaper
Pond water culture
Prepared slides of fly wing

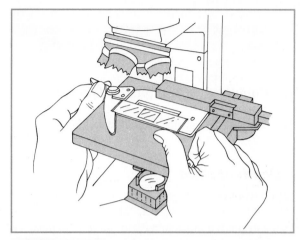

Figure 2.6. Placing the slide in the mechanical stage. The retainer lever is pulled back to place the slide against the stationary arm. Then, the retainer lever is released to secure the slide.

Assignment 2

1. Obtain a microscope slide and cover glass. If they are not clean, wash them with soap and water, rinse, and dry them. Use a paper towel to dry the slide, but use Kimwipes to blot the water from the fragile cover glass.
2. Use scissors to cut three letters from a newspaper with the letter *i* as the middle letter.
3. Prepare a wet-mount slide of the letters as shown in Figure 2.4. Use a paper towel to soak up any excess water. If too little water is present, add a drop at the edge of the cover glass and it will flow under the cover glass.
4. Place the slide on the stage with the letters over the **stage aperture,** the circular opening in the stage. Secure it with either a mechanical stage or stage clamps. See Figure 2.6. The slide should be parallel to the edge of the stage nearest you with the letters oriented so that they may be read with the naked eye.
5. Rotate the 10× objective into viewing position and bring the letters into focus using

the focusing procedure described above that is appropriate for your microscope.
6. Rotate the 4× objective into the viewing position. Center the letter *i* and bring it into sharp focus. Can you see all of the *i*? Can you see the other letters? What is different about the orientation of the letters when viewed with the microscope instead of the naked eye?
7. Move the slide to the left while looking through the ocular. Which way does the image move? Practice moving the slide while viewing through the ocular until you can quickly place a given letter in the center of the field.
8. Center the *i* and bring it into sharp focus. Rotate the 10× objective into position. Is the *i* centered and in focus? If not, center it and bring it into sharp focus. How much of the *i* can you see?
9. Rotate the 40× objective into position. Is the *i* centered and in focus? All that you can see at this magnification are "ink blotches" that compose the *i*. If you do not see this, center the *i* and bring it into focus with the *fine-focusing knob. Never use the coarse-focusing knob with the high-power objective.*
10. Practice steps 1 through 4 until you can quickly center the dot of the letter *i* and

bring it into focus with each objective. **Remember,** you are *never* to start observations with the high-power objective. Instead, start at a lower power and work up to the 40× objective.

11. ***Complete item 2 on the laboratory report.***
12. Remove the slide and set it aside.

Depth of Field

When you view objects with a microscope, you obviously are viewing the objects from above. The vertical distance within which structures are in sharp focus is called the **depth of field,** and it decreases as magnification increases. You will learn more about depth of field by performing the observations that follow.

Assignment 3

1. Obtain a prepared slide of a fly wing. Observe the tiny spines on the wing membrane with each objective starting with the 4× objective. Can you see all of a spine at each magnification?
2. Using the 4× objective, locate a large spine at the base of the wing where the veins converge and center it in the field. Can you see all of it?
3. Rotate the 10× objective into position and observe the spine. Can you see all of it? Practice focusing up and down the length of the spine, and note that you can see only a portion of the spine at each focusing position.
4. Rotate the 40× objective into position and observe the spine. At each focusing position, you can see only a thin "slice" of the spine. To determine the spine's shape, you have to focus up and down the spine using the *fine-focusing knob.*
5. ***Complete item 3 on the laboratory report.***

The preceding observations demonstrate that when viewing objects with a greater depth (thickness) than the depth of field, you see only a two-dimensional plane "optically cut" through the object. To discern an object's three-dimensional shape, a series of these images must be "stacked up" in your mind as you focus through the depth of the object.

Diameter of Field

When using each objective, you must know the diameter of the field to estimate the size of observed objects. Estimate the diameter of field for each magnification of your microscope as described in the section that follows.

Assignment 4

1. Place the clear, plastic ruler on the microscope stage as shown in Figure 2.7. The edge of the ruler should extend across the diameter of the field. Focus on the metric scale with the 4× objective, and adjust the ruler so that one of the millimeter marks is at the left edge of the field. Estimate and record the diameter of field at 40× by counting the spaces and

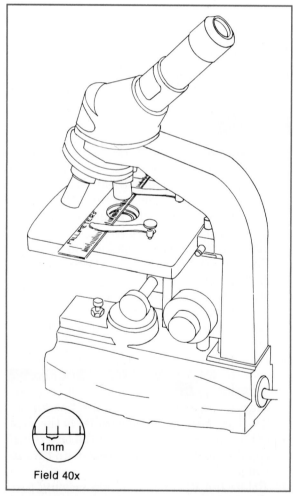

Figure 2.7. Estimating the diameter of field

portions thereof between the millimeter marks.

2. Use these equations to calculate the diameter of field at 100× and 400×:

$$\text{Diam. (mm)} \atop \text{at } 100\times = \frac{40\times}{100\times} \times \frac{\text{diam. (mm) at } 40\times}{1}$$

$$\text{Diam. (mm)} \atop \text{at } 100\times = \frac{40\times}{400\times} \times \frac{\text{diam. (mm) at } 40\times}{1}$$

3. Return to your slide of newspaper letters, and estimate the diameter of the dot of the letter *i* and the length of the letter *i* including the dot.

4. ***Complete item 4 on the laboratory report.***

Application of Microscopy Skills

In this section, you will use the skills and knowledge gained in the preceding portions of the exercise.

Assignment 5

1. Prepare a wet-mount slide of two crossed hairs, one blond and the other brunette. Ob-tain 1-cm lengths of hair from cooperative classmates.

2. Using the 4× objective, center the crossing point of the hairs in the field and observe. Are both hairs in sharp focus?

3. Examine the crossed hairs at 100× magnification. Are both hairs in sharp focus? Determine which hair is on top by using focusing technique. If you have trouble with this, see your instructor.

4. Examine the crossed hairs at 400× magnification. Are both hairs in focus? Move the crossing point to one side and focus on the blond hair. Using focusing technique, observe surface and optical midsection views of the hair as shown in Figure 2.8. At 400× magnification, the depth of field is less than half of diameter of a hair.

5. ***Complete items 5a to 5d on the laboratory report.***

6. Make slides of the pond water samples and examine them microscopically. The object is to sharpen your microscopy skills rather than to identify the organisms. However, you may see organisms like those shown in Figure 2.9. Note the size, color, shape, and motility of the

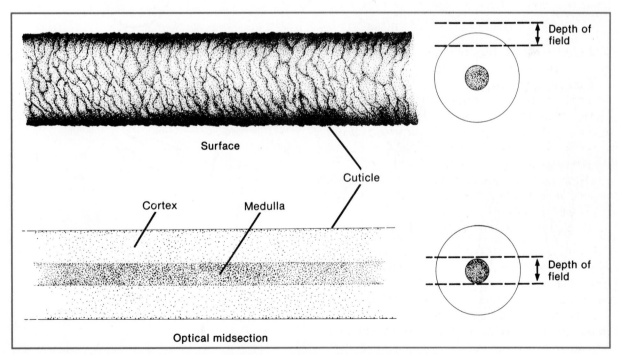

Figure 2.8. Surface and optical midsection views of a human hair, 400×

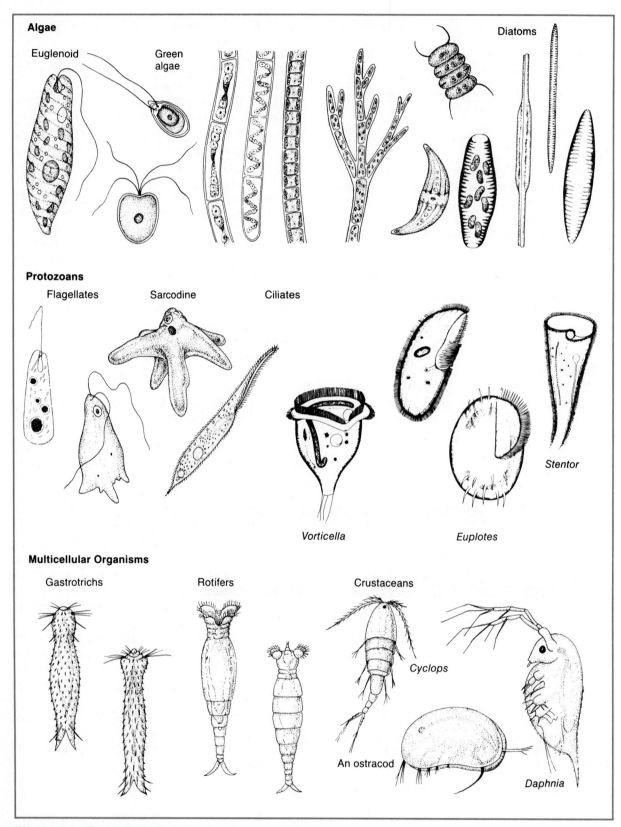

Figure 2.9. Representative pond water organisms

organisms. ***Draw a few of the organisms in the space for item 5e on the laboratory report.***

7. Prepare your microscope for return to the cabinet as described previously, and return it.

THE DISSECTING MICROSCOPE

A **dissecting microscope** is used to view objects that are too large or too opaque to observe with a compound microscope. The two oculars enable stereoscopic observations and usually are 10× in magnification. Most student models have two objectives that provide 2× and 4× magnification so that total magnification is 20× and 40×. Some models have a zoom feature that enables observations at intermediate magnifications. Objects are usually viewed with reflected light instead of transmitted light, although some dissecting microscopes provide both types of light sources.

The parts of a dissecting microscope are shown in Figure 2.10. Note the single focusing knob and the two oculars. The oculars may be moved inward or outward to adjust for the distance between the pupils of your eyes. One ocular has a focusing ring that may be adjusted to accommodate for differences in visual acuity in your eyes.

Materials

Per student
Coin
Desk lamp
Dissecting microscope
Metric ruler, clear plastic

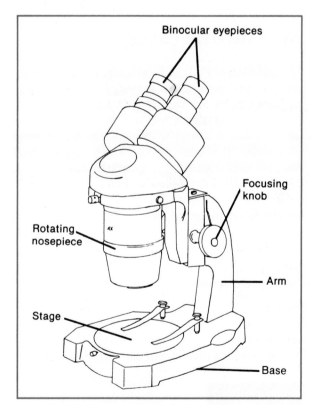

Figure 2.10. Dissecting microscope

Assignment 6

1. Obtain a dissecting microscope from the cabinet, and locate the parts shown in Figure 2.10.
2. Place a coin or other object on the stage, illuminate it with a desk lamp, and examine it with both objectives. To accommodate for differences in acuity in your eyes, focus first with the focusing knob while viewing through the ocular that *cannot* be individually adjusted. Then use the **focusing ring** on the other ocular to bring the object into sharp focus for that eye.
3. Practice focusing until you are good at it. Move the coin on the stage, noting the direction in which the image moves. Examine the grooves and ridges that produce your unique fingerprint.
4. Determine the diameter of field at each magnification.
5. ***Complete items 6 and 7 on the laboratory report.***

3

THE CELL

Living organisms exhibit two fundamental characteristics that are absent in nonliving things: **self-maintenance** and **self-replication.** The smallest unit of life that exhibits these characteristics is a single living cell. Thus, a single cell is the *structural and functional unit of life*:

1. All organisms are composed of cells.
2. All cells arise from preexisting cells.
3. All hereditary components of organisms occur in cells.

Two different types of cells occur in the biotic world: prokaryotic cells and eukaryotic cells. Prokaryotic cells are rather primitive and occur only in bacteria and cyanobacteria. Eukaryotic cells compose all other organisms. Your study in this exercise will emphasize the structure of eukaryotic cells.

PROKARYOTIC CELLS

Prokaryotic cells lack a nucleus and membrane-bound organelles. A single circular chromosome is located in an irregular **DNA (deoxyribonucleic acid) region** in the interior of the cell. A small amount of **cytoplasm** containing **ribosomes** lies between the DNA region and the **cell membrane.** A supportive and protective **cell wall** is secreted just exterior to the cell membrane. See Figure 3.1 and Plate 1, which is located following page 216.

Materials

Per student
Compound microscope

Per lab
Kimwipes
Lens paper
Prepared slides of bacteria and cyanobacteria

Assignment 1

1. Examine the prepared slides of bacteria and cyanobacteria at 400× total magnification. Can you detect the cellular components?
2. *Complete item 1 on Laboratory Report 3 that begins on page 307.*

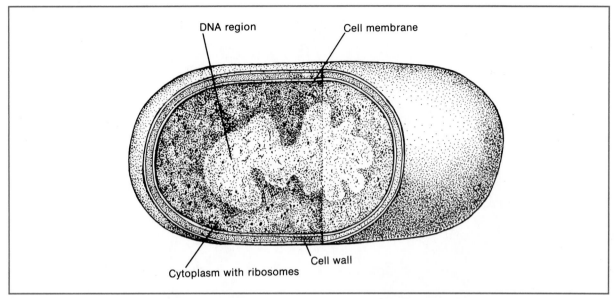

DNA region   Cell membrane

Cell wall

Cytoplasm with ribosomes

Figure 3.1. A bacterial cell

EUKARYOTIC CELLS

As you read the following descriptions of cell anatomy, locate and label the organelles on Figures 3.2 and 3.3 and note their functions in Table 3.1. The structures of the cellular organelles are shown as viewed with an electron microscope; most organelles are too small to be seen with your microscope.

The Animal Cell

Animal cells are surrounded by a **cell (plasma) membrane** that is composed of two back-to-back phospholipid layers and associated proteins. Intracellular membranes have the same structure. The bulk of the cell consists of **cytoplasm,** the semifluid or gel-like substance in which the nucleus and other **organelles** are embedded. It is supported by a lattice formed of very fine **microfilaments** (label 13) and slightly larger **microtubules** that form the cytoskeleton.

The **nucleus** is a large, spherical organelle that contains the **chromosomes.** In nondividing cells, the chromosomes are uncoiled and elongated so that only bits and pieces of them may be seen as **chromatin granules** at each focal plane when viewing the nucleus with a microscope. In dividing cells, the chromosomes coil tightly and appear as dark-staining rod-shaped structures. The spherical, dark-staining structure in the nucleus is the **nucleolus,** which is composed of ribonucleic acid (RNA) and protein. The nucleus is surrounded by a **nuclear envelope** that is composed of two adjacent membranes perforated by pores. The pores enable the movement of materials between the nucleus and cytoplasm.

The outer membrane of the nuclear envelope is continuous with the **endoplasmic reticulum (ER),** a series of folded membranes that permeate the cytoplasm. **Smooth ER** lacks ribosomes; **rough ER** is studded with ribosomes (shown as dots in the figures). **Ribosomes** (label 14) are tiny organelles consisting of RNA and protein that may occur singly, in clusters, or in chains. They are located either free in the cytoplasm or on the rough ER. The **Golgi apparatus** is a stack of membranes associated with the ER, usually near the nucleus.

A pair of **centrioles,** short cylindrical bodies composed of microtubules, are oriented perpendicular to each other near the nucleus. **Mitochondria** (label 17) are elongated structures formed of a larger, folded, inner membrane surrounded by a smaller, nonfolded membrane. Mitochondria are more abundant in cells with a high metabolic rate. **Vacuoles** (label 9), small fluid-filled spaces enveloped by a membrane,

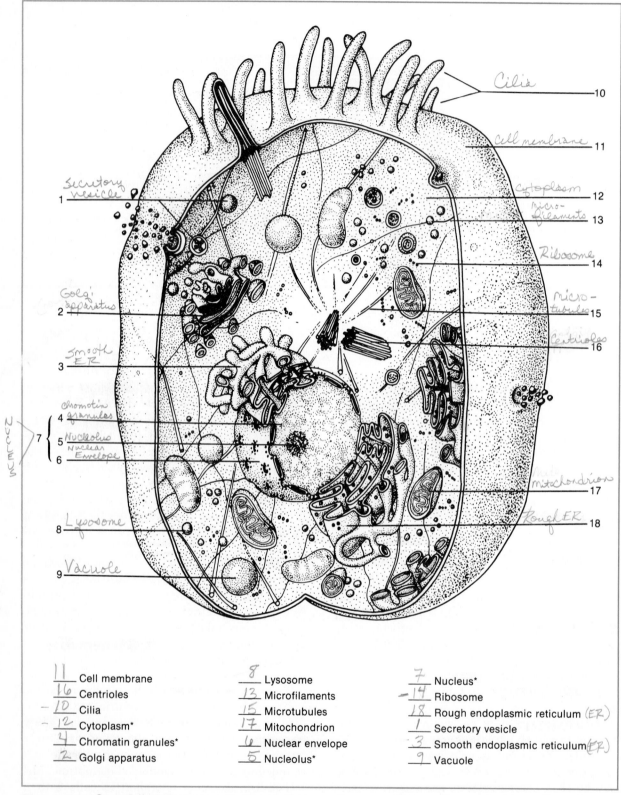

Handwritten labels on figure:
- Cilia — 10
- Cell membrane — 11
- Cytoplasm — 12
- Microfilaments — 13
- Ribosome — 14
- Microtubules — 15
- Centrioles — 16
- Mitochondrion — 17
- Rough ER — 18
- 1 — Secretory vesicle
- 2 — Golgi apparatus
- 3 — Smooth ER
- 4 — Chromatin granules
- 5 — Nucleolus
- 6 — Nuclear Envelope
- 7 — Nucleus
- 8 — Lysosome
- 9 — Vacuole

11 Cell membrane	8 Lysosome	7 Nucleus*
16 Centrioles	13 Microfilaments	14 Ribosome
10 Cilia	15 Microtubules	18 Rough endoplasmic reticulum (ER)
12 Cytoplasm*	17 Mitochondrion	1 Secretory vesicle
4 Chromatin granules*	6 Nuclear envelope	3 Smooth endoplasmic reticulum (ER)
2 Golgi apparatus	5 Nucleolus*	9 Vacuole

Figure 3.2. Generalized animal cell. Organelles are shown as seen with an electron microscope. Organelles visible with your microscope are noted with asterisks.

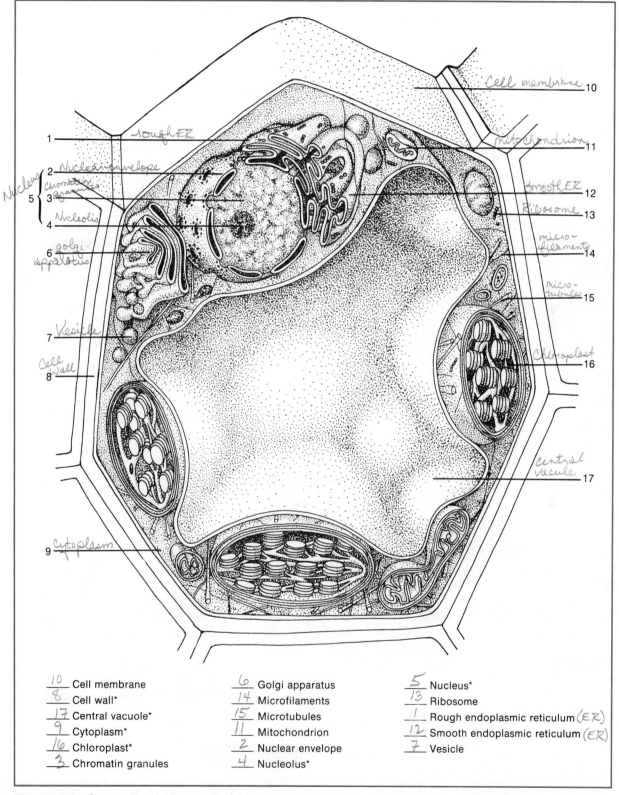

- Cell membrane — 10
- Mitochondrion — 11
- Smooth ER — 12
- Ribosome — 13
- micro-filaments — 14
- micro-tubules — 15
- Chloroplast — 16
- Central Vacuole — 17
- rough ER — 1
- Nuclear envelope — 2
- Chromatin granules — 3
- Nucleolus — 4
- Nucleus — 5
- golgi apparatus — 6
- Vesicle — 7
- Cell Wall — 8
- Cytoplasm — 9

Figure 3.3. Generalized plant cell. Organelles are shown as seen with an electron microscope. Organelles visible with your microscope are noted with asterisks.

Label key:

_10_ Cell membrane	_6_ Golgi apparatus	_5_ Nucleus*
_8_ Cell wall*	_14_ Microfilaments	_13_ Ribosome
_17_ Central vacuole*	_15_ Microtubules	_1_ Rough endoplasmic reticulum (ER)
_9_ Cytoplasm*	_11_ Mitochondrion	_12_ Smooth endoplasmic reticulum (ER)
_16_ Chloroplast*	_2_ Nuclear envelope	_7_ Vesicle
_3_ Chromatin granules	_4_ Nucleolus*	

TABLE 3.1
Functions of the Major Cellular Organelles

Organelle	Function
Cell membrane	Controls the passage of materials into and out of the cell: maintains the integrity of the cell.
Chromosomes	Contain the genetic code, in DNA molecules, that controls the life processes of the cell.
Centrioles	Short cylinders that form microtubule systems (absent in plant cells).
Endoplasmic reticulum	Forms channels for the movement of materials throughout the cell and surfaces for chemical reactions.
Golgi apparatus	Stores, modifies, and packages materials for export from the cell.
Lysosomes	Tiny sacs containing strong digestive enzymes that are released to digest old and damaged cells (absent in plant cells).
Microfilaments	Provide support for the cytoplasm.
Microtubules	Provide support for the cytoplasm; form the spindle in dividing animal cells, flagella in flagellated cells, and cilia in ciliated cells.
Mitochondria	Sites of aerobic cellular respiration that releases energy from nutrients to form adenosine triphosphate (ATP).
Nuclear envelope	Controls the passage of materials between the nucleus and the cytoplasm.
Nucleolus	Assembles precursors (RNA and protein) of ribosomes prior to their export to the cytoplasm.
Nucleus	Control center of the cell because it contains chromosomal DNA that controls cellular functions and inheritance.
Plastids	Chloroplasts are sites of photosynthesis. Chromoplasts contain pigments that give coloration to flowers and fruits. Leukoplasts are often sites of starch storage. (Only in plant cells.)
Secretory vesicles	Tiny sacs that carry substances to the cell membrane for release from the cell.
Vacuoles	Fluid-filled sacs containing a variety of substances.

may be scattered in the cytoplasm. **Lysosomes** (label 8) are smaller than vacuoles and contain powerful digestive enzymes. **Secretory vesicles** (label 1) are tiny sacs that carry materials to the cell membrane for export.

Some animal cells possess **cilia,** hairlike projections that move particles over the cell surfaces in multicellular forms. Sperm cells have a **flagellum,** a whiplike structure, that provides cell movement.

The Plant Cell

Plant cells contain all of the organelles found in animal cells, except centrioles and lysosomes. Mature plant cells are characterized by the presence of cell walls, plastids, and a central vacuole. A rigid **cell wall,** formed of cellulose, is located just exterior to the cell membrane.

Plastids are enveloped by a double membrane and are classified according to the pigments that they contain. **Chloroplasts** contain chlorophyll and carotenes and are the only plastids shown in Figure 3.3. **Chromoplasts** (not shown) contain various red, orange, or yellow pigments. **Leukoplasts** (not shown) contain no pigments and are colorless.

Immature plant cells have numerous small vacuoles, but in mature cells they combine to form a large **central vacuole** that constitutes much of the cell volume.

Assignment 2

1. Label Figures 3.2 and 3.3 and color-code the organelles.
2. *Complete item 2 on the laboratory report.*

MICROSCOPIC STUDY

In this section, you will prepare wet-mount slides for the study of cell structure.

Materials

Per student
Compound microscope
Dissecting instruments

Per lab
Dropping bottles of:
    iodine solution (IKI)
    sodium chloride, 0.9%
Kimwipes
Lens paper
Microscope slides and cover glasses
Toothpicks, flat
Medicine droppers
Amoeba proteus culture
Elodea shoots
Onion bulb scales, red

Onion Epidermal Cells

Epidermal cells of an onion scale show many of the features found in nongreen plant cells and are excellent subjects to use in beginning your study of cells.

Assignment 3

1. Label Figure 3.4 using information from the previous section.
2. Prepare a wet-mount slide of the inner epidermis of an onion scale as shown in Figure

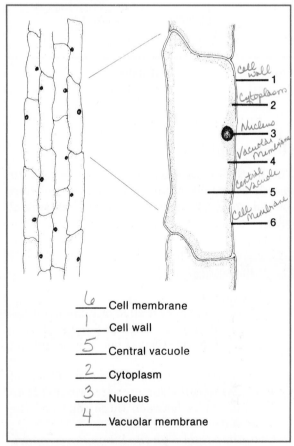

_6_ Cell membrane
_1_ Cell wall
_5_ Central vacuole
_2_ Cytoplasm
_3_ Nucleus
_4_ Vacuolar membrane

Figure 3.4. Onion epidermal cell

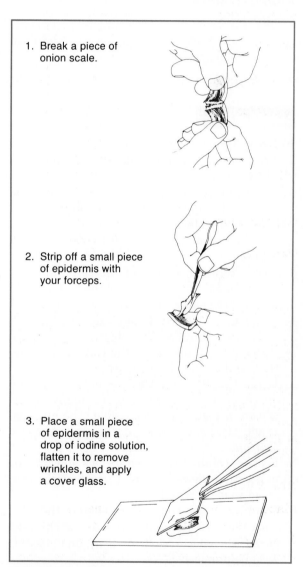

1. Break a piece of onion scale.

2. Strip off a small piece of epidermis with your forceps.

3. Place a small piece of epidermis in a drop of iodine solution, flatten it to remove wrinkles, and apply a cover glass.

Figure 3.5. Preparation of a slide of onion epidermis

3.5. Use a drop of iodine solution as the mounting fluid and add a cover glass.

3. Examine the cells at 100× and 400×. Note the arrangement of the cells. Locate the parts shown in Figure 3.4.

4. Prepare a wet-mount slide of the red epidermis from the outer surface of the onion scale. The color is due to **anthocyanin,** a water-soluble pigment in the central vacuole. Note the size of the vacuole and the location of the cytoplasm and nucleus.

5. *Complete item 3 on the laboratory report.*

Elodea Leaf Cells

Elodea is a water plant with simple leaves composed of cells that are easy to observe and that exhibit the characteristics of green plant cells.

1. Label Figure 3.6.
2. Prepare a wet-mount slide of an *Elodea* leaf in this manner:
   a. Use forceps to remove a young leaf from near the tip of the shoot and mount it in a drop of water.
   b. Add a cover glass and observe at 40× and 100×.
3. Note the arrangement of the cells and the "spine" cells along the edge of the leaf. Locate a light green area for study. The thickness of the leaf is composed of more than one layer of cells. Switch to the 40× objective and focus through the thickness of the leaf. Determine the number of cell layers present.
4. Using the 40× objective, focus through the depth of a cell and locate as many parts shown in Figure 3.6 as possible. The nucleus is spherical and slightly darker than the cytoplasm.
5. Focus carefully to observe surface and optical midsection views. See Figure 3.6. Note the basic shape of a cell.
6. Examine a spine cell at the edge of the leaf with reduced illumination to locate the nucleus, vacuole, and cytoplasm uncluttered with chloroplasts.
7. *Complete item 4 on the laboratory report.*

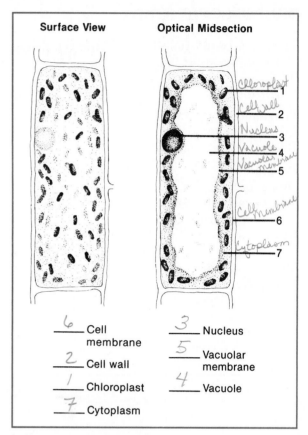

Figure 3.6. *Elodea* leaf cells

Human Epithelial Cells

The epithelial cells lining the inside of your mouth are easily obtained for study, and they exhibit some of the characteristics of animal cells.

1. Label Figure 3.7.
2. Prepare a wet-mount slide of human epithelial cells as shown in Figure 3.8, using 0.9% sodium chloride (NaCl) as the mounting fluid.
3. Add a cover glass and observe at 100× and 400×. What cell structures are present? Note the differences between these cells and the plant cells.
4. *Complete item 5 on the laboratory report.*

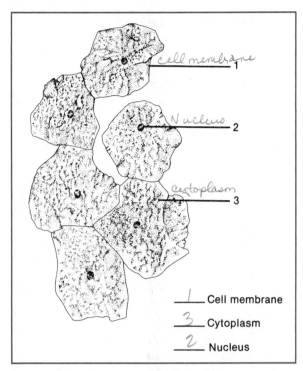

Figure 3.7. Human epithelial cells

The *Amoeba*

The common freshwater protozoan *Amoeba* exhibits many characteristics of animal cells, and it is large enough for you to see the cellular structure rather well. The nucleus, cytoplasm, vacuoles, and cytoplasmic granules are readily visible. The flowing movement of the *Amoeba* allows it to capture minute organisms that are digested in **food vacuoles.** The *Amoeba* also has **contractile vacuoles** that maintain its water balance by collecting and pumping out excess water.

Ectoplasm, the clear outer portion of the cytoplasm, is located just interior to the cell membrane. Most of the cytoplasm consists of the granular **endoplasm.** It contains the organelles and may be either gel-like, the **plasmagel,** or fluid, the **plasmasol.** Reversible changes between plasmagel and plasmasol result in the flowing **amoeboid movement** of the *Amoeba.* Your white blood cells also exhibit amoeboid movement as they slip through capillary walls and wander among the body tissues engulfing disease-causing organisms and cellular debris.

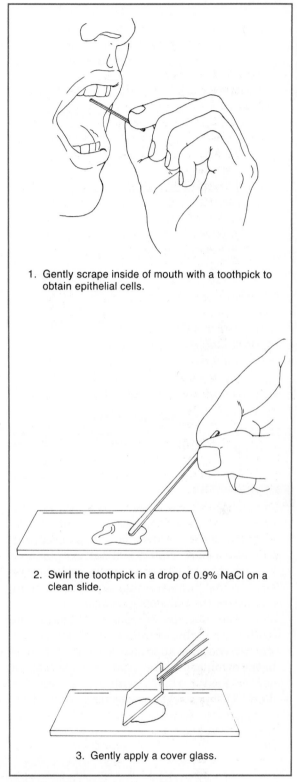

1. Gently scrape inside of mouth with a toothpick to obtain epithelial cells.

2. Swirl the toothpick in a drop of 0.9% NaCl on a clean slide.

3. Gently apply a cover glass.

Figure 3.8. Preparation of a slide of human epithelial cells

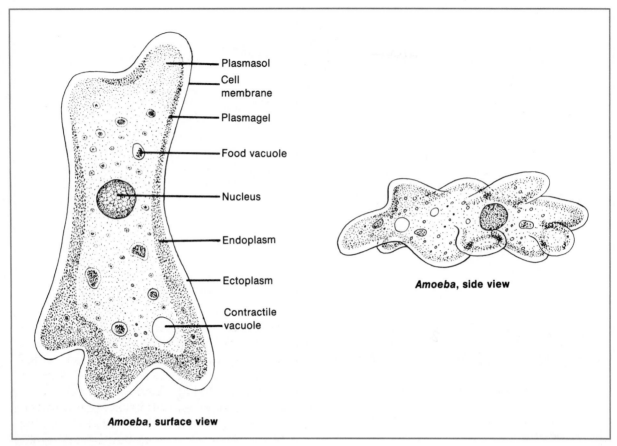

Labels (surface view): Plasmasol, Cell membrane, Plasmagel, Food vacuole, Nucleus, Endoplasm, Ectoplasm, Contractile vacuole

Amoeba, surface view

Amoeba, side view

Figure 3.9. An *Amoeba*

Assignment 6

1. Use a medicine dropper to obtain fluid *from the bottom* of the culture jar and place a drop on a clean slide. Do *not* add a cover glass.
2. Use the 10× objective to locate and observe an *Amoeba*. Note the manner of movement and locate the nucleus, cytoplasm, and vacuoles. The nucleus is spherical and a bit darker than the granular cytoplasm. Contractile vacuoles appear as spherical bubbles in the cytoplasm, while food vacuoles contain darker food particles within the vacuolar fluid. Compare your specimen with that shown in Figure 3.9 and Plate 2.5. Observe the way the *Amoeba* moves while remembering that some of your white blood cells move about in a similar way.
3. Note the appearance of the cytoplasm. Does the appearance of the inner and outer portions differ? Observe the nucleus and vacuoles. Gently add a cover glass to your slide and observe an *Amoeba* at 400×. When the pressure of the cover glass becomes too great, the cell membrane will rupture, and the contents of the cell will spill out, killing the *Amoeba*.
4. ***Complete item 6 on the laboratory report.***

4

CHEMISTRY OF CELLS

OBJECTIVES

After completion of the laboratory session, you should be able to:
1. Determine an acid or base by using pH test strips and recognize an acid or base by its pH value.
2. Explain the action of a buffer.
3. Perform chemical tests to identify carbohydrates, lipids, and proteins.
4. Explain the mode of enzyme action and the basis of enzyme specificity.
5. Define all terms in bold print.

Life at its most fundamental level consists of complex chemical reactions, so having some understanding of the chemicals in living organisms and the nature of chemical reactions is important in your study of biology.

Chemical substances are classified into two groups: elements and compounds. An **element** is a substance that cannot be broken down by chemical means into any simpler substance. Table 4.1 lists the most common elements found in living organisms. An **atom** is the smallest unit of an element that retains the properties (characteristics) of the element.

Two or more elements may combine to form a **compound.** Water (H_2O) and table salt (NaCl)

are simple compounds; carbohydrates, fats, and proteins are complex compounds. The smallest unit of a compound that retains the properties of a compound is a **molecule.** It is formed of two or more atoms joined by **chemical bonds.**

In this exercise, you will investigate acids and bases, the major biological molecules composing cells, and the action of enzymes. These investigations are best done by working in groups of two to four students.

Assignment 1

Complete item 1 on Laboratory Report 4 that begins on page 311.

ACIDS, BASES, AND pH

An **acid** is a substance that releases **hydrogen ions** (H^+) when dissolved in water. The greater the degree of dissociation, the greater is the strength of the acid. Hydrogen chloride is a strong acid.

$$HCl \rightarrow H^+ + Cl^+$$

A **base** is a substance that releases **hydroxide ions** (OH^-) when dissolved in water. The greater the degree of dissociation, the stronger is the base. Sodium hydroxide is a strong base.

$$NaOH \rightarrow Na^+ + OH^-$$

TABLE 4.1
Common Elements in the Human Body

Element	Symbol	Percentage*
Oxygen	O	65
Carbon	C	18
Hydrogen	H	10
Nitrogen	N	3
Calcium	Ca	2
Phosphorus	P	1
Potassium	K	0.35
Sulfur	S	0.25
Sodium	Na	0.15
Chlorine	Cl	0.15
Magnesium	Mg	0.05
Iron	Fe	0.004
Iodine	I	0.0004

* By weight.

Chemists use a **pH scale** to indicate the strength of acids and bases. Figure 4.1 shows the scale, which ranges from 0 to 14 and indicates the proportionate concentrations of hydrogen and hydroxide ions at the various pH values. When the concentration of hydrogen ions increases, the concentration of hydroxide ions decreases, and vice versa. A change of one (1.0) in the pH number indicates a 10-fold change in the hydrogen ion concentration because the pH scale is a logarithmic scale.

Pure water has a pH of 7, the point where the concentrations of hydrogen and hydroxide ions are equal. Observe this in Figure 4.1. Acids have a pH less than 7; bases have a pH greater than 7. The greater the concentration of hydrogen ions, the stronger the acid and the *lower* the pH number. The greater the concentration of hydroxide ions, the stronger the base and the *higher* the pH number.

Living organisms are sensitive to the concentrations of hydrogen and hydroxide ions and must maintain the pH of their cells within narrow limits. For example, your blood is kept very close to pH 7.4. Organisms control the pH of cellular and body fluids by buffers. A **buffer** is a compound or a combination of compounds that can combine with or release hydrogen ions to keep the pH of a solution relatively constant.

Materials

Per student group
Beakers, 100 ml, 2
Dropping bottles of:
 bromthymol blue
 hydrochloric acid (HCl), 1.0%
Graduated cylinder, 50 ml
pH test papers, wide and narrow ranges

Per lab
Buffer solution, pH 7, 1,000 ml
Distilled water, pH 7, 1,000 ml
Dropping bottles of:
 Alka-Seltzer solution
 detergent solution
 household ammonia
 lemon juice
 mouthwash
 white vinegar
 unknowns

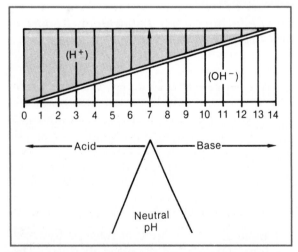

Figure 4.1. The pH scale. The diagonal line indicates the proportionate amount of hydrogen (H^+) and hydroxide (OH^-) ions at each pH value.

Assignment 2

1. *Complete items 2a and 2b on the laboratory report.*
2. Use pH papers to determine the pH of the solutions that your instructor has placed in dropping bottles. Place 1 drop of the solution being tested on a small strip of pH test paper over a paper towel. Compare the color of the

wet pH test paper with the color code on the dispenser to determine the pH. First, use a wide-range pH test paper to determine the approximate pH. Then use a narrow-range test paper for the final pH determination. ***Record your results in item 2c on the laboratory report.***

3. Now you will observe the effect of a buffer. Place 25 ml of (a) distilled water at pH 7 and (b) buffer solution at pH 7 in separate labeled beakers. Use pH papers to verify the pH of each.
4. Add 6 drops of bromthymol blue to each beaker and mix by swirling the liquids. Place the beakers on a white sheet of paper. Each liquid should be a pale blue. Bromthymol blue is a pH indicator that is blue at pH 7.6 and yellow at pH 6.0.
5. Starting with the distilled water, add 1.0% HCl to the water, drop by drop, and mix thoroughly by swirling the water after each drop. *Count and record* the number of drops required to turn the distilled water yellow (pH 6).
6. Use the same procedure to determine the number of drops of 1.0% HCl required to turn the buffer solution yellow (pH 6).
7. ***Complete item 2d on the laboratory report.***

IDENTIFYING BIOLOGICAL MOLECULES

Carbohydrates, lipids, and proteins are complex organic compounds found in living organisms. Carbon atoms form the basic framework of their molecules. In this section, you will analyze plant and animal materials for the presence of these compounds, which constitute the basic organic foods of heterotrophs. Work in groups of two to four students. If time is limited, your instructor will assign certain portions to separate groups, with all groups sharing the data.

Carbohydrates

Carbohydrates are formed of carbon, hydrogen, and oxygen, and the ratio of hydrogen to oxygen is 2:1. **Monosaccharides** are simple sugars containing three to seven carbon atoms, and they serve as building blocks for more complex carbohydrates. The combination of two monosaccharides forms a **disaccharide** sugar, and the union of many monosaccharides forms a **polysaccharide.** Six-carbon sugars, such as glucose, are primary energy sources for living organisms. Starch in plants and glycogen in animals are polysaccharides used for nutrient storage. See Figure 4.2.

Use the iodine test for starch and Benedict's test for reducing sugars (most 6-carbon and some 12-carbon sugars) to identify the presence of these carbohydrates.

Iodine Test for Starch

1. Place one dropper (1 ml) of the liquid to be tested in a clean test tube. (A "dropper" means *one dropper full* of liquid.)
2. Add 3 drops of iodine solution to the liquid in the test tube and shake gently to mix. If the substance being tested is not a liquid (e.g., a potato), crush a small piece in a plastic Petri dish and add 3 drops of iodine solution.
3. A gray to blue-black coloration of the liquid indicates the presence of starch in that order of increasing concentration.

Benedict's Test for Reducing Sugars

1. Place 1 dropper (1 ml) of the liquid to be tested in a clean test tube.
2. Add 3 drops of Benedict's solution to the liquid in the test tube and shake gently to mix. If the substance being tested is not a liquid (e.g., a potato), macerate (grind to a pulp) a small piece with a mortar and pestle in a dropper of water. Then use a medicine dropper to transfer the liquid to a test tube. Add 3 drops of Benedict's solution and shake to mix.
3. Heat the tube to near boiling in a water bath for 2–3 min. (A water bath consists of a beaker that is half-full of water, heated on a hot plate or over a burner. The test tube is placed in the beaker so that it is heated by the water in the beaker rather than directly by the heat source. Add 3 or 4 boiling chips to the beaker to prevent spatter of hot water.)
4. A light green, yellow, orange, or brick-red coloration of the liquid indicates the presence of reducing sugars in that order of increasing concentration.

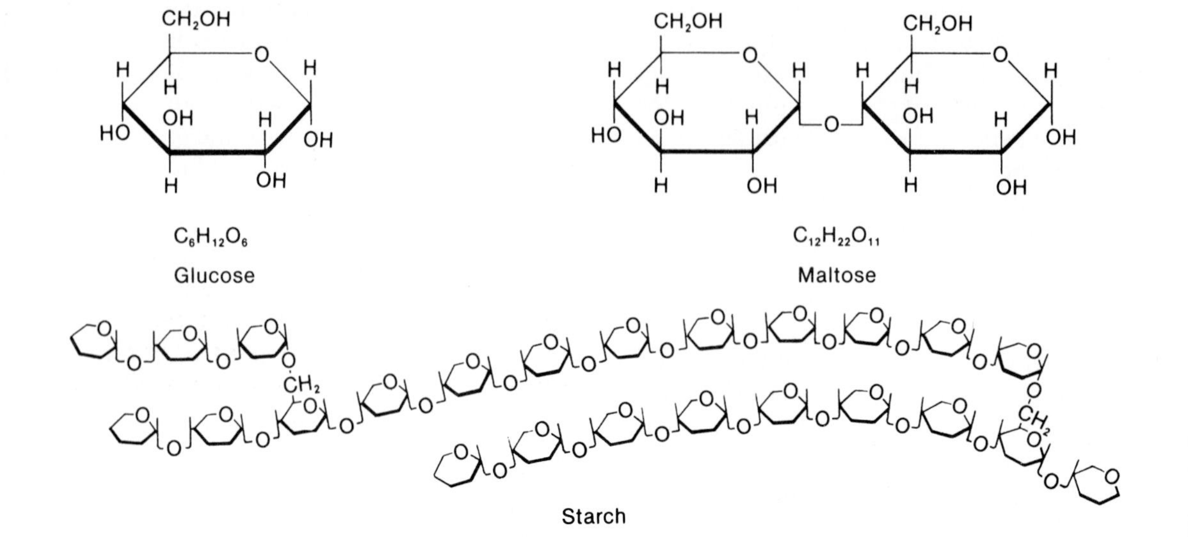

Figure 4.2. Carbohydrates

Lipids

Lipids include an array of oily and waxy substances, but the most familiar are the neutral fats (triglycerides). They are composed of carbon, oxygen, and hydrogen and have fewer oxygen atoms than carbohydrates. A fat molecule consists of three long fatty acid chains joined to a single glycerol molecule. See Figure 4.3. Note the difference in the carbon-to-carbon bonding between saturated and unsaturated fats. Fats serve as an important means of nutrient storage in organisms and are not soluble in water.

The presence of lipids may be determined by a paper spot test and Sudan IV, a dye that selectively stains lipids.

Paper Sport Test

1. Place a drop of the liquid to be tested on a piece of paper. Blot it with a paper towel and let it dry. If the substance to be tested is a solid, it can be rubbed on the paper. Remove the excess with a paper towel. Animal samples may need to be gently heated.
2. After the spot has dried, hold the paper up to the light and look for a permanent translucent spot on the paper indicating the presence of lipids.

Sudan IV Test

1. Place a dropper of distilled water in a clean test tube and add 3 drops of Sudan IV. Shake the tube from side to side to mix well.
2. Add 10 drops of the substance to be tested and mix again. Let the mixture stand for 5 min.
3. Lipids will be stained red and will rise to the top of the water or be suspended in tiny globules.

Proteins

Proteins are usually large, complex molecules composed of many **amino acids** joined together by **peptide bonds**. Their molecules contain nitrogen in addition to carbon, oxygen, and hydrogen. Proteins play important roles by forming structural components of cells and tissues and by serving as enzymes. See Figure 4.4.

Proteins may be identified by a Biuret test that is specific for peptide bonds.

Biuret Test

1. Place a dropper of the solution to be tested in a clean test tube.
2. Add a dropper of 15% sodium hydroxide (NaOH) and mix by shaking the tube from

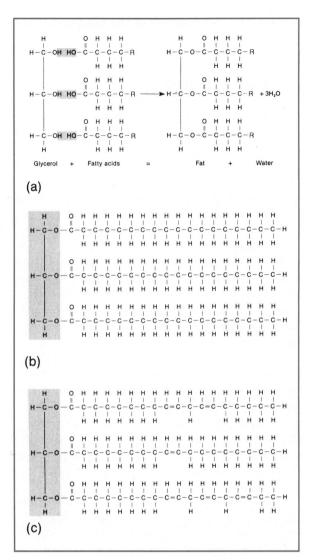

Figure 4.3. Fats. (a) The bonding of fatty acids to glycerol to form a fat. (b) Glycerol tristearate, a saturated fat. (c) Linseed oil, an unsaturated fat. R represents the remainder of the molecule.

Figure 4.4. Formation of peptide bonds. Proteins are formed of a large number of amino acids joined together by peptide bonds. R represents the remainder of the amino acids.

Materials

Per student group
Beaker, 250 ml
Boiling chips
Brown paper, 1 sheet
Dissecting instruments
Hot plate
Medicine droppers, 3
Mortar and pestle
Petri dish, plastic
Test tubes, 12
Test-tube holder
Test-tube rack
Dropping bottles of:
  Copper sulfate, 0.5%
  Benedict's solution
  Iodine solution (IKI)
  Sodium hydroxide, 15%
  Sudan IV dye

Per lab
Apple
Onion
Potato
Dropping bottles of:
  albumin, 0.5%
  corn oil
  egg white, 0.5%
  glucose, 0.5%
  soluble starch, 0.5%

side to side. *Caution:* NaOH is a caustic substance that can cause burns. If you get it on your skin or clothing, wash it off *immediately* with lots of water.

3. Add five drops of 0.5% copper sulfate ($CuSO_4$) to the tube and shake from side to side to mix. Observe after 3 min.

4. A violet color indicates the presence of proteins.

Assignment 3

1. Be sure that you understand the tests to be performed.

2. Prepare a set of standards that yield positive and negative results for each test that you will use. The standards will help you interpret your results when testing substances with unknown organic compounds. Prepare the standards as follows.

   a. Obtain 8 test tubes and number them 1A through 4A and 1B through 4B and place them in a test tube rack. Add the following substances to the test tubes. You will use *droppers* (about 1 ml) of the test materials.

      1A: 1 dropper 0.5% starch
      1B: 1 dropper water
      2A: 1 dropper 0.5% glucose
      2B: 1 dropper water
      3A: 1 dropper water plus 5 drops corn oil
      3B: 1 dropper water
      4A: 1 dropper 0.5% albumin
      4B: 1 dropper water

   b. Add the reagents to the test tubes and mix the contents by shaking the tubes from side to side.

      1A and 1B: Add 3 drops iodine solution to each tube.
      2A and 2B: Add 3 drops Benedict's solution to each tube and heat both tubes in a boiling water bath for 3 minutes.
      3A and 3B: Add 3 drops of Sudan IV to each tube. Read your results after 5 minutes.
      4A and 4B: Add 1 dropper 15% sodium hydroxide and 5 drops 0.5% copper sulfate to each tube. (*Caution:* NaOH can cause burns. If spilled on skin or clothing, wash it off *immediately* with lots of water.)

   c. For the paper spot test for lipids, apply a drop of corn oil and a drop of water to different parts of a brown paper. Blot with a paper towel and write the name of each substance near its spot. Read your results after the spots have dried.

3. ***Record your results in item 3a on the laboratory report.*** Refer to the description of the tests and positive results earlier in this section.

4. Test the substances provided by your instructor to see if they contain starch, sugar, lipid, or protein. ***Record your results in item 3b and complete item 3 on the laboratory report.***

ENZYME ACTION

The chemical reactions in cells would not occur fast enough to support life without the action of enzymes. **Enzymes** are organic catalysts that greatly accelerate the rate of chemical reactions in cells by reducing the required activation energy. All chemical reactions require a certain amount of activation energy to start. For example, energy (heat) is added by a lighted match to start a fire.

Figure 4.5 contrasts the required activation energy for a chemical reaction with and without an enzyme. By lowering the required activation energy, an enzyme greatly increases the rate of a chemical reaction—up to a million times a second in some cases.

Enzymes are proteins, so each enzyme consists of a specific sequence of amino acids. Weak hydrogen bonds that form between some of the amino acids help to determine the three-dimensional shape of the enzyme, and it is this shape that allows the enzyme to fit onto a specific **substrate molecule** (substance acted upon). The enzyme and the substrate molecule must fit together like a lock and key.

The interaction of an enzyme and substrate is shown in Figure 4.6. In this reaction, the substrate is split into two products. Another way to express an enzymatic reaction is as follows:

$$E + S \rightarrow ES \rightarrow E + P$$

The *enzyme* (E) combines with the *substrate molecule* (S) to form a temporary *enzyme-substrate complex* (ES), where the specific reaction occurs. Then the *product molecules* (P) separate from the enzyme, and the unchanged enzyme is recycled to combine with another substrate molecule. Note that the enzyme is not altered in the reaction, so a few enzyme molecules can catalyze a great number of reactions.

An enzyme is inactivated by a change in shape, and its shape is altered by anything that disrupts the pattern of hydrogen bonding. For example, many enzymes function best within

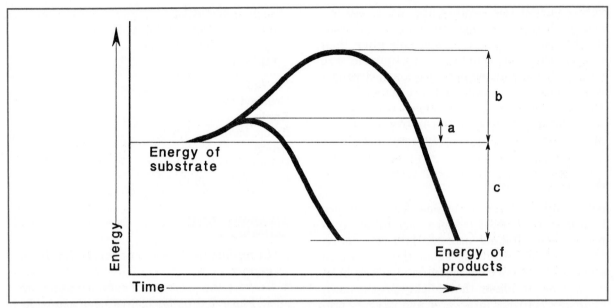

Figure 4.5. Activation energy and enzymes. (a) Activation energy required with enzyme. (b) Activation energy required without enzyme. (c) Net energy released by the reaction.

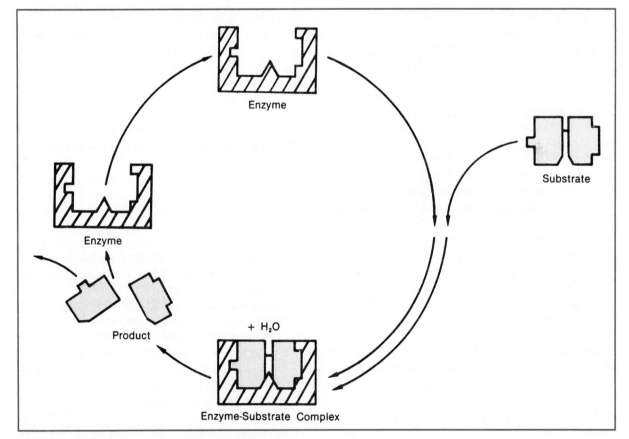

Figure 4.6. The mechanism of enzyme action

rather narrow temperature and pH ranges, because substantial changes in temperature or pH disrupt their hydrogen bonds and alter their shapes. However, some enzymes function well over rather broad temperature and pH ranges because their hydrogen bonds are not easily disrupted. It is the unique hydrogen bonding pattern that determines the sensitivity of each enzyme to changes in temperature and pH.

Action of Catalase

All organisms using molecular oxygen produce hydrogen peroxide (H_2O_2) as a harmful by-product of some cellular reactions. Hydrogen peroxide is a strong oxidizing agent that can cause serious damage to cells. Fortunately, cells have an enzyme, **catalase,** that quickly breaks down hydrogen peroxide into water and oxygen, preventing cellular damage. Catalase is especially abundant in the liver cells of humans and other vertebrates, because the liver is an organ that detoxifies many harmful substances, including peroxides.

$$\text{Hydrogen peroxide} \xrightarrow{\text{(catalase)}} \text{Water} + \text{Oxygen}$$
$$2H_2O_2 \qquad\qquad H_2O \qquad O_2$$

In a cell, oxygen released by the above reaction is used for other cellular processes, but when the reaction occurs in a test tube, oxygen gas bubbles to the surface, producing a layer of foam on the surface of the peroxide. The amount of foam produced and the speed with which it forms are measures of catalase activity. In the following experiments, you will determine the degree of catalase action by measuring the thickness of the foam layer.

Materials

Per student group
Beakers, 250 ml, 3
Celsius thermometer
Glass-marking pen
Hot plate
Medicine dropper
Metric ruler
Mortar and pestle
Scalpel or knife
Test tubes, 6
Test-tube rack

Dropping bottles of:
buffer solutions, pH 4, 6, 8, 10
hydrogen peroxide, 3%
tap water

Per lab
Crushed ice
Apple
Liver, 1/8 lb
Onion
Potato
Ground beef, 1/8 lb

Assignment 4

1. ***Complete items 4a to 4e on the laboratory report.***
2. Test for the presence of catalase as follows.
   a. Place 1 dropper of hydrogen peroxide in each of 6 numbered test tubes.
   b. Use a mortar and pestle to macerate (grind to a pulp) a small amount (0.5 cm³) of beef liver in 2 droppers of tap water.
   c. Use a medicine dropper to add 1 drop of tap water to Tube 1. Use a different medicine dropper to add 1 drop of liver extract to Tube 2. (The drop should land directly in the peroxide rather than running down the side of the tube.) What happens?
   d. Measure the thickness of the foam layer (mm) in each tube after 1 minute and ***record its thickness in item 4f on the laboratory report.***
   e. Use the same procedure to see if these tissues contain catalase: ground beef, apple, onion, and potato. Measure the thickness of the foam layers after 1 minute. Wash the dropper and mortar and pestle thoroughly after macerating each tissue.
   f. ***Complete item 4f on the laboratory report.***
3. Determine the effect of cold and hot temperatures on liver catalase action as follows.
   a. Macerate a cube (0.5 cm³) of liver in 2 droppers of tap water using a mortar and pestle as before. Place 1 drop of the liver extract and 1 dropper of tap water in each of 4 test tubes numbered 1A, 2A, 3A, and 4A.
   b. Place 1 dropper of hydrogen peroxide in each of 4 test tubes numbered 1B, 2B, 3B, and 4B.

c. Place all of the tubes for 5 minutes as follows.

> Tubes 1A and 1B: in a beaker of crushed ice.
>
> Tubes 2A and 2B: in a test tube rack at room temperature.
>
> Tubes 3A and 3B: in a beaker of 70 °C water.
>
> Tubes 4A and 4B: in a boiling water bath.

Measure the temperature of exposure for each tube.

d. After 5 minutes, pour the peroxide from Tubes 1B, 2B, 3B, and 4B into corresponding Tubes 1A, 2A, 3A, and 4A. Measure the thickness of the foam layer after 1 minute.

e. ***Complete item 4g on the laboratory report.***

4. Determine the effect of pH on liver catalase activity as follows:

a. Macerate a cube (0.5 cm³) of liver in 2 droppers of tap water using a mortar and pestle as before. Place 1 drop of the liver extract in each of 4 test tubes numbered 1, 2, 3, and 4.

b. Add to these tubes 1 dropper of the following buffers.

> Tube 1: pH 4
>
> Tube 2: pH 6
>
> Tube 3: pH 8
>
> Tube 4: pH 10

Place the tubes in a test tube rack for 5 minutes.

c. After 5 minutes, add 1 dropper of hydrogen peroxide to each tube. Measure the thickness of the foam layer after 1 minute.

d. ***Complete the laboratory report.***

5

DIFFUSION AND OSMOSIS

Materials constantly move into and out of cells. The **selective permeability** of the cell membrane permits some materials to pass through it, but prevents others. Furthermore, the materials that can pass through the membrane may change from moment to moment.

Materials move through a cell membrane by two different processes. **Passive transport,** commonly called diffusion, results from the normal, random motion of molecules and does not require an expenditure of energy by the cell. **Active transport** requires the use of energy by the cell, does not depend on molecular motion, and may move materials either with or against a **concentration gradient.** In a similar way, materials that would normally diffuse out of a cell may be prevented from doing so by **active retention.**

You may wish to refresh your understanding of the scientific method in Exercise 1 since you will be using it in this exercise. A general hypothesis, prediction, and null hypothesis are provided for each experiment to help you think through the steps of the scientific method.

BROWNIAN MOVEMENT

Molecules of liquids and gases are in constant, random motion. A molecule moves in a straight-line path until it bumps into another molecule, and then it bounces off into a different straight-line course. Since this motion is temperature dependent, the higher the temperature, the greater the molecular movement. This movement cannot be observed directly. However, tiny particles suspended in a liquid may be observed with a microscope as they are moved in a random fashion due to bombardment by molecules composing the liquid. This vibratory movement is indirect evidence of molecular motion and is called **Brownian movement** after Robert Brown, who first described it in 1827.

Materials

Per student
Compound microscope

Per lab
Dissecting needles
Microscope slides and cover glasses
Powdered carmine dye
Dropping bottles of detergent-water solution

1. Place a drop of water-detergent solution on a clean slide.
2. Dip a *dry* tip of a dissecting needle into powdered carmine, and, while holding the tip above the drop of fluid, tap the needle with your finger to shake only a few dry particles into the drop. Add a cover glass.
3. Examine the dye particles at 400× to observe Brownian movement.
4. ***Complete item 1 on Laboratory Report 5 that begins on page 315.***

DIFFUSION

The constant, random motion of molecules is what enables diffusion to occur. **Diffusion** is the net movement of the same kind of molecules from an area of their higher concentration to an area of their lower concentration. Thus, the molecules move down a **concentration gradient.** Molecules that initially are unequally distributed in a liquid or gas tend to move by diffusion throughout the medium until they are equally distributed. Diffusion is an important means of distributing materials within cells and of passively moving substances through cell membranes.

Diffusion and Temperature

General hypothesis: The rate of diffusion is affected by temperature.

Prediction: If temperature affects the rate of diffusion, then the same water-soluble substance placed in water at different temperatures will diffuse at different rates.

Null hypothesis: There is no difference in the rates of diffusion of potassium permanganate

molecules in water when exposed to two different temperatures.

You are to test the null hypothesis by performing a controlled experiment. The temperature of water in each beaker will each serve as the control for the other. There is only one independent variable, temperature. All other variables are controlled.

Materials

Per student group
Hot plate
Beakers, 250 ml, 2
Beaker tongs
Celsius thermometer

Per lab
Forceps
Granules of potassium permanganate
Ice water

1. ***Complete items 2a to 2c on the laboratory report.***
2. Fill a beaker two-thirds full with water and heat it on a hot plate to about 50 °C. Use beaker tongs to place the beaker on a pad of paper towels.
3. Place an equal amount of ice water (but no ice) in another beaker. Record the temperature of the water in each beaker.
4. Keeping both beakers motionless so that the water is also motionless, add a granule of potassium permanganate to each. Record the time. Observe the rate of diffusion by the potassium permanganate molecules during a 15-min period. Is the rate identical in each beaker?
5. ***Complete item 2 on the laboratory report.***

Diffusion and Molecular Mass

General hypothesis: The rate of diffusion is affected by the molecular mass of the substance.

Prediction: If molecular mass affects the rate of diffusion, then heavier molecules will diffuse more slowly than lighter molecules.

Null hypothesis: There is no difference in diffusion rates of molecules of potassium permanganate (mol. mass 158) and methylene blue (mol. mass 320) when placed in water.

You are to test the null hypothesis by measuring the rate of diffusion of potassium permanganate and methylene blue through an agar gel that is about 98% water.

Materials

Per student group
Petri dish of agar gel

Per lab
Forceps
Granules of:
  methylene blue
  potassium permanganate

Assignment 3

1. Place equal-sized granules of potassium permanganate and methylene blue about 5 cm apart on an agar plate. Gently press the granules into the agar with forceps to assure good contact. Record the time.
2. After 1 hr, determine the distance each substance has diffused by measuring the diameter of each colored circle on the agar plate.
3. ***Complete item 3 on the laboratory report.***

Diffusion and Molecular Size

At times, materials seem to diffuse through cell membranes if their molecules are sufficiently small while larger molecules are prevented from passing through. This selectivity based on molecular size can be observed in an experiment in which a cellulose membrane simulates a cell membrane. The cellulose membrane has many microscopic pores scattered over its surface, and molecules smaller than the pores are able to pass through the membrane.

You will use the iodine test for starch and Benedict's test for reducing sugars to determine the results of the experiment.

Starch Test

1. Place one full dropper of the substance to be tested in a clean test tube.
2. Add 3 drops of iodine solution to the test tube, and shake to mix. A gray to blue-black coloration indicates the presence of starch in that order of increasing concentration.

Benedict's Test

1. Place one full dropper of the substance to be tested in a clean test tube.
2. Add 5 drops of Benedict's solution to the test tube, and mix by shaking.
3. Heat to near boiling for 2–3 min in a boiling water bath. A water bath is set up by placing the test tube in a beaker, which is about half full of water, and by heating the beaker on a hot plate. Add 3 or 4 boiling chips to the beaker to prevent spatter of hot water. In this way the test tube is heated by the water in the beaker rather than directly by the hot plate. A light green, yellow, orange, or brick-red coloration of the liquid in the test tube indicates the presence of reducing sugars in that order of increasing concentration.

Materials

Per student group
Hot plate
Beaker, 250 ml
Boiling chips
Rubber band
Test tube, 24 × 200 mm
Test tubes, 14 × 150 mm, 2
Test-tube holder
Test-tube rack
Dropping bottle of:
  Benedict's solution
  iodine solution (IKI)

Per lab
Cellulose tubing, 1″ width
Flasks or squeeze bottles of:
  glucose, 20%
  soluble starch, 0.1%

Assignment 4

1. Set up the experiment as shown in Figure 5.1. *Be certain* to rinse the outside of the sac before inserting it into the test tube.
2. Place the test tube in a beaker or rack for 20 min.
3. Examine your test tube and record any color change that has occurred.
4. Test the solution outside the sac to see if glucose diffused from the sac. Place one dropper of the solution in a clean test tube and perform Benedict's test.

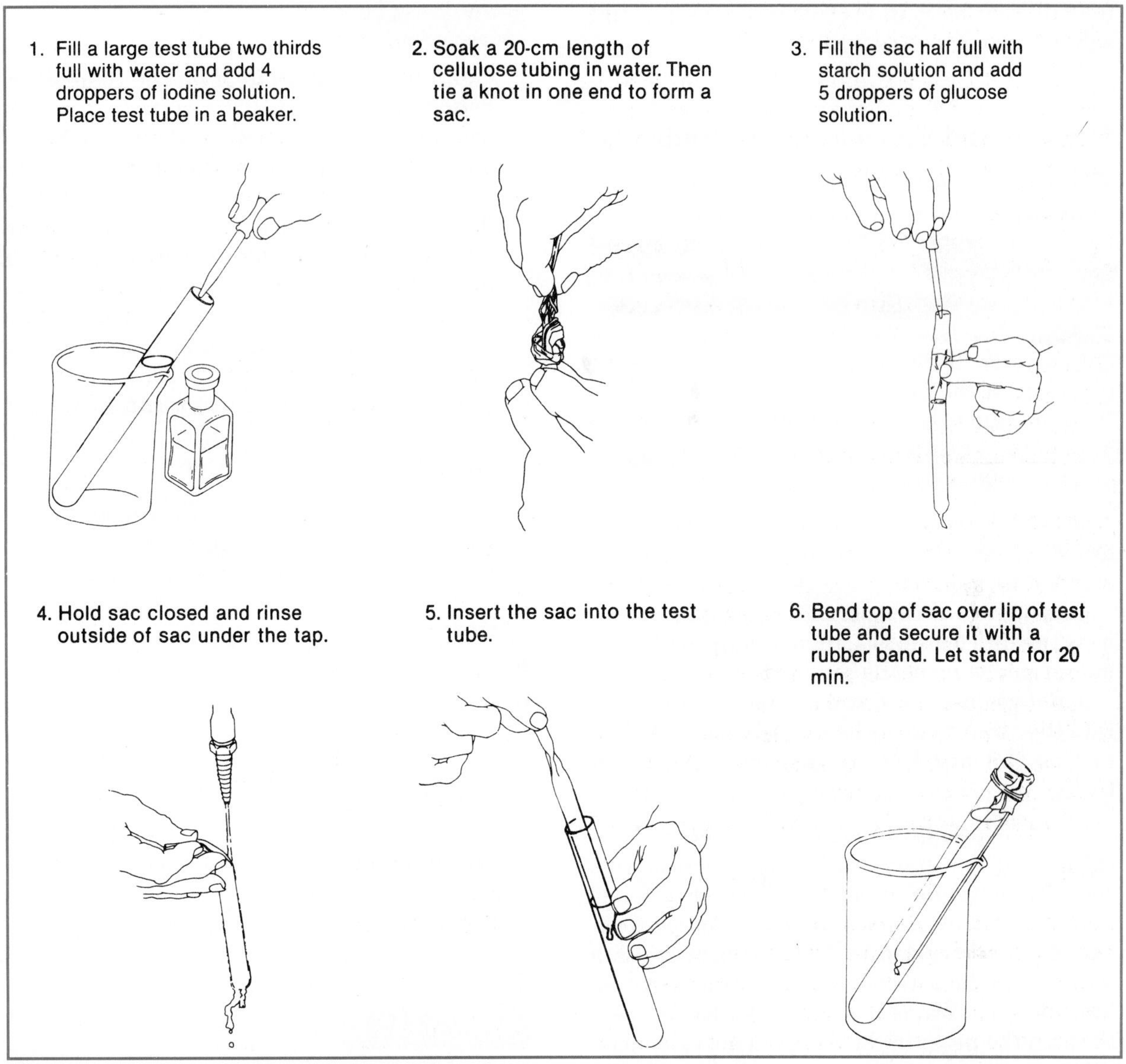

1. Fill a large test tube two thirds full with water and add 4 droppers of iodine solution. Place test tube in a beaker.

2. Soak a 20-cm length of cellulose tubing in water. Then tie a knot in one end to form a sac.

3. Fill the sac half full with starch solution and add 5 droppers of glucose solution.

4. Hold sac closed and rinse outside of sac under the tap.

5. Insert the sac into the test tube.

6. Bend top of sac over lip of test tube and secure it with a rubber band. Let stand for 20 min.

Figure 5.1. Preparation of an experiment to study the effect of molecular size on the diffusion of molecules through a cellulose membrane

5. ***Complete item 4 on the laboratory report.***
6. Clean the test tubes thoroughly using a test-tube brush.

OSMOSIS

Water is essential for life. It is the **solvent** of living systems and is the aqueous medium in which the chemical reactions of life occur. Substances dissolved in a solvent are called **solutes.** Water is the most abundant substance in cells, and it contains both organic and inorganic solutes. Water is a small molecule that diffuses freely in and out of cells in accordance with the concentration of water inside and outside the cells. **Osmosis** is simply the diffusion of water through a semipermeable or selectively permeable membrane. Like all substances, water diffuses down a concentration gradient. It moves from an area of higher concentration of water to an area of lower concentration of water. Remember that the concentration of water increases as the concentration of solutes decreases. Thus, a

5% salt solution contains 95% water while a 10% salt solution contains 90% water.

Osmosis and Concentration Gradients

When unequal solute concentrations in aqueous solutions are on each side of a semipermeable membrane, water always moves from the **hypotonic solution** (lower solute concentration) into the **hypertonic solution** (higher solute concentration), that is, down the concentration gradient of water. If the concentrations of solutes on each side of the membrane are equal, the solutions are said to be **isotonic solutions.** They are at **osmotic equilibrium,** and no net movement of water occurs through the membrane.

General hypothesis: The rate of osmosis is affected by the concentration gradient of water across a membrane.

Prediction: If the concentration gradient affects the rate of osmosis, then osmosis will occur more rapidly when the gradient is greater.

Null hypothesis: There is no difference in the rate that water moves by osmosis into sacs containing 20% and 10% sucrose solutions, respectively.

In this experiment, you will test the null hypothesis by immersing sacs of 20% sucrose, 10% sucrose, and water in beakers of water to determine their gain in mass (weight). The percentage of mass increase in 1 hr for each sac is a measure of the rate of osmosis as water moves into the sacs. Water molecules can move freely through the pores in the sacs, but sucrose molecules are too large to pass through.

Materials

Per student group
Beakers, 400 ml, 3
Cellulose tubing, 1″ diameter
Marking pen, felt-tip, waterproof
String

Per lab
Squeeze bottles of:
    20% sucrose, 4 per lab
    10% sucrose, 4 per lab
Triple-beam balance

Assignment 5

1. *Complete items 5a to 5f on the laboratory report.*
2. Set up the experiment as shown in Figure 5.2. Using a felt-tip marker pen, mark the string of each sac to code for its contents. Blot the sacs and measure their mass to the nearest 0.1 g. *Record your measurements in item 5g on the laboratory report.*
3. Set the beakers with their sacs aside for 1 hr. Then blot the sacs dry and measure their mass again to the nearest 0.1 g.
4. *Complete item 5 on the laboratory report.*

Osmosis and Living Cells

In this section, you will observe the movement of water into and out of living cells that are exposed to hypertonic and hypotonic solutions.

Materials

Per student group
Compound microscope
Dropping bottles of:
    10% NaCl solution
    distilled water

Per lab
Microscope slides and cover glasses
Elodea shoots
Osmosis demonstration using celery sticks

Assignment 6

1. *Complete items 6a to 6c on the laboratory report.*
2. Examine the osmosis demonstration using celery sticks. Celery sticks were placed in (a) distilled water and (b) 10% salt solution at the beginning of the lab session. Evaluate the crispness and flexibility of the celery sticks. *Complete item 6d on the laboratory report.*
3. Mount an *Elodea* leaf in tap water. Note the size of the central vacuoles and the location of the protoplasm in the cells.
4. *In the space for item 6e on the laboratory report, draw an optical midsection view of a normal* **Elodea** *cell.* Show the

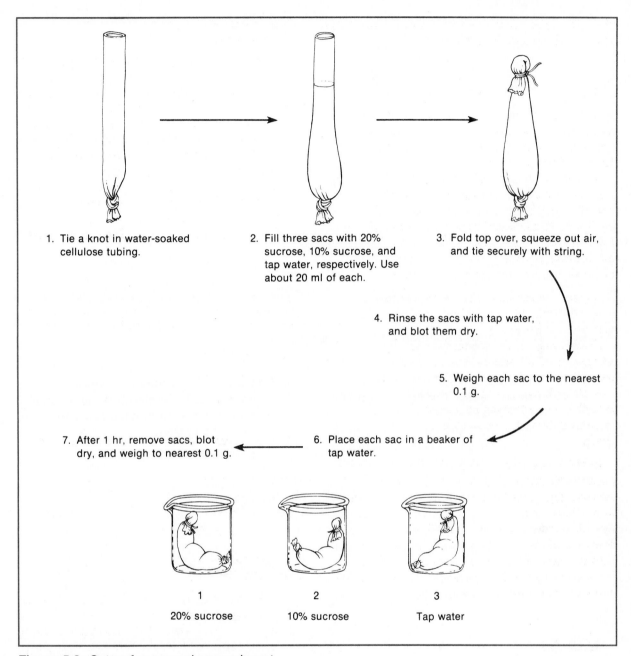

1. Tie a knot in water-soaked cellulose tubing.

2. Fill three sacs with 20% sucrose, 10% sucrose, and tap water, respectively. Use about 20 ml of each.

3. Fold top over, squeeze out air, and tie securely with string.

4. Rinse the sacs with tap water, and blot them dry.

5. Weigh each sac to the nearest 0.1 g.

7. After 1 hr, remove sacs, blot dry, and weigh to nearest 0.1 g.

6. Place each sac in a beaker of tap water.

1
20% sucrose

2
10% sucrose

3
Tap water

Figure 5.2. Setup for osmosis experiment

distribution of the protoplasm and the central vacuole.

5. Place 1 drop of 10% salt solution at the edge of the cover glass, and draw it under the cover glass as shown in Figure 5.3. Repeat the process. Now the leaf is mounted in salt solution. This experiment allows you to observe at the cellular level the effect of water moving into or out of plant cells. What do you think will happen? *Central vacuole size decrease as water moves from plant to mount solution. Salt will move to plant*

6. Examine the cells with your microscope. Observe the protoplasm as it shrinks. The shrinking of protoplasm caused by the loss of water is called **plasmolysis.**

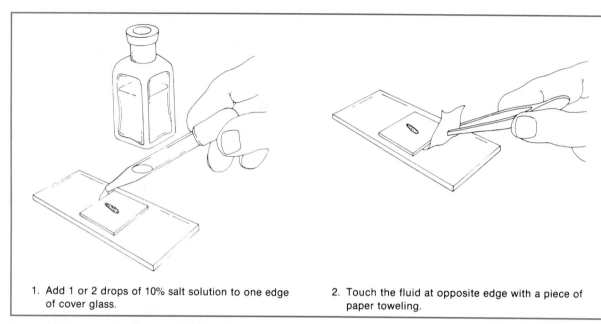

1. Add 1 or 2 drops of 10% salt solution to one edge of cover glass.

2. Touch the fluid at opposite edge with a piece of paper toweling.

Figure 5.3. How to draw a solution under a cover glass

7. ***Make a drawing of a cell mounted in salt solution in the space for item 6e on the laboratory report.***

8. Remove the leaf from the salt solution, rinse it, and remount it on a clean slide in tap water. Quickly observe it with your microscope. What happens?

9. ***Complete the laboratory report.***

10. Clean and dry the objectives and stage of your microscope to remove any salt water that may be present.

6

PHOTOSYNTHESIS

Living organisms require a constant source of energy to operate their metabolic (living) processes. The ultimate source of energy is the sun. Except for the few chemosynthetic bacteria, all organisms depend on the conversion of **light energy** into the **chemical energy** of organic nutrients by **photosynthesis.** The photosynthesizers are plants, plantlike protists, and cyanobacteria. All of these organisms contain **chlorophyll,** the green pigment that captures light energy and converts it into the chemical energy of organic molecules. Photosynthetic organisms are called **autotrophs** (self-feeders) since they are able to synthesize organic nutrients from inorganic materials. **Heterotrophs** (feeding on others) must obtain their organic nutrients by feeding on other organisms.

The summary equation for photosynthesis is shown in Figure 6.1. The process is not as simple as the summary equation suggests, however. It actually is a complex series of chemical reactions that may be separated into two parts: (1) a light reaction and (2) a dark reaction.

During the **light reaction,** chlorophyll absorbs light energy and converts it into chemical energy, and in the process, water is split, releasing oxygen molecules. The **dark reaction** normally occurs in the presence of light, but it does not *require* light energy, hence its name. In this reaction, chemical energy, which was formed in the light reaction, is used to synthesize glucose by combining (1) carbon dioxide from the atmosphere and (2) hydrogen split from water in the light reaction. Consult your text for a complete discussion of photosynthesis.

When **glucose,** a six-carbon sugar, is formed in a photosynthesizing cell, much of it is converted into **starch** for temporary storage. Starch is a polysaccharide composed of many glucose units. Using glucose as the base material, plants can synthesize all other organic compounds required for their metabolic and structural needs.

Although photosynthesis occurs in all cells containing chlorophyll, the leaf is the primary organ of photosynthesis in vascular plants. In this exercise, you will use the presence of starch in leaf cells as indirect evidence for photosynthe-

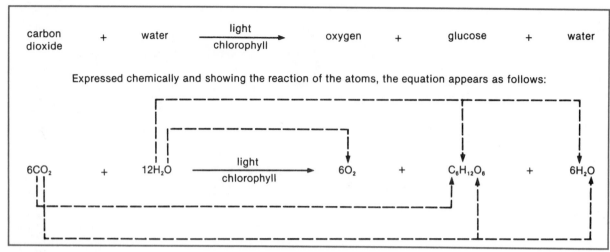

Figure 6.1. Summary equation of photosynthesis

sis. The presence of sugars cannot be used as evidence of photosynthesis because their presence may be due to transport from other parts of a plant.

Materials

Per student group
Alcohol
Beakers, 100 ml, 250 ml
Boiling chips
Hot plate
Petri dish
Scissors
Test tubes, 2
Test-tube rack
Dropping bottles of:
  Benedict's solution
  iodine (IKI) solution

Per lab
Setup for photosynthesis and CO_2 experiment
  (Figure 6.2)
  bell jars, 2
  glass funnels, 2
  fluorescent lamps, 2
  Petri dish
  rubber stoppers for bell jars, 1-hole
  soda lime granules
  fairy primrose (*Primula malacoides*), 4″ pots
Setup for photosynthesis and light experiment
  leaf shields
  fluorescent lamps
  fairy primrose (*Primula malacoides*), 4″ pots

Setup for photosynthesis and chlorophyll experiment
  fluorescent lamps
  Coleus plants, green and white, 4″ pots

Assignment 1

Complete item 1 on Laboratory Report 6 that begins on page 319.

CARBON DIOXIDE AND PHOTOSYNTHESIS

General hypothesis: Carbon dioxide is necessary for photosynthesis.

Prediction: If carbon dioxide is necessary for photosynthesis, then photosynthesis will not occur in the absence of CO_2.

Null hypothesis: Photosynthesis will occur in the absence of carbon dioxide.

Two healthy plants where placed in darkness for 48 hr and then placed under separate bell jars. See Figure 6.2. Plant A has been exposed to air containing CO_2. Plant B has been exposed to air from which CO_2 has been removed by soda lime. Both plants have been exposed to light for 24 hr prior to the laboratory session. You will test the null hypothesis by determining the presence or absence of starch in a leaf from each plant, since the presence of starch is an indicator of photosynthesis.

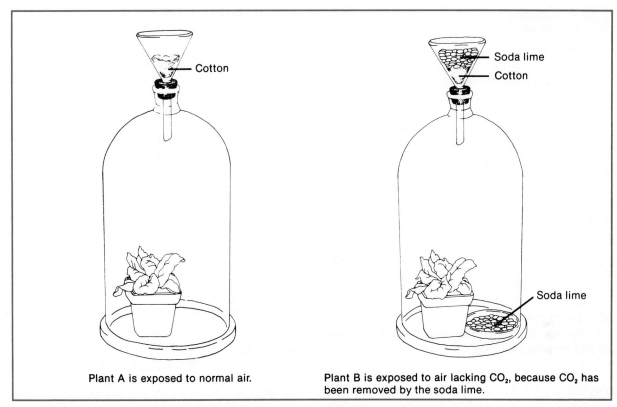

Figure 6.2. Setup for the carbon dioxide and photosynthesis experiment

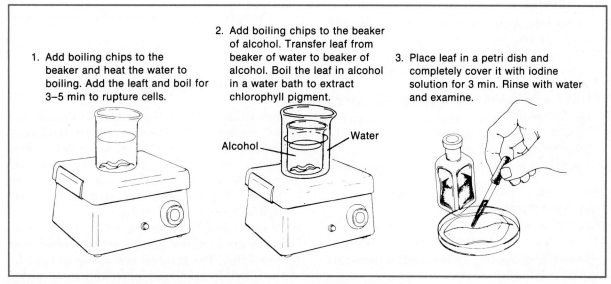

Figure 6.3. Procedure for testing a leaf for the presence of starch

Assignment 2

1. Remove a leaf from each plant. Cut off the petiole of the leaf from plant A so that you can distinguish the leaves.
2. Carefully follow the procedures in Figure 6.3 to test each leaf for the presence of starch. A gray to blue-black color indicates the presence of starch. *Caution: Alcohol is flammable; do not get it near an open flame.*
3. *Complete item 2 on the laboratory report.*

LIGHT AND PHOTOSYNTHESIS

General hypothesis: Light is necessary for photosynthesis.
Prediction: If light is necessary for photosynthesis, then photosynthesis will not occur in the absence of light.
Null hypothesis: Photosynthesis will occur in the absence of light.

Healthy green plants were placed in darkness for 48 hr. Leaf shields, that screen part of a leaf from light, were placed on the leaves, and the plants have been exposed to light for 24 hr prior to the laboratory session. You will test the null hypothesis by comparing the position of the light shield (area receiving no light) with the distribution of starch in the leaves.

Assignment 3

1. *Complete items 3a and 3b on the laboratory report.*
2. Remove a leaf from one of the plants. *Trace or sketch the leaf in item 3c, showing the position of the leaf shield.*
3. Remove the leaf shield and test the leaf for the presence of starch as shown in Figure 6.3.
4. Compare the distribution of starch and the position of the leaf shield.
5. *Complete item 3 on the laboratory report.*

CHLOROPHYLL AND PHOTOSYNTHESIS

General hypothesis: Chlorophyll is necessary for photosynthesis.
Prediction: If chlorophyll is necessary for

photosynthesis, then photosynthesis will not occur in the absence of chlorophyll.
Null hypothesis: Photosynthesis will occur in the absence of chlorophyll.

Some plants have variegated leaves that have an unequal distribution of chlorophyll. Some parts of variegated leaves lack chlorophyll. Such leaves are good subjects for investigating whether chlorophyll is necessary for photosynthesis. You will test the null hypothesis by comparing the distribution of chlorophyll and starch in variegated leaves of *Coleus* plants that have been exposed to light for 24 hr prior to the laboratory session.

Assignment 4

1. Remove a variegated leaf from a *Coleus* plant on the stock table. Cut the leaf in half along the midrib. Sketch half of the leaf, showing the distribution of chlorophyll.
2. Test one half of the leaf for the presence of starch as shown in Figure 6.3, and compare the starch and chlorophyll distributions.
3. Use scissors to separate the green and non-green portions of the other half of the leaf, and test *each portion* separately for the presence of sugar as shown in Figure 6.4. A light green, yellow, or orange coloration indicates the presence of glucose. Compare your results with the distribution of chlorophyll.
4. *Complete item 4 on the laboratory report.*

CHLOROPLAST PIGMENTS

When white light is passed through a prism, the component wavelengths are separated to form a rainbowlike spectrum that we perceive as colors ranging from violet (380 nanometers, nm) through blue, green, yellow, and orange to red (760 nm). Light energy decreases as the wavelengths increase.

When white light strikes a leaf, some wavelengths are absorbed and others are reflected or transmitted. The greatest absorption of light by chlorophyll is in the red, blue, and violet wavelengths. See Figure 6.5. Most leaves appear

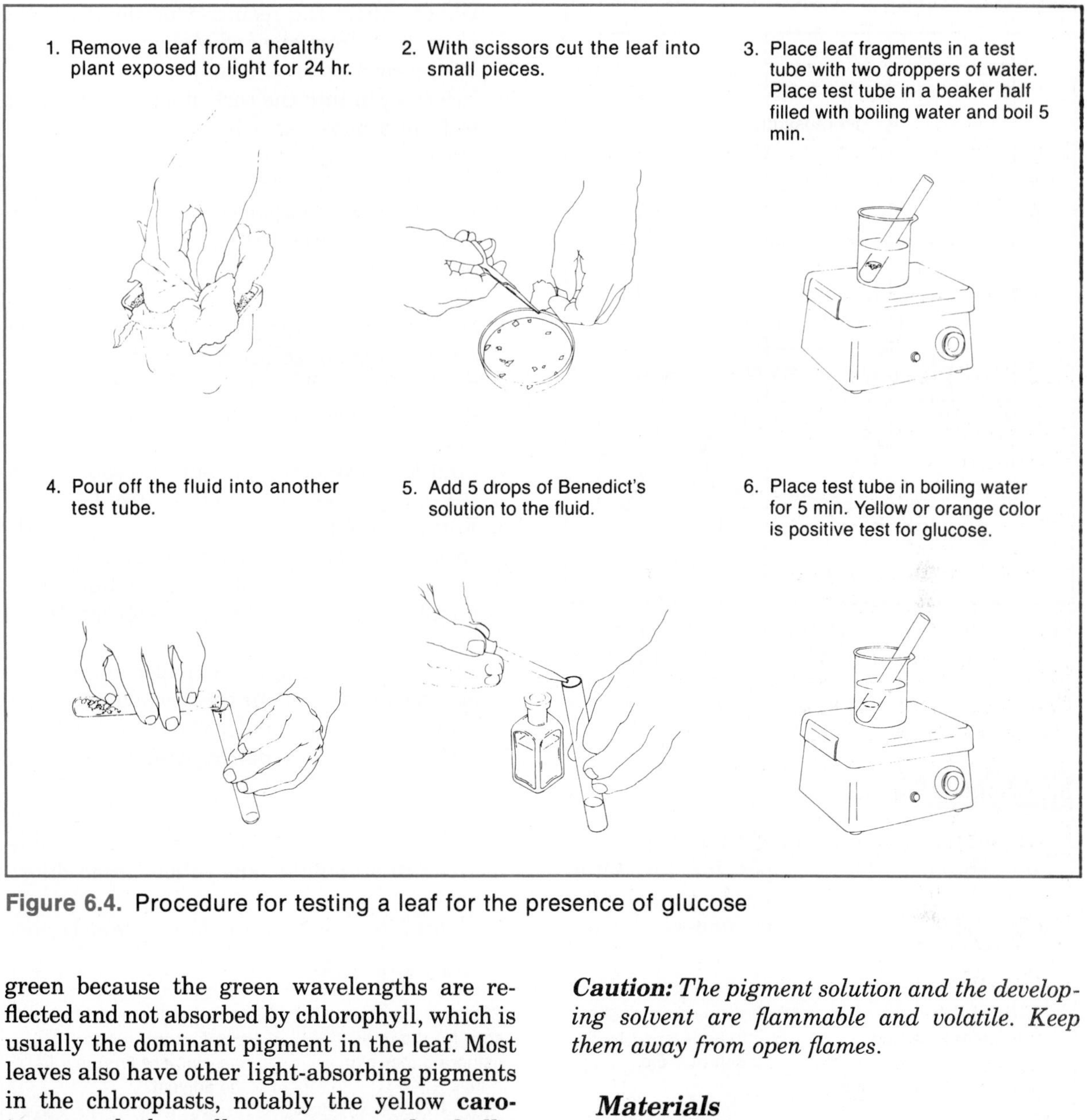

1. Remove a leaf from a healthy plant exposed to light for 24 hr.

2. With scissors cut the leaf into small pieces.

3. Place leaf fragments in a test tube with two droppers of water. Place test tube in a beaker half filled with boiling water and boil 5 min.

4. Pour off the fluid into another test tube.

5. Add 5 drops of Benedict's solution to the fluid.

6. Place test tube in boiling water for 5 min. Yellow or orange color is positive test for glucose.

Figure 6.4. Procedure for testing a leaf for the presence of glucose

green because the green wavelengths are reflected and not absorbed by chlorophyll, which is usually the dominant pigment in the leaf. Most leaves also have other light-absorbing pigments in the chloroplasts, notably the yellow **carotenes** and the yellow-orange **xanthophylls.** These pigments absorb different wavelengths of light than chlorophyll and pass the captured energy to chlorophyll.

Chloroplast pigments may be separated by **paper chromatography,** a technique that takes advantage of slight differences in solubility of the pigments. The most soluble pigment will be carried farthest by the solvent, and the least soluble pigment will be carried the shortest distance.

Your instructor has prepared a concentrated solution of chloroplast pigments for your use.

Caution: The pigment solution and the developing solvent are flammable and volatile. Keep them away from open flames.

Materials

Per student group
Chromatography jar with cork stopper
Chromatography paper, 1″ width
Paper clip
Scissors
Straight pin

Per lab
Chloroplast pigment solution
Chromatography developing solvent
Chromatography waste collecting jar
Paintbrushes, watercolor, fine-tipped
Pliers, pointed nose

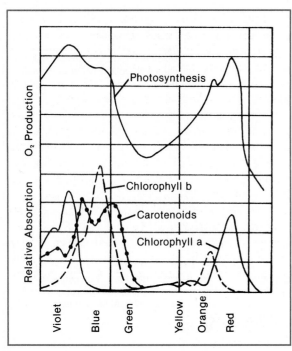

Figure 6.5. Absorption spectrum of chloroplast pigments and action spectrum of photosynthesis

1. Obtain the jar, cork, straight pin, paper clip, and chromatography paper for the setup shown in Figure 6.6. Handle the paper by the edges since oils from your fingers can ad-

versely affect the results. Cut small notches about 2 cm from one end of the paper, and cut off the corners of this end.

2. Insert a pin into the cork stopper and bend it to form a hook. Attach the paper clip to the paper and suspend it in the jar *before adding the solvent.* The end of the paper should be *just above the bottom of the jar.* Adjust the paper clip or cut off the paper to achieve this length. Then remove the paper and clip for step 3.

3. Use a small paintbrush to paint a line of chloroplast pigment solution across the paper between the notches. Let it dry. Repeat this step six times to obtain a dark pigment line.

4. Pour the developing solvent into your jar *under a fume hood* to a depth of about 1 cm.

5. Return to your workstation, and insert the cork and suspended chromatography paper into the jar. The end of the paper, but not the pigment line, should be in the solvent. *Do not move the jar for 30 min.*

6. After 30 min, remove the paper and place it on a paper towel. Pour the developing solvent into the waste jar under the fume hood.

7. ***Attach your chromatogram to your lab report*** and label the separated pigments: bright yellow carotene at the top of the paper, 1 or 2 bands of yellow xanthophylls, blue-green chlorophyll a, and yellow-green chlorophyll b.

8. ***Complete item 5 on the laboratory report.***

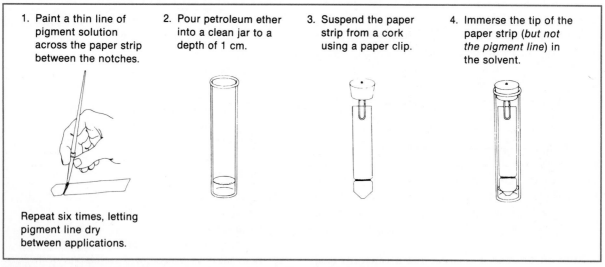

1. Paint a thin line of pigment solution across the paper strip between the notches.

Repeat six times, letting pigment line dry between applications.

2. Pour petroleum ether into a clean jar to a depth of 1 cm.

3. Suspend the paper strip from a cork using a paper clip.

4. Immerse the tip of the paper strip (*but not the pigment line*) in the solvent.

Figure 6.6. Procedure for paper chromatography of chloroplast pigments

LIGHT INTENSITY AND THE RATE OF PHOTOSYNTHESIS

You have shown that light is necessary for photosynthesis. Of course, plants are exposed to different light intensities throughout the day. In addition, some plants grow mostly in shade while others thrive in direct sunlight. Do you think that the rate of photosynthesis is affected by light intensity? Let's find out.

General hypothesis: Light intensity affects the rate of photosynthesis.

Prediction: If light intensity affects the rate of photosynthesis, then the rate of photosynthesis will be faster when a plant is exposed to bright light and slower when it is exposed to dim light.

Null hypothesis: The rate of photosynthesis will not be affected by differences in light intensity.

As noted in the summary equation of photosynthesis (Figure 6.1), oxygen molecules are one of the products. For each molecule of oxygen produced, a molecule of carbon dioxide is used in the formation of glucose. Some of the oxygen is used by the plant cells, and the excess is released into the atmosphere. This excess oxygen forms the oxygen in the atmosphere.

Since the rate of oxygen production is an indicator of the rate of photosynthesis, you will test the null hypothesis by measuring the rate of oxygen production by an *Elodea* shoot exposed to different light intensities. Light intensity decreases with distance from a light source, so you will vary the light intensity by changing the distance between the *Elodea* shoot and the light source. See Figure 6.7. A good range of light intensities is obtained by placing the *Elodea* shoot at 30 cm, 60 cm, 90 cm, and 120 cm from a 150-watt spot light. If you use a fluorescent lamp, decrease the distance by half at each setting.

Materials

Per student group
Heat filter (rectangular glass container of water)
Light source: 150-watt spot lamp or fluorescent lamp

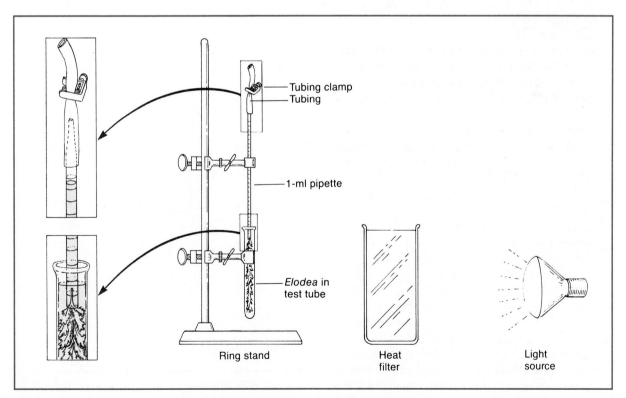

Figure 6.7. Setup for studying the effect of light intensity on the rate of photosynthesis

Meter stick
Pipette, 1 ml
Plastic tubing, 7 cm
Ring stand
Ring stand test-tube clamps
Scalpel or razor blade
Syringe and needle, 5 or 10 ml
Test tube
Tubing clamp, screw type

Per lab
Sodium bicarbonate solution, 2%
Elodea shoots

Assignment 6

1. ***Complete items 6a and 6b on the laboratory report.***
2. Fill a test tube about three-fourths full of 2% sodium bicarbonate ($NaHCO_3$) solution, which will provide an adequate concentration of CO_2.
3. Make a diagonal cut through the stem of a leafy shoot of *Elodea* about 10–13 cm from its tip. Use a sharp scalpel, being careful not to crush the stem. Insert the *Elodea* shoot, tip down, into the test tube so that the cut end is 1–2 cm below the surface of the solution. Place the tube in a test-tube holder on a ring stand. If bubbles of oxygen are released too slowly from the cut end of the stem, recut the stem at an angle to obtain a good production of bubbles.
4. Place a short piece of plastic tubing snugly over the tip (pointed end) of a 1-ml pipette as shown in Figure 6.7. Place a screw-type tubing clamp on the tubing and tighten it until it is *almost closed*. Place the pipette in a ring-stand clamp with the tip up, and lower the base of the pipette into the test tube so that the cut end of the *Elodea* shoot is inserted into the pipette. Secure the pipette in this position. See Figure 6.7. Study the 0.01-ml graduations on the pipette to ensure that you known how to read them.
5. If using a spot incandescent light source, fill a rectangular glass container with water to serve as a heat filter. The heat filter must be placed between the *Elodea* shoot and the spot lamp, about 10 cm in front of the *Elo-*

dea. A heat filter is not necessary if using a fluorescent lamp.

6. You will vary the light intensity by placing the light source at 30 cm, 60 cm, 90 cm, and 120 cm from the tube containing the *Elodea* shoot. Start at the 30-cm distance and arrange your setup as shown in Figure 6.7. Turn off the room lights.
7. After allowing 10 min for equilibration, tighten the screw clamp to close the tubing on the tip of the pipette. Insert the needle of a 5- or 10-ml syringe into the tubing and gently pull out the syringe plunger to raise the level of the fluid in the pipette to the 0.9-ml mark. Then remove the syringe.
8. Bubbles of oxygen should start forming at the cut end of the *Elodea* stem and rise into the pipette, displacing the solution.
9. Read and record the fluid level in the pipette, and record the time of the reading. Be sure to take your readings at the bottom of the meniscus, as shown in Figure 6.8. ***Record your data in the chart in item 6c on the laboratory report.***
10. *Exactly* 3 min later, record another reading. Then determine the volume of oxygen pro-

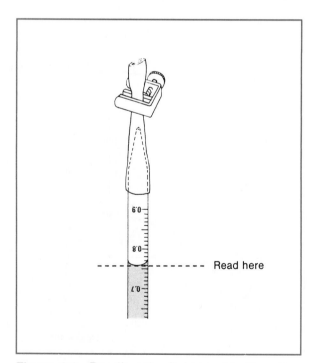

Figure 6.8. Reading the fluid level in the pipette at the meniscus

duced within the 3-min interval. Repeat the process two more times to determine oxygen production during three separate 3-min intervals at this light intensity. Then calculate the average volume of oxygen produced per minute milliliters of oxygen per minute (ml O_2/min), and record this figure in the chart in item 6c on the laboratory report.

11. Use the same procedure as in 9 and 10 to determine the rate of oxygen production at 60 cm, 90 cm, and 120 cm. Move the light source to obtain the new distance and change the water in the heat filter at each new distance. Allow 10 min for equilibration to the new distance before starting your readings.

12. ***Complete the laboratory report.***

7

CELLULAR RESPIRATION

All organisms must have a continuous supply of energy from an external source to operate their metabolic functions. The sun is the ultimate energy source, and photosynthesis converts light energy into the **chemical bond energy** of organic nutrients. The three major classes of organic nutrients are **carbohydrates, fats,** and **proteins.** For the energy stored in these molecules to be used for cellular work, the chemical bonds must be broken by **cellular respiration.**

Cellular respiration and combustion are both oxidation processes. Combustion is an *uncontrolled oxidation* that releases large amounts of heat energy due to the simultaneous breaking of many chemical bonds. For burning to occur, the substance must be heated (the addition of energy) to its combustion temperature.

In contrast, cellular respiration is an *enzymatically controlled oxidation* that breaks bonds sequentially and releases energy in small amounts so that energy may be "captured" in high-energy phosphate bonds ($\sim$P). The captured energy ($\sim$P) combines with **adenosine diphosphate (ADP)** to form **adenosine triphosphate (ATP).** ATP is the immediate source of energy for cellular work. It transfers $\sim$P to power the chemical reactions within the cell. Usable energy is always transferred as $\sim$P. Study Figure 7.1.

The two types of cellular respiration are aerobic and anaerobic. Compare the summary equation in Figure 7.2.

Most organisms depend on the **aerobic respiration** of organic nutrients for ATP production. Aerobic respiration requires oxygen and yields a net of 36 ATP molecules for each molecule of glucose respired. About 40% of the released energy is transferred to ATP while the remainder is "lost" as heat. Aerobic respiration is about twice as efficient as an automobile engine.

Anaerobic respiration does not require oxygen, but it produces only a net of 2 ATP molecules for each glucose molecule respired. Only a few bacteria and yeasts can survive on the low ATP output of anaerobic respiration, and some of these organisms are used commercially to pro-

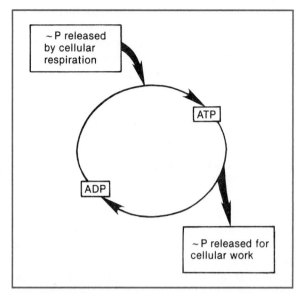

Figure 7.1. Transfer of ~P from respiration to cellular work

duce alcohol and industrial solvents. *Fermentation* is a synonym for anaerobic respiration in bacteria and fungi.

Anaerobic respiration occurs in humans, but for very brief periods of time when energy (ATP) needs for muscle contraction exceed the oxygen supply to the muscles, such as when a sprinter runs the 200-meter dash. Anaerobic respiration provides ATP for muscle contraction but results in an accumulation of lactic acid in the muscles and blood. The rapid breathing and heart rate after such exertion is the body's way of providing an increased amount of oxygen to muscles and liver to metabolize the accumulated lactic acid and to remove the excess carbon dioxide. Once this has been accomplished, both breathing and heart rate return to normal.

This exercise focuses on aerobic cellular respiration. It will be detected by the production of CO_2 and measured by the consumption of O_2.

Assignment 1

Complete item 1 on Laboratory Report 7 that begins on page 323.

Materials

Per student group
Drinking straws, 2
Glass-marking pen
Glass tubing, 2-cm lengths, 9
Medicine dropper
Rubber-bulbed air syringe
Test tubes, 3
Test-tube rack
Bromthymol blue, 0.004%, in dropping bottle

Per lab
Crickets, live
Pea seeds, germinating
Respiration and heat demonstration
    Celsius thermometers, 3
    vacuum bottles, 3
    cotton plugs for vacuum bottles

The equation for aerobic respiration (Figure 7.2) indicates that CO_2 is a product of the reaction. Therefore, an accumulation of CO_2 may be used as an indicator of cellular respiration. A dilute solution of bromthymol blue, a pH indicator, may be used to detect an increase in CO_2 concentration. Carbon dioxide easily dissolves in water, and, as shown in Figure 7.3, it combines with water to form carbonic acid. Carbonic acid then dissociates, releasing hydrogen ions (H^+) that, in turn, react with bromthymol blue, causing it to turn yellow.

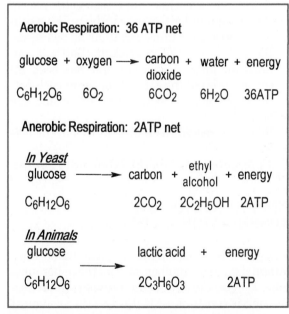

Aerobic Respiration: 36 ATP net

glucose + oxygen $\longrightarrow$ carbon dioxide + water + energy

$C_6H_{12}O_6$ $6O_2$ $6CO_2$ $6H_2O$ 36ATP

Anerobic Respiration: 2ATP net

In Yeast
glucose $\longrightarrow$ carbon dioxide + ethyl alcohol + energy

$C_6H_{12}O_6$ $2CO_2$ $2C_2H_5OH$ 2ATP

In Animals
glucose $\longrightarrow$ lactic acid + energy

$C_6H_{12}O_6$ $2C_3H_6O_3$ 2ATP

Figure 7.2. Summary equations for aerobic and anaerobic respiration

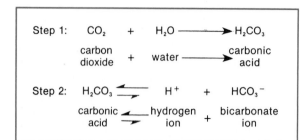

Figure 7.3. Reaction of carbon dioxide and water

Using the equation for aerobic respiration as the general hypothesis, let's see if living organisms actually release CO_2 at concentrations greater than that in atmospheric air.

Prediction: If CO_2 is a product of cellular respiration, then respiratory gases released by living organisms should turn a bromthymol blue solution yellow.

Null hypothesis: Respiratory gases released by living organisms will not turn a bromthymol blue solution yellow.

Assignment 2

1. Experiment 1 will determine if your exhaled breath contains a greater concentration of CO_2 than atmospheric air.
   a. Place 3 drops of 0.004% bromthymol blue into each of two numbered test tubes.
   b. Exhale your breath through a drinking straw into tube 1 for 3 min.
   c. Use a rubber-bulbed syringe to pump atmospheric air into tube 2 for 3 min.
   d. ***Record any color change in the bromthymol blue solutions in the table in item 2 on the laboratory report.***
2. Experiment 2 will determine if germinating pea seeds and crickets produce CO_2 concentrations greater than that in the air.
   a. Place 3 drops of 0.004% bromthymol blue into each of 3 numbered test tubes.
   b. Place several short segments of glass tubing into each tube so that pea seeds and crickets will be kept out of the solution.
   c. Place 6–10 germinating pea seeds into tube 1, place 3–6 crickets into tube 2, and place nothing else in tube 3.

d. Gently (to prevent a sudden increase in air pressure from injuring the crickets) insert a rubber stopper or cotton plug into each tube. Place the tubes in a test-tube rack for 20 min. Then observe any color change in the bromthymol blue.
   e. ***Record your results on the laboratory report and complete item 2.***

RESPIRATION AND HEAT PRODUCTION

Since about 40% of the energy released by aerobic respiration is captured in high-energy phosphate bonds of ATP, the remainder is lost as heat. Heat produced by aerobic respiration maintains normal body temperatures in humans and other homeothermic animals. Do you think simpler organisms like germinating seeds and crickets produce heat by aerobic respiration?

Assignment 3

1. Using the general hypothesis that *simple organisms produce heat by aerobic respiration,* ***write a prediction and null hypothesis to be tested in the following experiment in items 3a and 3b on the laboratory report.***
2. A few hours earlier, your instructor set up three vacuum bottles to test the null hypothesis. Bottle 1 contains germinating pea seeds. Bottle 2 contains live crickets. Bottle 3 contains air only. A thermometer has been inserted through the cotton stopper of each bottle to measure the temperature inside the bottle.
3. Read and record the temperatures in the bottles.
4. ***Complete item 3 on the laboratory report.***

TEMPERATURE AND RESPIRATION RATE

Now you will consider the *rate* of aerobic respiration, i.e., the number of reactions per unit time. Do you think that the rate of cellular respiration is constant, or is it, like other chemical reactions, affected by factors such as temperature? Let's find out.

General hypothesis: The rate of cellular respiration is affected by changes in temperature.

Prediction: If temperature affects the rate of cellular respiration, then the rate will increase as temperature increases within a normal range.

Null hypothesis: There will be no difference in the rate of cellular respiration when organisms are exposed to different temperatures within a normal range.

Germinating Peas and Crickets

The rate of oxygen consumption is a good measure of the rate of aerobic respiration, because six molecules of oxygen are consumed for each glucose molecule respired. You will test the null hypothesis by measuring the rate of oxygen consumption by germinating peas and crickets at three different temperatures. In order to allow comparisons among the organisms, it is necessary to determine the oxygen consumption per hour per gram ($O_2/hr/g$) of body mass for each organism.

These experiments are best done by groups of four students, with each group assigned to a particular temperature: 10 °C, room temperature, or 40 °C. If time is limited, your instructor may assign a different experiment to each group, with all groups sharing the data.

When your group has been formed, read through the experiments and establish a division of labor to cover the tasks. Once the experiments have been set up, one person should keep the time, one or two persons should take the readings, and one person should record the data.

Materials

Per student group
Celsius thermometer
Dropping bottle of 10% NaCl, colored
Glass marking pen
Respirometer (Figure 7.4)
    beaker, 500 or 1,000 ml
    pipettes, 1 ml, 3
    rubber stoppers for test tubes, 1-hole, 3
    test tubes, 3
Test-tube rack

Per lab
Cotton, absorbent
Crushed ice

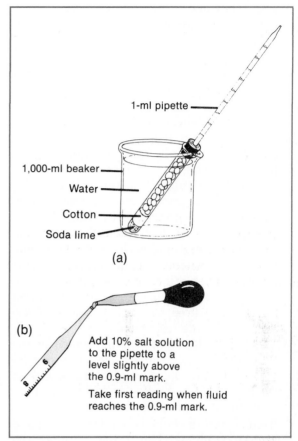

Figure 7.4. Respirometer for pea seeds and small invertebrates. (a) Using beaker as a water bath. (b) Method of adding salt solution to pipette.

Soda lime
Triple-beam balance
Water bath, 40 °C
Crickets, live
Germinating pea seeds

Assignment 4

1. Set up three numbered test tubes at the assigned temperature. The interior of each tube must be dry. Place soda lime in each tube to a depth of about 2 cm. Then insert a loose pad of cotton to separate the specimens from the soda lime. Place the tubes in a test-tube rack.

2. Obtain 10–12 pea seeds and 3–6 crickets, which are to be placed in the test tubes. Blot any water from them and measure their mass to the nearest 0.1 g on the balance.

3. Add specimens to the tubes as follows:
   Tube 1: germinating pea seeds
   Tube 2: crickets
   Tube 3: nothing
4. Insert the base of a 1-ml pipette into the one-hole stopper so that it is flush with the inner surface of the stopper. Insert the stopper into a test tube. Repeat for each tube.
5. Place the respirometer at the assigned temperature for 10 min for temperature equilibration. It is important that the respirometer is placed at an angle (not vertical) as shown in Figure 7.4 to prevent the salt solution from running down into the pipette. After 10 min, place a small drop of 10% sodium chloride solution at the tip of each pipette, and note how it is drawn into the pipettes of the experimental tubes. Since carbon dioxide is absorbed by the soda lime, oxygen consumption can be measured by the movement of the fluid toward the test chamber.
6. You are to determine the movement of the fluid for each tube during five 3-min test intervals. Record the reading on the pipette at the beginning and end of each 3-min test period. Be sure to take your readings at the front edge of the fluid. Discard the lowest and highest readings (Why?) and calculate the average of the remaining three readings. Subtract the average movement in the control tube (tube 3) from the average movement in the experimental tubes. Calculate the average respiration rate (ml O_2/hr/g) for each specimen as shown below. **Record your data in item 4a on the laboratory report.**

$$\frac{\text{ml } O_2}{3 \text{ min}} \times \frac{60 \text{ min}}{1 \text{ hr}} = \text{ml } O_2/\text{hr}$$

$$\frac{\text{ml } O_2/\text{hr}}{\text{mass in grams}} = \text{ml } O_2/\text{hr/g}$$

7. Clean the apparatus and your workstation.
8. Exchange data with groups doing the experiment at different temperatures.
9. **Complete item 4 on the laboratory report.**

Frog and Mouse

Now you will investigate the effect of temperature on the rate of cellular respiration in a frog, a poikilothermic animal whose body temperature varies directly with ambient temperature, and a mouse, a homeothermic animal whose body temperature is constant in spite of moderate changes in ambient temperature. Do you think that exposure to temperatures of 10 °C, room temperature, and 40 °C will have the same effect on aerobic respiration in each animal? Let's find out.

General hypothesis: The rate of cellular respiration in poikilothermic and homeothermic animals increases with an increase in the temperature of exposure.

Prediction: If the rate of cellular respiration in poikilothermic and homeothermic animals increases with an increase in temperature, then the rate of cellular respiration in both a frog and a mouse will be slowest at 10 °C and fastest at 40 °C.

Null hypothesis: The rate of cellular respiration in a frog and a mouse is not slowest at 10 °C and fastest at 40 °C.

The null hypothesis will be tested by measuring oxygen consumption at 10 °C, room temperature, and 40 °C. Your group will be assigned to do part of the experiment at one temperature by your lab instructor. Exchange results with student groups doing part of the experiment at other temperatures.

The manometer (U-tube) of the respirometer (Figure 7.5) connects the experimental and control chambers. A pressure change in one will cause the movement of the manometer fluid toward the chamber with the lowest pressure. Since carbon dioxide is absorbed by the soda lime, oxygen consumption can be measured by the movement of fluid toward the experimental chamber. Note that any change in pressure in the control chamber is automatically reflected in the level of the fluid.

Perform the experiment as described below. Record your results and those of other student groups in item 5a on the laboratory report.

Materials

Per student group
Celsius thermometer
Respirometer (Figure 7.5)
   colored water for manometer
   glass jars, wide-mouth
   glass tubing
   pipette, 5 ml (manometer)
   plastic tubing

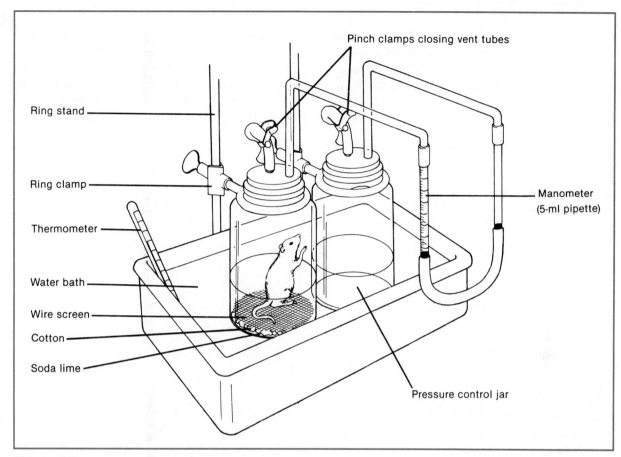

Figure 7.5. Respirometer setup

rubber stoppers for glass jars, 2-hole
ring stands with ring clamps, 2
tubing clamps, pinch type
water bath at assigned temperature
wire screen

Per lab
Cotton, absorbent
Crushed ice
Soda lime
Triple-beam balance

Assignment 5

1. Set up a respirometer as shown in Figure
   7.5. Place soda lime in the bottom of each
   chamber to a depth of about 2 cm. Cover the
   soda lime with about 1 cm of cotton and add
   a wire screen. Place the respirometer in a
   water bath at the assigned temperature.

2. Measure the mass of the animal assigned to
   you to the nearest 0.1 g. (Mice should be
   picked up by their tails.)

3. Place the animal in the experimental chamber. *With the vent tubes open,* loosely replace
   the stopper. ***Caution:*** *Failure to keep the
   vent tubes open when inserting the stopper
   may injure the test animal due to a sudden
   increase in air pressure.*

4. After 5–10 min for temperature equilibration, *with the vent tubes open,* insert the
   stopper snugly into the jar. Close the vent
   tubes, record the time, and take the first
   reading from the manometer. Exactly 3 min
   later, take the second reading. *Open the vent
   tubes, remove the stopper, and place it loosely
   on top of the jar.*

5. After 3–5 min, insert the stopper, close the
   vent tubes, and take the first reading of the
   second replica. Proceed as before. Repeat to

make at least five replicates. ***Record your data in item 5a on the laboratory report.***

6. Discard the lowest and highest values, and calculate the average oxygen consumption (milliliters per 3-min interval) and the ml O_2/hr/g of body weight.

7. Return the animal to its cage, and clean the respirometer and your workstation.

8. Exchange data with groups doing the experiment at different temperatures.

9. ***Plot the respiration level (ml O_2/hr/g) for each of the organisms studied in item 5d on the laboratory report.***

10. ***Complete the laboratory report.***

8

CELL DIVISION

All new cells are formed by the division of preexisting cells. In **prokaryotic cells,** cell division is relatively simple. The process is known as **binary fission,** and it occurs by (1) replication and separation of the circular DNA molecules and (2) the formation of additional cell membrane and cell wall material to separate the original cell into two new cells. Figure 8.1 depicts the process of binary fission. The cells are too small for you to observe this process in the laboratory, however.

In **eukaryotic cells,** two processes of cell division produce distinctly different types of cells. Study Table 8.1. Cells formed by **mitotic cell division** contain the same number and composition of chromosomes as the parent cell. In contrast, cells formed by **meiotic cell division** have only one-half the number of chromosomes as the parent cell. Thus, these two types of cell division differ in the way the chromosomes are dispersed to the new cells that are formed. The terms **mitosis** and **meiosis** refer to the orderly process of separating and distributing the replicated chromosomes to the new cells. **Cytokinesis** (division of the cytoplasm) is the process of actually forming the **daughter cells.**

Each organism has a characteristic number of chromosomes in the nuclei of its cells. If a single set of chromosomes is present, the cell is **haploid (n),** and it contains only one chromosome of each chromosome pair. If two sets are present, the cell is **diploid (2n),** and each chromosome pair is composed of **homologous chromosomes.**

The body cells of animals and most higher plants are diploid. For example, fruit flies have 8 chromosomes (4 pairs), onions have 16 (8 pairs), and humans have 46 (23 pairs). Gametes (eggs and sperm) of these organisms are always haploid and contain 4, 8, and 23 chromosomes, respectively. The body cells of most simple organisms are haploid.

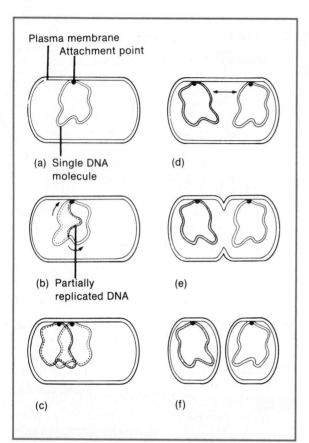

Figure 8.1. Cell division in a prokaryotic cell. (a) Cell before replication of DNA. Note attachment of DNA to cell membrane. (b) Replication of DNA moving in both directions from starting point. (c) DNA replication completed. (d) Growth of cell membrane and cell wall occurs between the points of DNA attachment. (e) Cleavage furrow of cell membrane and cell wall begins. (f) New cells formed.

MITOTIC CELL DIVISION

In unicellular organisms, mitotic cell division serves as a means of reproduction. In multicellular organisms, it serves as a means of growth and repair. As worn-out or damaged cells die, they are replaced by new cells formed by mitotic division in the normal maintenance and healing processes. Millions of new cells are formed in the human body each day in this manner.

Mitotic cell division is an orderly, controlled process, but it sometimes breaks out of control to form massive numbers of nonfunctional, rapidly dividing cells that constitute either a benign tumor or a cancer. Seeking the causes of uncontrolled mitotic cell division is one of the major efforts of current biomedical research.

The Cell Cycle

A cell passes through several recognizable stages during its life span. These stages constitute the **cell cycle.** There are two major stages. **Mitosis,** the M stage, accounts for only 5–10% of the cell cycle. The **interphase** forms the remainder. See Figure 8.2.

Interphase has three subdivisions. Immediately after mitosis is a growth period, the G_1 **stage.** Next is the **synthesis (S) stage,** when chromosome and centriole (if present) replication occurs. Each replicated chromosome consists of two **sister chromatids** joined at the **centromere.** See Figure 8.3. A second growth stage, the G_2 **stage,** follows and prepares the cell for the next mitotic division. Cells that will not divide again remain in the G_1 stage and carry out their normal functions.

Mitotic Phases in Animal Cells

The process of mitosis is arbitrarily divided into recognizable stages or phases to facilitate understanding, although the process is a continuous

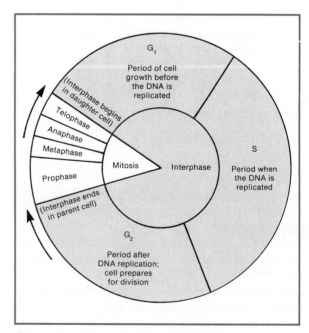

Figure 8.2. Eukaryotic cell cycle

TABLE 8.1
Significant Differences in Mitotic and Meiotic Cell Divisions

Mitotic Cell Division	*Meiotic Cell Division*
1. Occurs in both haploid (n) and diploid (2n) cells.	1. Occurs in diploid (2n) cells, but not in haploid (n) cells.
2. Completed when one cell divides to form two cells.	2. Requires two successive cell divisions to produce four cells from the single parent cell.
3. Duplicated chromosomes do not align themselves in homologous pairs during division.	3. Duplicated chromosomes arrange themselves in homologous pairs during the first cell division.
4. The two daughter cells contain (a) the same genetic composition as the parent cell and (b) the same chromosome number as the parent cell.	4. The four daughter cells contain (a) different genetic compositions and (b) one-half the chromosome number of the parent cell.

one. These phases are **prophase, metaphase, anaphase,** and **telophase.** The characteristics of each phase as observed in animal cells are noted here to aid your study. Interphase is also included for comparative purposes. Compare these descriptions with Figure 8.4.

Interphase

Cells in interphase have a distinct nucleus and two pairs of **centrioles.** The chromosomes are uncoiled and are visible only as **chromatin granules.**

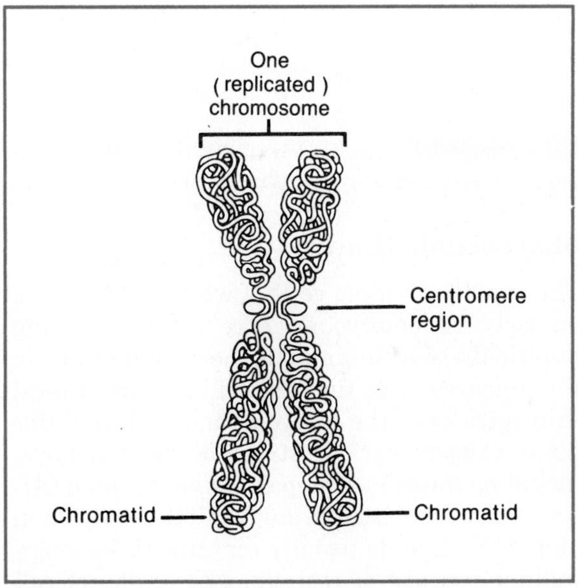

Prophase

During prophase, (1) the nuclear membrane and nucleolus disappear, (2) the chromosomes coil tightly to appear as rod-shaped structures, (3) each pair of centrioles migrates to opposite ends of the cell, and (4) the **spindle** forms. Each pair of centrioles and its radiating **astral rays** constitute an **aster** at each end (pole) of the spindle.

Metaphase

This brief phase is characterized by the chromosomes lining up at the equator of the spindle. The sister chromatids of each replicated chromosome are attached to separate spindle fibers by their **centromeres.**

Anaphase

Anaphase begins with the separation of the centromeres of the sister chromatids, which migrate toward opposite poles of the spindle. Once the sister chromatids separate, they are called **daughter chromosomes.** Thus, a cell in anaphase contains two complete sets of chromosomes.

Telophase

In telophase, (1) a new nuclear membrane forms around each set of chromosomes to form two new

Figure 8.3. A replicated chromosome

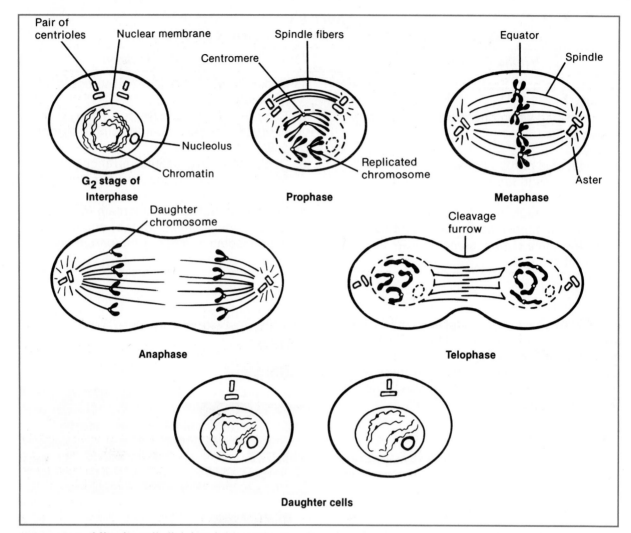

Figure 8.4. Mitotic cell division in an animal cell

nuclei, (2) the nucleolus reappears, (3) the chromosomes start to uncoil, and (4) a **cleavage furrow** forms to divide the parent cell into two **daughter cells.** Recall that the division of the parent cell is called cytokinesis.

Mitotic Division in Plant Cells

Mitotic division in plants follows the same basic pattern that occurs in animals, with some notable exceptions. Most plants do not possess centrioles, although a spindle of fibers is present. The rigid cell wall prevents the formation of a cleavage furrow during cytokinesis; instead, a **cell plate** forms to separate the parent cell into two daughters cells, and a new cell wall forms along the cell plate. Cytokinesis usually, but not always, occurs during telophase. See Figure 8.5.

← in plants

Microscopic Study

The rapidly dividing cells of whitefish blastula, an early fish embryo, are excellent for studying mitotic division in animals. See Figure 8.6. On the prepared slide that you will use are several thin sections of the blastula, and each contains many cells in various stages of the cell cycle, including mitosis. A prepared slide of onion (*Allium*) root tip is used to study mitotic division in plants. Each slide usually contains three longitudinal sections of root tip. The region of cell division is near the pointed tip.

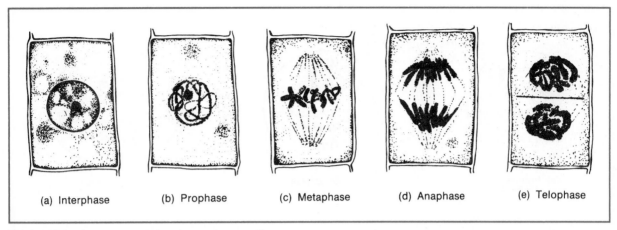

Figure 8.5. Mitotic division in a plant cell

Materials

Per student
Compound microscope

Per lab
Prepared slides of:
  whitefish blastula, x.s.
  onion root tip, l.s.

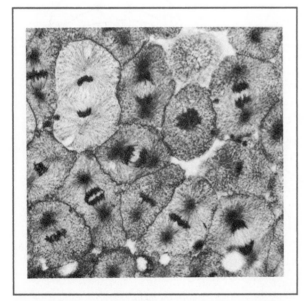

Figure 8.6. Various stages of mitotic division in whitefish blastula cells. Courtesy of Turtox/Cambosco, Macmillan Science Co., Inc., Chicago.

Assignment 1

Complete items 1, 2a, and 2b on Laboratory Report 8 that begins on page 327.

Assignment 2

1. Obtain a prepared slide of whitefish blastula. Locate a section for study with the 4× objective. Then switch to the 10× objective to find mitotic phases for observation with the 40× objective. ***Locate cells in each phase of mitosis, and draw them in the space for item 2c on the laboratory report.*** You may have to examine all the sections on your slide, or even additional slides, to observe each phase of mitosis.
2. Examine a prepared slide of onion root tip. Locate the cells in mitotic phases near the tip of the section. Observe cells in metaphase, anaphase, and telophase. Note the cell plate. Note how mitotic division in an onion root tip differs from that observed in a whitefish blastula.
3. ***Complete item 2 on the laboratory report.***

Meiotic Cell Division

In contrast to mitotic cell division, meiotic cell division consists of *two* successive divisions but only *one* chromosome replication. This results in the formation of four cells that have only half the number of chromosomes of the diploid (2n)

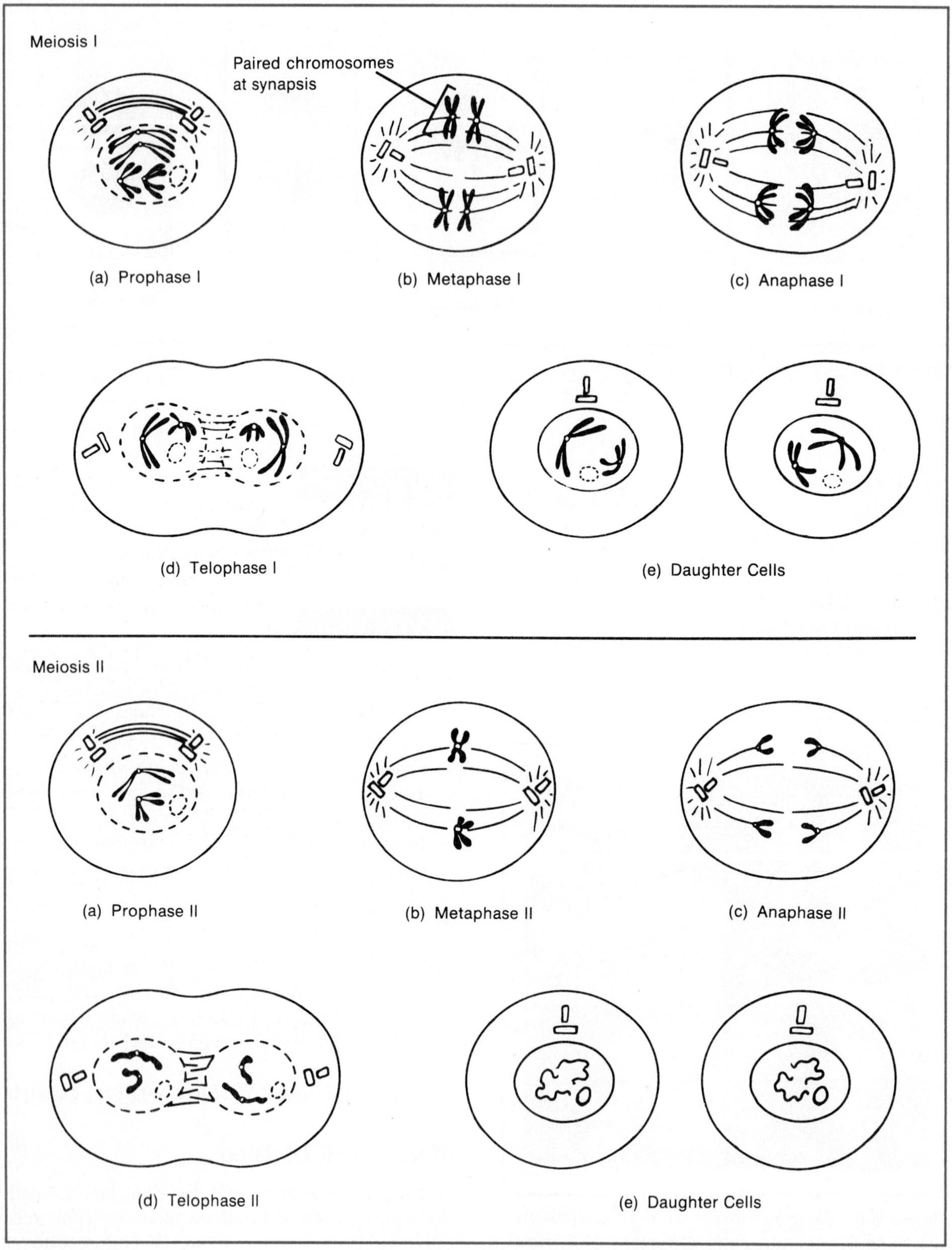

Meiosis I

Paired chromosomes
at synapsis

(a) Prophase I

(b) Metaphase I

(c) Anaphase I

(d) Telophase I

(e) Daughter Cells

Meiosis II

(a) Prophase II

(b) Metaphase II

(c) Anaphase II

(d) Telophase II

(e) Daughter Cells

Figure 8.7. Meiotic cell division. For simplicity, only one of the cells formed in meiosis I is shown in meiosis II.

parent cell. Thus, the daughter cells have a haploid (n) number of chromosomes since they each contain only *one member of each chromosome pair*. In addition to reducing the chromosome number in the daughter cells, meiosis also reshuffles the genes, hereditary units formed of small segments of DNA, and this greatly increases the genetic variability among the daughter cells.

In humans and most animals, cells formed by meiotic division become either sperm or eggs. In plants, meiotic cell division results in the formation of meiospores that do not fuse like gametes but grow into haploid gametophytes which, in turn, produce gametes by mitotic division. In either case, the basic result of meiosis is the same: haploid cells with increased genetic variation.

Meiotic Phases in Animal Cells

Study Figure 8.7 as you read the following description of meiotic cell division in an animal cell. Chromosome and centriole replication occur in the S stage of interphase prior to the start of meiosis.

Meiosis I

Prophase I exhibits the following characteristics. Each chromosome is composed of two sister chromatids joined together at the centromere. The replicated members of each chromosome pair join together in a side-by-side pairing called **synapsis.** Chromosomes in synapsis are often called **tetrads** since they consist of four chromatids. An exchange of chromosome segments (crossover) frequently occurs between members of the tetrad and increases the genetic variability of the cells produced by meiotic division. The chromosomes coil tightly to appear as rod-shaped structures, the nuclear membrane and nucleolus disappear, and a spindle forms.

Metaphase I is characterized by the synapsed chromosomes lining up at the equatorial plane, where they attach to spindle fibers by their centromeres.

Anaphase I begins with the separation of the members of each chromosome pair. The centromeres do *not* separate, so each chromosome still consists of two chromatids joined at their centro-

meres. Members of each chromosome pair migrate to opposite poles of the spindle in the replicated state.

Telophase I proceeds to form a nuclear membrane around each set of chromosomes. The chromosomes untwist and the nucleolus reappears. Cytokinesis separates the mother cell into two daughter cells. Keep in mind that the nucleus of each daughter cell contains only *one member of each chromosome pair* in a replicated state. Thus, each daughter cell is haploid (n).

Meiosis II

Both cells formed by meiosis I divide again in meiosis II, but for discussion purposes we will follow only one of these cells in the second division. In interphase between meiosis I and II, the centrioles replicate but chromosomes do *not* replicate again. Recall that they are already replicated.

Prophase II is characterized by the usual loss of the nuclear membrane and nucleolus, spindle formation, and the appearance of rod-shaped chromosomes.

Metaphase II is characterized by the chromosomes lining up at the equator of the spindle. Each chromosome consists of two sister chromatids joined together at the centromere that is attached to a spindle fiber.

Anaphase II begins with the separation of the centromeres. The sister chromatids, now called daughter chromosomes, move toward opposite poles of the spindle.

Telophase II proceeds as usual to form the new nuclei, and cytokinesis divides the cell to form two haploid (n) daughter cells.

Since each cell entering meiosis II forms two daughter cells, a total of four haploid (n) cells are produced from the original diploid (2n) parent cell entering meiosis I. Thus, meiotic cell division may be summarized as:

$$1 \text{ cell } (2n) \xrightarrow{\text{M I}} 2 \text{ cells } (n) \xrightarrow{\text{M II}} 4 \text{ cells } (n)$$

Materials

Per student group

Chromosome simulation kits or colored pipe cleaners

Assignment 3

1. Study Figure 8.7.
2. Using colored pipe cleaners to represent chromosomes or a chromosome simulation kit, simulate the replication and distribution of chromosomes in both mitosis and meiosis where 2n = 4.
3. ***Complete items 3 and 4 on the laboratory report.***

PART II

INHERITANCE

9

HEREDITY

OBJECTIVES

After completion of the laboratory session, you should be able to:
1. Explain Mendel's principle of segregation and principle of independent assortment and give examples of each.
2. Solve simple genetic problems involving dominance, recessiveness, codominance, and sex linkage.
3. Determine gametes from genotypes where genes are linked or nonlinked.
4. Perform a chi-square analysis.
5. Define all terms in bold print.

A human baby begins with the fusion of a haploid (n) egg and a haploid (n) sperm to form a diploid (2n) zygote (fertilized egg). The zygote contains one haploid set of chromosomes from the mother and one haploid set from the father. At the moment of egg and sperm fusion, the baby's **inherited characteristics** (traits) are determined. All that is left is for it to grow and develop in accordance with the inherited instructions received from mom and dad. So it is in all sexually reproducing organisms: inherited traits are determined at the moment of zygote formation. Subsequent growth and development enable the expression of those traits.

The genetic information that determines hereditary traits is found in the structure of the **DNA molecules** in the **chromosomes.** A short segment of DNA that codes for a particular protein constitutes a **gene,** a hereditary unit. Both genes and chromosomes occur in homologous pairs in diploid organisms. See Figure 9.1.

In the simplest situation, an inherited trait, such as flower color, is determined by a single pair of genes. The members of a **gene pair** may be identical (e.g., each codes for purple flowers) or may code for a different variation of the trait (e.g., one codes for purple flowers, and the other codes for white flowers). Again, in the simplest case, only two forms of a gene exist. Alternate forms of a gene are called **alleles.**

Biologists use symbols (usually letters like "P" or "p") to represent alleles when solving genetic problems. When both members of a gene pair consist of the same allele, such as PP or pp, the individual is **homozygous** for the expressed trait. When the members of the gene pair consist of unlike alleles, such as Pp, the individual is **heterozygous** (hybrid) for the expressed trait.

The genetic composition of the gene pair (e.g., PP, Pp, or pp) is known as the **genotype** of the individual. The observable (expressed) form of a trait (e.g., purple flowers or white flowers) is called the **phenotype.**

An understanding of inheritance patterns enables the prediction of an **expected ratio** for the occurrence of a trait in the progeny (off-spring) of parents of known genotypes.

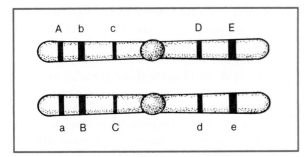

Figure 9.1. Diagrammatic representation of homologous chromosomes and genes

MENDEL'S PRINCIPLES

Gregor Mendel, an Austrian monk, worked out the basic patterns of simple inheritance in 1860, long before chromosomes or genes were associated with inheritance. Mendel's work correctly identified the existence of the units of inheritance now known as genes.

Mendel proposed two principles concerning the activity of genes, and these principles form the basis for the study of inheritance. Look for evidence of these principles as you work through this exercise. In modern terms, these principles may be stated as follows:

1. The **principle of segregation** states that (1) genes occur in pairs and exist unchanged in the heterozygous state and (2) members of a gene pair are segregated (separated) from each other during gametogenesis, ending up in separate gametes.
2. The **principle of independent assortment** states that genes for one trait are assorted (segregated into the gametes) independently from genes for other traits. This principle applies *only* to traits whose genes are located on different chromosome pairs (i.e., the genes are not linked).

DOMINANT-RECESSIVE TRAITS

When a gene pair consists of two alleles, and one is expressed and the other is not, the expressed allele is **dominant.** The unexpressed allele is **recessive.** Many traits are inherited in this manner. For example, the following traits in garden peas exhibit a dominant/recessive pat-

tern of inheritance. In each case, the dominant trait is in italics, and the dominant allele is capitalized in the genotype.

> Flower color: *Purple flowers* (PP, Pp) or white flowers (pp)
> Plant height: *Tall plants* (TT, Tt) or dwarf plants (tt)

Table 9.1 shows the genotypes and phenotypes that are possible for purple or white flowers in peas. Note that the dominant allele is assigned an uppercase P, while the recessive allele is represented by a lowercase p. Only one dominant allele is required for the expression of purple flowers. In contrast, both recessive alleles must be present for white flowers to be expressed in the phenotype. *This relationship is true for all dominant and recessive alleles.*

Solving Genetic Problems

Consider this genetic problem: What are the expected genotype and phenotype ratios (probabilities) in the progeny of a cross between purple-flowering and white-flowering pea plants when each parent is homozygous? Steps used to solve genetic problems such as this are listed in Table 9.2. Note how the steps are used in Figure 9.2 to solve this problem.

The determination of the gametes is a critical step. Recall that meiosis separates homologous chromosomes (and genes) into different gametes. Thus, the members of the gene pair are separated into different gametes. In Figure 9.2, the gametes formed by each parent are identical since each parent is homozygous.

In setting up the Punnett square, the gametes of one parent are placed on the vertical axis, and the gametes of the other parent are placed on the horizontal axis. The number of squares composing a Punnett square depends on the number

TABLE 9.1
Genotypes and Phenotypes for Flower Color in Garden Peas

Genotype	Phenotype
PP	Purple
Pp	Purple
pp	White

TABLE 9.2
Steps Used to Solve Standard Genetics Problems

1. Be sure that you understand what you are to solve. Write down what is known.
2. Write out the cross using genotypes of the parents.
3. Determine the possible gametes that may be formed.
4. Use a Punnett square to establish the genotypes of all possible progeny.
5. Determine the genotype ratio of the progeny. Count the number of identical genotypes and express them as a ratio of the total genotypes (e.g., $\frac{1}{4}$ PP : $\frac{2}{4}$ Pp : $\frac{1}{4}$ pp).
6. Use the information obtained in step 1 to determine the phenotype ratio from the genotypes (e.g., $\frac{3}{4}$ purple flowers to $\frac{1}{4}$ white flowers).

of different gametes formed by the parents. Four squares are used in Figure 9.2 to enable you to understand the setup, although only one square is actually needed.

All possible combinations of gametes are simulated by recording the gametes on the vertical axis into each square to their right and those on the horizontal axis into each square below them. Note that uppercase letters (dominant alleles) always compose the first letter in each gene pair.

When the Punnett square is complete, the individual squares contain the expected genotypes of progeny in the F_1 (first filial) generation. The genotype and phenotype ratios may then be determined.

Test Cross

It is usually not possible to distinguish between homozygous and heterozygous phenotypes exhibiting a dominant trait, but they may be determined by using a **test cross.** In a test cross, the individual exhibiting the dominant phenotype is crossed with an individual exhibiting the

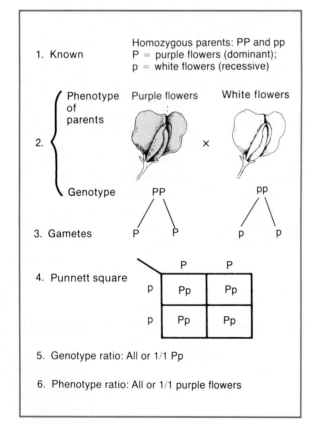

Figure 9.2. Method of solving genetic crosses

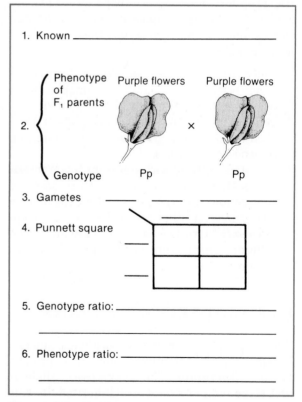

Figure 9.3. Determination of the progeny of monohybrid parents

recessive phenotype. Recall that an individual exhibiting a recessive phenotype is *always* homozygous for that trait. If all progeny exhibit the dominant trait, the parent with the dominant phenotype is homozygous. If half the progeny exhibit the dominant trait and half exhibit the recessive trait, the parent with the dominant phenotype is heterozygous.

Assignment 1

Complete item 1 on Laboratory Report 9 that begins on page 331.

Materials

Per lab
Trays of tall-dwarf corn seedlings from monohybrid crosses

Assignment 2

1. Using Figure 9.2 as a guide, determine the expected progeny in the F_2 generation by crossing two members of the F_1 generation, both of which are monohybrids (Pp). A monohybrid is heterozygous for one trait. Complete the Punnett square in Figure 9.3.
2. Once you have completed the Punnett square, determine the genotypes by counting the identical genotypes and recording them as fractions of the total number of genotypes. The different types of genotypes are then expressed as a proportion to establish the expected ratios.

$$\text{¼ PP : ² ⁄₄ Pp : ¼ pp}$$
or
$$\text{1 PP : 2 Pp : 1 pp}$$

The phenotype ratio may then be determined from the genotype ratio. Remember that the presence of a single dominant allele in a genotype produces a dominant phenotype.

¾ purple-flowering plants : ¼ white-flowering plants
or
3 purple-flowering plants : 1 white-flowering plant

These ratios are always obtained in a monohybrid cross where the gene consists of only two alleles and one allele is dominant.

3. **Complete items 2a–2d on the laboratory report.**
4. Now that you know how to predict the genotype and phenotype ratios of progeny when the genotypes of the parents are known, examine the tray of corn seedlings. These are progeny of a monohybrid cross. Count the number of tall and dwarf plants and calculate the ratio of tall to dwarf plants. **Complete items 2e–2g on the laboratory report.**
5. Now use the knowledge you have gained to determine the phenotype and genotype of parents when the progeny are known. **Complete item 2h on the laboratory report.**
6. Examine Table 9.3, which shows several human traits that are inherited in a dominant/recessive manner. Using this information, **complete items 2i–2n on the laboratory report.**

CODOMINANCE

The alleles of some genes are always expressed in the phenotype and are never recessive. Such alleles exhibit **codominance.** The inheritance of color in snapdragons is an example. When a homozygous red-flowering snapdragon (RR) is crossed with a homozygous white-flowering snapdragon (rr), the F_1 progeny always have pink flowers since both alleles are expressed.

Sickle-cell anemia in humans is inherited in this manner. The structure of the hemoglobin molecule is controlled by a single gene pair consisting of two alleles: Hb^A for normal hemoglobin and Hb^S for sickle-cell hemoglobin. A person who is homozygous for sickle cell is afflicted with the disease and usually dies early in life. The heterozygote has relatively few abnormal hemoglobin molecules, shows no ill effects, and has an increased resistance to malaria.

Assignment 3

Complete item 3 on the laboratory report.

MULTIPLE ALLELES

Some traits are controlled by genes with more than two alleles. The inheritance of ABO blood groups in humans is an example. Three alleles

TABLE 9.3
Dominant and Recessive Phenotypes for a Few Human Traits

Trait	Dominant Phenotype	Recessive Phenotype
Ear lobes	Free	Attached
Pigment distribution	Freckles	No Freckles
Hairline	Widow's peak	Straight
Little finger	Bent	Straight
Tongue roller	Yes	No

are involved: I^A codes for type A blood, I^B codes for type B blood, and i codes for type O blood. Table 9.4 shows the relationship between genotypes and phenotypes. Note that alleles I^A and I^B are both dominant over i, but that they are codominant to each other.

TABLE 9.4
Phenotypes and Genotypes of the A, B, O Blood Types

Blood Type	Genotype
O	ii
A	$I^A I^A$ or $I^A i$
B	$I^B I^B$ or $I^B i$
AB	$I^A I^B$

Assignment 4

Complete item 4 on the laboratory report.

DIHYBRID CROSS

Let's see how to predict progeny ratios when considering two traits at the same time. In humans, handedness and earlobe type are controlled by genes located on separate (nonhomologous) chromosomes. Right-handed (H) is dominant over left-handed (h), and free earlobes (E) are dominant over attached earlobes (e). Therefore, the genotypes of persons homozygous for these dominant and recessive traits are:

Phenotype	Genotype
Right-handed–Free earlobes	HHEE
Left-handed–Attached earlobes	hhee

What is the predicted progeny ratio for these traits in children of parents with these genotypes?

Mom   **Dad**
HHEE × hhee

The alleles for handedness are located on one chromosome pair, and the alleles for earlobes are located on a different chromosome pair. Recall that the members of each gene pair are separated into different gametes when the chromosomes are segregated by meiotic division, so the only possible genotypes of gametes produced by the parents are:

Parents' genotypes:	HHEE	hhee
Gametes' genotypes:	HE	he

Figure 9.4 shows the relationships between homologous chromosomes and genes in the parents, their gametes, and the offspring. A Punnett square may be used to predict the ratio in the offspring as shown below. Note that when combining genotypes of gametes in the square, the alleles of homologous genes are grouped together in the resulting genotype. The alleles for handedness have arbitrarily been placed first.

Therefore, all children of these parents will have a genotype of HhEe for these traits and a phenotype of right-handed–free earlobes. Each child is heterozygous for each trait, and since we are considering two traits, each child is a **dihy-**

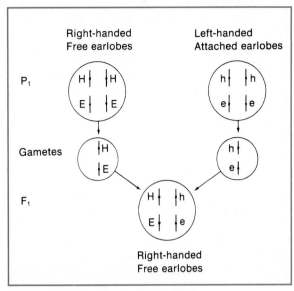

Figure 9.4. Formation of dihybrid

brid. (Actually, all people are hybrid for many traits.)

Now consider a dihybrid cross. What is the predicted ratio for these traits in children of parents that are both dihybrid (HhEe)? This can be solved with a Punnett square. The most difficult part of this problem is determining the genotypes of the gametes, but it is rather simple if you keep in mind Mendel's Principle of Independent Assortment.

Figure 9.5 shows the distribution of homologous genes and chromosomes in gamete formation. Table 9.5 shows an easier way to determine the genotypes of the gametes in such crosses. Once you set up the table with the first three columns filled in, multiply algebraically each of

TABLE 9.5
Gamete Determination in a Dihybrid with Nonlinked Genes

Parent Genotype	First Gene Pair	Second Gene Pair	Possible Gametes
HhEe	H	E	HE
		e	He
	h	E	hE
		e	he

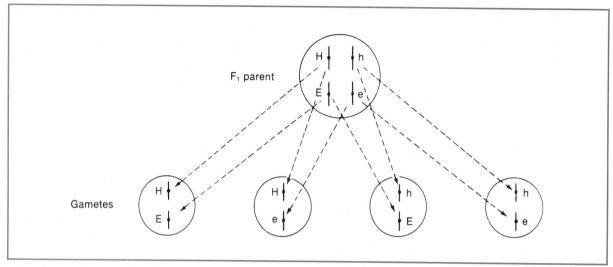

Figure 9.5. Gamete formation in a dihybrid with nonlinked genes

the first pair of alleles by both of the second pair of alleles as shown to yield the four types of genotypes found in the gametes.

Assignment 5

1. **Work out this cross in item 5a on the laboratory report.** In writing the genotypes of the offspring in the squares, group the homologous genes together and place the alleles for handedness first.
2. The predicted phenotype ratio of progeny from dihybrid crosses is always $9:3:3:1$ when each trait is determined by two alleles and a dominant/recessive mode of inheritance, and when genes for the two traits are located on different chromosome pairs. **Complete item 5 on the laboratory report.**

LINKED GENES

Each chromosome contains many genes that are linked together in a definite sequence. When members of a chromosome pair are separated in gamete formation, the genes of each chromosome tend to remain linked together as a unit. See Figure 9.6.

Sex-Linked Traits

In humans, sex is determined by a single pair of sex chromosomes. Females possess two X chromosomes (XX), and males possess an X and a Y (XY). Sex in humans is inherited as shown in Figure 9.7.

The Y chromosome is shorter than the X and lacks some of the genes present on the X chromosome. Those genes that are present on the X chromosome but absent on the Y chromosome are the **sex-linked genes** that control inheritance of sex-linked traits. See Figure 9.8.

For the sex-linked genes, a female is diploid and a male is haploid. Therefore, a recessive allele on the X chromosome of a male will be expressed, whereas the recessive allele must be present on both X chromosomes of a female to be expressed. Common sex-linked traits in humans are red-green color blindness and hemophilia.

Assignment 6

Complete item 6 on the laboratory report.

Pedigree Analysis

Now that you understand the fundamentals of simple inheritance patterns, it is possible to trace a trait in a pedigree (family tree) to determine if it is inherited in a simple dominant/recessive or a sex-linked pattern of inheritance.

Assignment 7

Complete item 7 on the laboratory report.

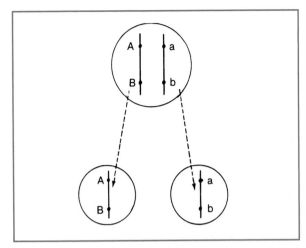

Figure 9.6. Gametogenesis in linked genes

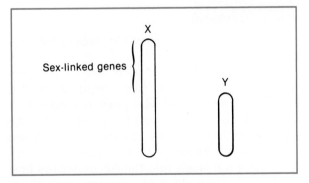

	♀		♂
P₁	XX	x	XY
Gametes	X		X Y
F₁	XX		XY

Figure 9.7. Sex inheritance in humans

Figure 9.8. The sex chromosome

POLYGENIC INHERITANCE

Traits inherited as dominants or recessives are qualitative traits. For example, people are either left-handed or right-handed, and they either have freckles or not. General observations suggest that some traits are not inherited in this manner, that is, some are quantitative in nature. For example, people are not either short or tall but show a gradation of heights typical of a normal (bell-shaped) curve. Such traits exhibit **polygenic inheritance** in which (1) several genes control the same trait and (2) codominance is evident among the alleles.

Assignment 8

Complete item 8 on the laboratory report.

CHI-SQUARE ANALYSIS

To this point in the exercise, you have learned how to predict the expected genotype and phenotype ratios of progeny. However, biologists must verify the expected ratio of a cross to establish the pattern of inheritance. This is done by using the **chi-square** (χ^2) test. This statistical test indicates the probability (p) that differences between the expected ratio and the actual ratio are due to chance alone or whether use of a different hypothesis (expected ratio) would be more appropriate to explain the results (observed ratio). The formula for the chi-square test is $\chi^2 = \Sigma(d^2/e)$, where

χ^2 = chi square
Σ = sum of
d = deviation (difference) between expected and observed results
e = expected results

Consider a monohybrid cross involving flower color in garden peas. The predicted phenotype ratio is 3 purple-flowering plants to 1 white-flowering plant. Thus, if 100 plants were produced from the cross, 75 should have purple flowers and 25 should have white flowers. Table 9.6 shows the results of such a cross and the calculation of chi square.

Comparing the calculated value of chi-square (χ^2) with the values in Table 9.7 is necessary to determine the probability (p) that the deviation from the expected ratio is either (1) by chance and verifies the predicted ratio or (2) greater than chance and does not support the predicted ratio. Note that the chi-square values are arranged in columns headed by probability values and in horizontal rows by phenotype classes minus one (C − 1).

The two classes of progeny in the example are purple flowers and white flowers. Since 2 classes

TABLE 9.6
Chi-Square Determination

Phenotype	Actual Results	Expected Results	Deviation (d)	(d^2)	(d^2/e)
Purple flowers	78	75	3	9	9/75 = 0.12
White flowers	22	25	3	9	9/25 = 0.36
					$\Sigma(d^2/e) = 0.48$
					$\chi^2 = 0.48$

TABLE 9.7
Chi-Square Values

	Probability (p)						
	Deviation Insignificant: Hypothesis Supported					Deviation Significant: Hypothesis Not Supported	
$C - 1$	.99	.80	.50	.20	.10	.05	.01
1	.00016	.064	.455	1.642	2.706	3.841	6.635
2	.0201	.446	1.386	3.219	4.605	5.991	9.210
3	.115	1.005	2.366	4.642	6.251	7.815	11.341
4	.297	1.649	3.357	5.989	7.779	9.488	13.277

$- 1 = 1$, you must look for the calculated chi-square value in the first horizontal row of values. A χ^2 value of 0.48 falls between the columns of 0.50 and 0.20 probability. This means that by random chance the deviation between the expected and actual results will occur between 20% and 50% of the time. Thus, the predicted ratio for the progeny is supported. Probabilities greater than 5% ($p \geq 0.05$) are generally accepted as supporting the hypothesis (expected ratio), while those of 5% or less indicate that the results could not be due to chance.

Materials

Per lab

Corn ears with purple and white kernels from a monohybrid cross

Assignment 9

1. **Complete item 9a on the laboratory report.**
2. Examine a corn ear with both purple and white kernels that have resulted from a monohybrid cross. The predicted ratio is 3 purple to 1 white. Count the purple and white kernels to determine the actual ratio. Mark the row of kernels where you start counting with a pin stuck into the cob under the first kernel. Then **do a chi-square analysis of the results in item 8b on the laboratory report.**
3. **Complete the laboratory report.**

MOLECULAR AND CHROMOSOMAL GENETICS

After completion of the laboratory session, you should be able to:

1. Describe the basic structure of DNA and RNA.
2. Describe the process of information transfer in (a) DNA replication, (b) RNA synthesis, and (c) protein synthesis.
3. Explain how mutations involving base substitution, addition, or deletion affect protein synthesis.
4. Prepare a karyotype from a metaphase smear of human chromosomes.
5. Describe the basis of the chromosomal abnormalities studied.
6. Define all terms in bold print.

Chromosomes are responsible for transmitting the hereditary material from cell to cell in cell division and from organism to progeny in reproduction. This is why the distribution of replicated chromosomes in mitotic and meiotic cell divisions is so important in eukaryotic cells. The genetic information is contained in the structure of **deoxyribonucleic acid (DNA),** which forms the hereditary portion of the chromosomes.

DNA AND THE GENETIC CODE

DNA is a long, thin molecule consisting of two strands twisted in a spiral arrangement to form a double helix somewhat like a twisted ladder. See Figure 10.1. The sides of the ladder are formed of sugar and phosphate molecules, and the rungs are formed by nitrogeneous bases joined together by hydrogen bonds.

Each strand of DNA consists of a series of **nucleotides** joined together to form a polymer of nucleotides. Each nucleotide of DNA is formed of three parts: (1) a deoxyribose (C_5) sugar, (2) a phosphate group, and (3) a nitrogenous base. Four kinds of nitrogenous bases are in DNA. The purine bases (double-ring structure) are **adenine** (A) and **guanine** (G). The pyrimidine bases (single-ring structure) are **thymine** (T) and **cytosine** (C). Note the **complementary pairing** of the bases in Figure 10.1. Can you discover a pattern to their pairing? It is the sequence of nucleotides with their respective purine or pyrimidine bases that contains the genetic information of the DNA molecule.

DNA Replication

Each DNA molecule is able to replicate itself during interphase of the cell cycle and thereby maintain the constancy of the genetic informa-

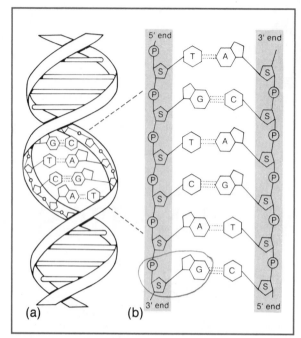

Figure 10.1. DNA structure. (a) The double helix of a DNA molecule. Complementary pairing of the nitrogenous bases joins the sides like rungs of a twisted ladder. A = adenine, G = guanine, C = cytosine, and T = thymine. (b) If a DNA molecule is untwisted, it would resemble a ladder in shape. The sides of the ladder are formed of deoxyribose sugar and phosphate, and the rungs consist of nitrogenous bases.

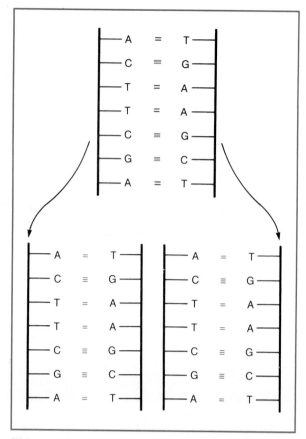

Figure 10.2. DNA replication. A = adenine, G = guanine, C = cytosine, and T = thymine

tion in new cells that are formed. **Replication begins with the breaking of the weak hydrogen bonds that join the nitrogen bases of the nucleotides.** This results in the separation of the DNA molecule into two strands of nucleotides. See Figure 10.2. Each strand then serves as a **template** for the synthesis of a complementary strand of nucleotides that is formed from nucleotides available in the nucleus. The complementary pairing of nitrogen bases determines the sequence of nucleotides in the new strands and results in the formation of two DNA molecules that are identical. Since each new DNA molecule contains one "old" strand and one "new" strand, replication is said to be **semiconservative.** Occasionally, errors are made during replication, and such errors are a type of **mutation.** Of course, replication is controlled by a series of enzymes that catalyze the process.

Materials

Per student group
Colored pencils
DNA, RNA, and protein synthesis kit

Assignment 1

1. Color-code the nitrogenous bases in Figure 10.1 and circle one nucleotide.
2. *Complete items 1a–1c on Laboratory Report 10 that begins on page 337.*
3. Use a DNA kit to construct a segment of a DNA molecule that matches the base sequence of the DNA segment in item 1c on the laboratory report.
4. *Complete item 1d on the laboratory report.*
5. Use a DNA kit to construct a segment of a DNA molecule that matches the "old" non-replicated DNA segment in item 1d on the laboratory report. Then separate the strands

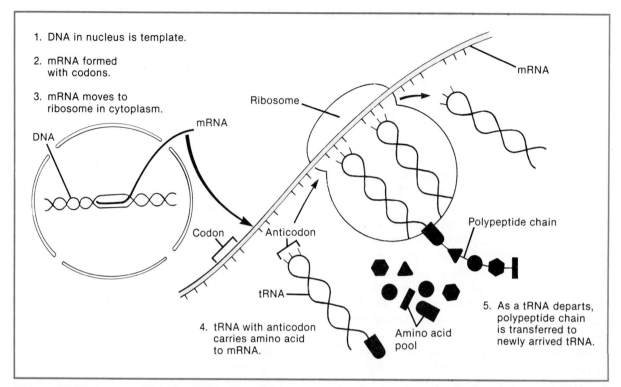

1. DNA in nucleus is template.

2. mRNA formed with codons.

3. mRNA moves to ribosome in cytoplasm.

DNA

mRNA

Ribosome

mRNA

Codon

Anticodon

Polypeptide chain

tRNA

4. tRNA with anticodon carries amino acid to mRNA.

Amino acid pool

5. As a tRNA departs, polypeptide chain is transferred to newly arrived tRNA.

Figure 10.3. A summary of protein synthesis

and construct the replicated strands as shown in item 1d.

RNA Synthesis

DNA serves as the template for the synthesis of **ribonucleic acid (RNA).** RNA differs from DNA in three important ways: (1) it consists of a single strand of nucleotides, (2) its nucleotides contain ribose sugar instead of deoxyribose sugar, and (3) **uracil** (U) is substituted for thymine as one of the four nitrogenous bases.

To synthesize RNA, a segment of a DNA molecule untwists and the hydrogen bonds between the nucleotides are broken. The nucleotides of one strand pair with complementary RNA nucleotides in the nucleus. When the RNA nucleotides are joined by sugar-phosphate bonds, the RNA strand is complete, and it separates from the DNA strand. Few or many RNA molecules may be formed before the DNA strands reunite.

Assignment 2

1. **Complete item 2 on the laboratory report.**
2. Use a DNA-RNA kit to synthesize an RNA molecule with a base sequence identical to the hypothetical RNA molecule in item 2b on the laboratory report.

Protein Synthesis

The genetic information of DNA functions by determining the kinds of protein molecules that are synthesized in the cell. A sequence of three bases—a base triplet—in a DNA molecule has been shown to code indirectly for an amino acid. By controlling the sequence of amino acids, DNA determines the kind of protein produced. Recall that enzymes are proteins and that the chemical reactions in a cell are controlled by enzymes. Thus, DNA indirectly controls cellular functions by controlling enzyme production.

Each of the three types of RNA molecules plays an important role in protein synthesis. **Messenger RNA (mRNA)** is a complement of the genetic information of DNA. A **transcription** of the genetic information in DNA is made when mRNA is synthesized. A base triplet of mRNA is called a **codon,** and the codons for the 20 amino acids composing proteins have been determined. The genetic information is carried by mRNA as it passes from the nucleus to the **ribosomes,** sites of protein synthesis in the cytoplasm.

Transfer RNA (tRNA) carries amino acids to the ribosomes. At one end of a tRNA molecule are three nitrogenous bases, an **anticodon,** which is complementary to a codon of the mRNA. At the other end is an attachment site for 1 of the 20 types of amino acids.

The codon of mRNA and anticodon of tRNA briefly join to place a specific amino acid in its position in a polypeptide chain. This interaction takes place on the surface of a ribosome containing the necessary enzymes for the reaction. **Ribosomal RNA (rRNA),** the third type of RNA, is an integral component of ribosomes, and it plays an important role in decoding the mRNA message to enable protein synthesis. The formation of an amino acid chain is the **translation** of the genetic information.

Figure 10.3 depicts the interaction of mRNA, tRNA, and rRNA in the formation of a polypeptide. Note how the sequence of the amino acids is controlled by the pairing of the codons and anticodons. It may be simplified as follows:

Transcription   **Translation**

Base triplets ⟶ Codons ⟶ Polypeptide
of DNA   of RNA

The Genetic Code

In protein synthesis, the genetic information inherent in the sequence of base triplets in DNA is transcribed into the sequence of codons in mRNA which, in turn, are translated into the sequence of amino acids in a polypeptide chain. In this way, DNA determines both the kinds of amino acids and their sequence in proteins that are synthesized.

There are 64 possible combinations of nucleotide bases in mRNA codons. Their translation is shown in Table 10.1. Note that AUG specifies the amino acid methionine and also is the start signal for protein synthesis. Three codons, UUA, UAG, and UGA, do not specify an amino acid, but they signal the ribosomes to stop assembling the polypeptide chain. Most amino acids are specified by more than one codon, but no codon specifies more than one amino acid. Thus, the code is **redundant** but not **ambiguous.**

TABLE 10.1
The Codons of mRNA and the Amino Acids That They Specify

AAU AAC	Asparagine	CAU CAC	Histidine	GAU GAC	Aspartic acid	UAU UAC	Tyrosine
AAA AAG	Lysine	CAA CAG	Glutamine	GAA GAG	Glutamic acid	UAA UAG	(Stop)*
ACU ACC ACA ACG	Threonine	CCU CCC CCA CCG	Proline	GCU GCC GCA GCG	Alanine	UCU UCC UCA UCG	Serine
AGU AGC	Serine	CGU CGC	Arginine	GGU GGC	Glycine	UGU UGC	Cysteine
AGA AGG	Arginine	CGA CGG		GGA GGG		UGA UGG	(Stop)* Tryptophan
AAU AUC AUA	Isoleucine	CUU CUC CUA	Leucine	GUU GUC GUA	Valine	UUU UUC	Phenylalanine
AUG	Methionine and start	CUG		GUG		UUA UUG	Leucine

* Signals the termination of the polypeptide chain.

Mutations

Base-pair substitution, deletion, or addition constitutes a mutation in a gene. The effect of such mutations is variable, depending on how the mutation is translated via the genetic code.

For example, if a base substitution mutation resulted in a codon change from GCU to GCC, there would be no effect since both specify the amino acid alanine. But if the change was from GCU to GUU, valine would be substituted for alanine in the polypeptide chain and may have a marked effect on the protein. Similarly, if UAU mutated to UAA, it would terminate the polypeptide chain at that point instead of adding tyrosine to the chain. This likely would form a nonfunctional protein.

The mutation involving the addition or deletion of one or two base pairs will cause a **frameshift** in the reading of the codons that may either terminate the polypeptide chain or insert different amino acids into the chain. Usually, addition or deletion mutations have a more disastrous effect than substitution mutations.

Assignment 3

1. Add the bases of the DNA template and the anticodons of tRNA in Figure 10.4.

2. **Complete items 3a–3c on the laboratory report.**
3. Use a DNA-RNA-protein synthesis kit to synthesize an amino acid sequence as shown in item 3c on the laboratory report, starting with the DNA template.
4. **Complete item 3 on the laboratory report.**
5. Use a DNA–RNA-protein synthesis kit to synthetize the amino acid sequences determined in items 3e and 3f on the laboratory report.

HUMAN CHROMOSOMAL DISORDERS

In the study of cell division, you learned that (1) chromosomes occur in pairs in diploid cells, (2) chromosomes are faithfully replicated and equally distributed in mitotic cell division, and (3) cells formed by meiotic cell division receive only one member of each chromosome pair. In both types of cell division, the distribution of the chromosomes is systematically controlled, but errors sometimes occur. In this section of the exercise, you will consider human genetic defects, **chromosomal aberrations,** in which whole chromosomes or large parts thereof are missing or added. Since you now understand the

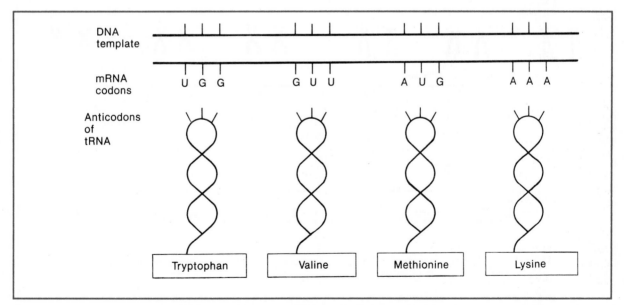

Figure 10.4. Interaction of DNA, mRNA, and tRNA in protein synthesis. Complete the figure by adding the bases in the anticodons of tRNA and the base triplets of DNA.

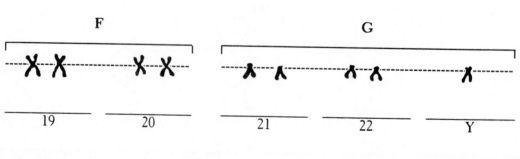

Figure 10.5. Karyotype of a normal male

ganisms. The evaporation of perspiration secreted by sweat glands helps to cool the body surface.

Skeletal System

The skeletal system consists of **bones, cartilages,** and **ligaments.** It provides support for the body and protection for vital organs like the heart, lungs, and brain.

Muscle System

Muscles of the body form the muscle system. **Skeletal muscles** are attached to bones by **tendons** to form levers that enable movement when the muscles contract. About half of the body weight consists of skeletal muscles.

Nervous System

The nervous system consists of the **brain, spinal cord, cranial nerves, spinal nerves,** and **sensory receptors.** The formation and transmission of neural impulses enable the nervous system to provide rapid perception of environmental changes and coordination of body functions. Self-awareness, intelligence, will, and emotions are functions of the human brain.

Cardiovascular System

The cardiovascular system consists of the **heart, blood vessels,** and **blood.** Blood is the transporting medium that carries materials throughout the body as it is pumped by the heart through blood vessels. Blood also provides a primary defense against invasion by disease-causing microorganisms.

Lymphatic System

Lymphatic vessels, spleen, tonsils, and **lymph nodes** compose the lymphatic system. Extracellular fluid is collected by lymphatic vessels; filtered through lymph nodes, where microorganisms and cellular debris are removed; and then returned to the blood. The lymphatic system plays an important role in immunity.

Respiratory System

The respiratory system consists of the lungs and the air passages. The **lungs** are gas exchange organs, where atmospheric air and blood are separated by thin membranes enabling oxygen in the air to diffuse into the blood and carbon dioxide in the blood to diffuse into the air in the lungs. The air passages through which air enters and leaves the lungs are the **nasal cavity, pharynx, larynx, trachea,** and **bronchi.**

Digestive System

The digestive system consists of the alimentary canal, through which food passes, and the accessory digestive glands. The parts of the alimentary canal are **mouth, pharynx, esophagus, stomach, small intestine, large intestine** or **colon, rectum,** and **anus.** The digestive glands are **salivary glands, pancreas,** and **liver.** The function of the digestive system is to convert large nonabsorbable food molecules into small absorbable nutrient molecules.

Urinary System

The urinary system removes metabolic wastes and excessive minerals from the body as urine. Wastes and excessive minerals are removed from the blood by the paired **kidneys,** carried through **ureters** to the **urinary bladder** for temporary storage, and voided from the body through the **urethra.**

Endocrine System

The endocrine system consists of hormone-producing endocrine glands: **pituitary, thyroid, parathyroid, thymus, adrenal, pancreas, pineal, ovaries,** and **testes.** Hormones are chemical messengers that provide a relatively slow-acting, but long-lasting, coordination of body functions.

Reproductive System

The function of male and female reproductive systems is the production of babies. The male reproductive system consists of sperm-producing **testes** located in the **scrotum,** the exterior pouch; two **vas deferens,** two tubes carrying sperm to the **urethra; accessory glands** providing fluid for sperm transport; and the **penis,** the male copulatory organ. The female reproductive system consists of egg-producing **ovaries;** two **oviducts** that carry eggs and, after egg–sperm fusion, early embryos to the **uterus,** where prenatal development occurs; **accessory**

glands that secrete lubricating fluids; and the **vagina,** the female copulatory organ and birth canal.

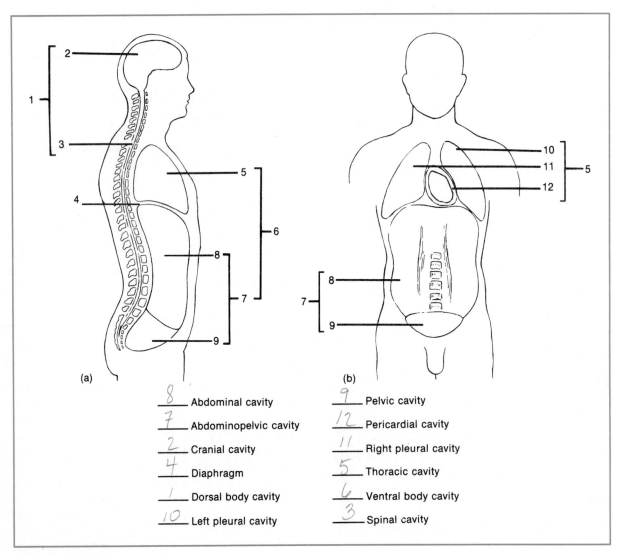

Assignment 1

Complete item 1 on Laboratory Report 11 that begins on page 345.

BODY CAVITIES

Many organs are located in body cavities. There are two body cavities that are named for their location. See Figure 11.1. The **ventral body** cavity is the larger cavity, and it is located near the anterior (front) of the body. It is divided by a dome-shaped sheet of muscle, the **diaphragm,** into the **thoracic cavity** and the **abdomino-pelvic cavity.** The thoracic cavity is further subdivided into left and right **pleural cavities** containing the lungs and the **pericardial cavity** containing the heart. The abdominopelvic cavity is further subdivided into an **abdominal cavity** and a **pelvic cavity.**

The **dorsal body cavity** is the smaller cavity, and it consists of the **cranial cavity** containing the brain and the **spinal cavity** containing the spinal cord. Note that the dorsal body cavity is surrounded by bones.

(a)   (b)

_8_ Abdominal cavity   _9_ Pelvic cavity

_7_ Abdominopelvic cavity   _12_ Pericardial cavity

_2_ Cranial cavity   _11_ Right pleural cavity

_4_ Diaphragm   _5_ Thoracic cavity

_1_ Dorsal body cavity   _6_ Ventral body cavity

_10_ Left pleural cavity   _3_ Spinal cavity

Figure 11.1. The body cavities

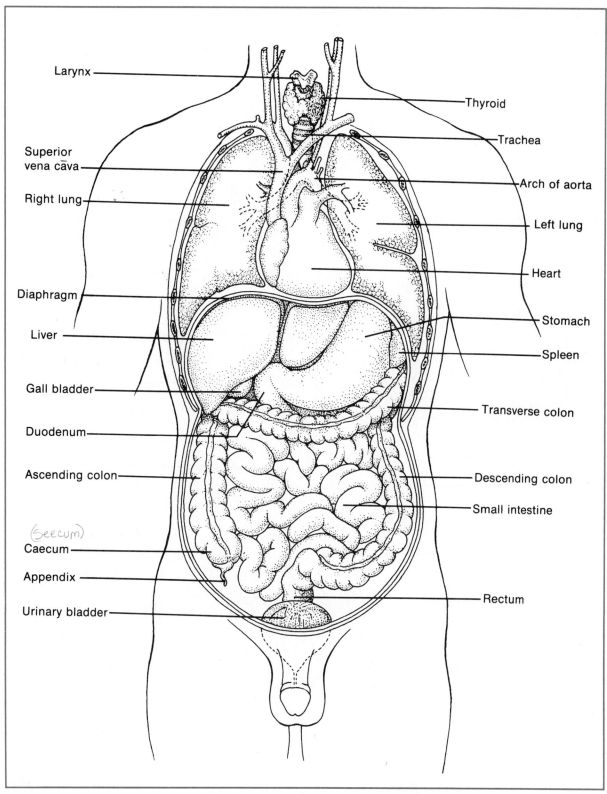

Figure 11.2. Male torso with the anterior body wall removed to expose the thoracic and abdominopelvic organs

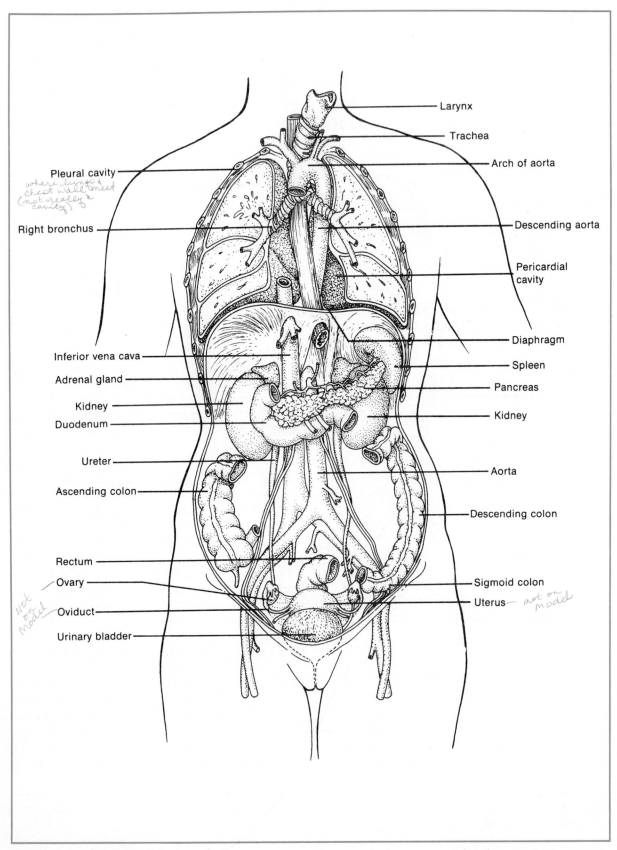

Labels on the figure:

Larynx
Trachea
Arch of aorta
Pleural cavity — *where lungs & chest wall meet (not really a cavity)*
Right bronchus
Descending aorta
Pericardial cavity
Diaphragm
Inferior vena cava
Spleen
Adrenal gland
Pancreas
Kidney
Kidney
Duodenum
Ureter
Aorta
Ascending colon
Descending colon
Rectum
Ovary — *Not on Model*
Sigmoid colon
Oviduct
Uterus — *not on model*
Urinary bladder

Figure 11.3. Female torso with some anterior organs removed to expose the deeper organs

Materials

Per lab
Colored pencils
Human torso models with removable parts

1. Label Figure 11.1 and **complete item 2a on the laboratory report.**
2. Study and color-code the organs in Figures 11.2 and 11.3. Note the location of the organs in the ventral cavity and identify the organ system to which each belongs.
3. Examine the human torso model. Identify the organs in the ventral cavity and the organ system to which each belongs. Remove the anterior organs as necessary to locate the more posterior organs. Note how the organs fit together in the available space. Replace the organs carefully when finished.
4. **Complete item 2 on the laboratory report.**

TISSUES

Organs of the body are formed of two or more tissues, and each tissue is formed of similar cells that are specialized to perform their particular functions. There are four basic types of tissues in the body.

Epithelial tissue	Covers surfaces and lines cavities; forms secretory portions of glands; functions include protection, secretion, and absorption.
Connective tissue	Binds tissues together; provides protection and support for organs and body.
Muscle tissue	Specialized for contraction enabling body movements.
Nerve tissue	Transmission of neural impulses enables rapid coordination of body functions.

In this section your task is to learn the recognition characteristics of body tissues and their general functions, and to correlate their microscopic structure with their functions.

Epithelial Tissues

The cells of epithelial tissues are tightly packed together, forming single or multiple sheetlike layers. The innermost layer is attached to underlying connective tissue by a very thin, noncellular **basement membrane.** The tissue surface opposite the basement membrane is always exposed, i.e., it is not attached to other tissues.

Epithelial tissues are named according to (1) whether there is only one layer (simple epithelium) or more than one layer (stratified epithelium) of cells, (2) the shape of the cells in the outermost cell layer, and (3) the presence or absence of cilia.

Simple Epithelium

Simple epithelium consists of a single layer of cells that form a thin layer covering underlying tissues. It is involved in diffusion, secretion, and absorption of materials.

Simple squamous epithelium consists of a single layer of thin, flat, tilelike cells arranged in a mosaic pattern. See Figure 11.4. It forms the tiny air sacs in the lungs and the capillaries (smallest blood vessels), where its thin, flat cells aid the diffusion of substances. Simple squamous epithelium also provides a smooth, friction-reducing interior lining of the heart, blood vessels, lymphatic vessels, and body cavities.

Simple cuboidal epithelium (Figure 11.4) occurs as a single layer of cubelike cells. It forms the secretory portions of glands, and it composes kidney tubules, where it is involved in both secretion and absorption.

Simple ciliated columnar epithelium lines the interior of the oviducts, and its beating cilia move eggs released from ovaries toward the uterus. Scattered goblet cells secrete a protective layer of mucus.

Simple nonciliated columnar epithelium (Figure 11.4) lines the interior of the digestive tract. It is involved in the secretion of digestive juices and the absorption of nutrients. Scattered goblet cells secrete a protective layer of mucus.

Pseudostratified ciliated columnar epithelium (Figure 11.4) looks as if it consists of more than one cell layer, but it does not. All of

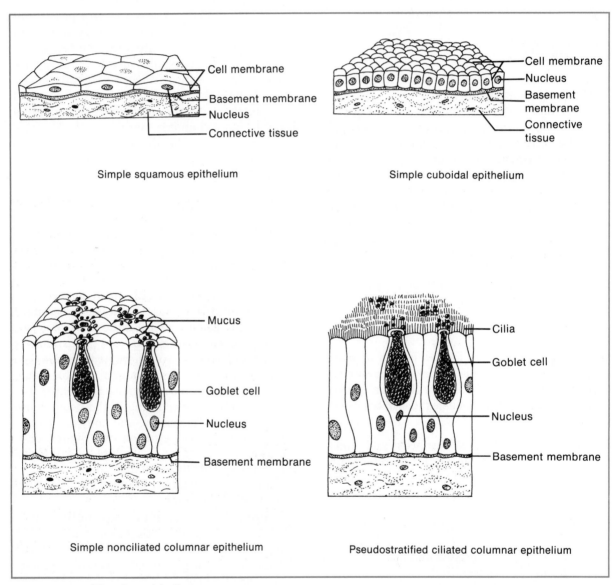

Figure 11.4. Simple epithelium

the cells extend from the basement membrane to the surface of the tissue. Scattered goblet cells secrete a protective layer of mucus. This epithelium lines the interior of the upper respiratory passages, where its beating cilia remove airborne particles trapped in mucus released from goblet cells.

Stratified Epithelium

Stratified epithelium consists of multiple layers of cells, which makes this epithelium resistant to abrasion. The innermost layer of cells continuously forms new cells by mitotic division, so as old cells are worn and lost from the tissue surface, new cells are pushed upward to replace them. As you might expect, protection of underlying tissues is an important function. We'll consider only one example.

Stratified squamous epithelium (Figure 11.5) occurs in two forms. The keratinized form composes the outer layer (epidermis) of the skin. As cells migrate to the surface they are filled with **keratin,** a waterproofing substance that prevents evaporative water loss directly through the skin. The nonkeratinized form lines the mouth, esophagus, and vagina. The cells from your mouth that you observed microscopically in Exercise 3 were cells from surface layers of stratified squamous epithelium.

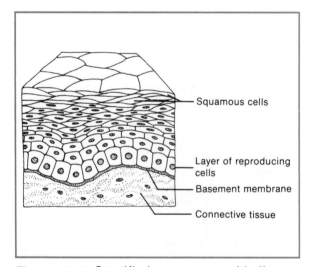

Figure 11.5. Stratified squamous epithelium

Materials

Per student
Compound microscope

Per lab
Prepared slides of epithelium:
   pseudostratified ciliated columnar
   simple columnar
   simple cuboidal
   stratified squamous

Assignment 3

1. Examine microscopically the prepared slides of simple columnar, simple cuboidal, pseudostratified ciliated columnar, and stratified squamous epithelia. These slides have been stained to make it easier to observe cellular structure. Start at 100× to locate the tissue, and then use 400× for observing cellular details. Compare your observations with the corresponding figures. ***Draw each of the tissues as they appear on your slide at 100× in item 3a on the laboratory report.***
2. ***Complete item 3 on the laboratory report.***

Connective Tissues

Connective tissues are characterized by the presence of relatively few cells embedded in a large amount of intercellular, nonliving material called **matrix.** The matrix may be jellylike, fibrous, solid, or combinations of these. Your task is to learn the characteristics of the connective tissues studied and correlate their functions with their structure.

Loose fibrous connective tissue (Figure 11.6) is the most widespread connective tissue in the body. It provides flexible support within and around internal organs, muscles, and nerves, and it attaches the skin to underlying muscles. It is characterized by having a jellylike matrix in which relatively few elastic and nonelastic (collagen) fibers are embedded along with a few cells (fibroblasts) that produce the fibers.

Adipose tissue (Figure 11.6) is a special type of loose fibrous connective tissue that contains a

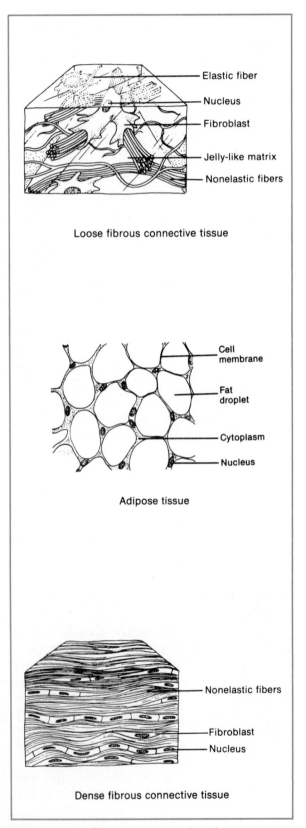

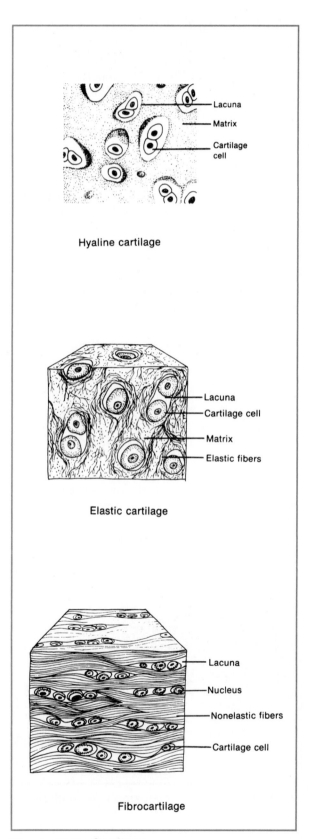

Figure 11.6. Fibrous connective tissue

Figure 11.7. Cartilage

large number of fat cells, where surplus energy is stored as fat. Vacuoles of fat cells are filled with fat droplets that push the cytoplasm and nucleus against the cell membrane. Adipose tissue is abundant under the skin, where it provides insulation, and around internal organs, where it provides a protective cushion.

Dense fibrous connective tissue (Figure 11.6) has a matrix filled with nonelastic (collagen) fibers that give it great tensile strength. Cells (fibroblasts) occur in rows among the fibers. It forms ligaments that attach bones to each other at joints and tendons that join muscles to bones. One form with interwoven fibers forms the dermis of the skin.

Hyaline cartilage (Figure 11.7) has a white, glassy, flexible matrix, and the scattered cartilage cells are located in tiny spaces in the matrix called **lacunae.** Most of the skeleton of a developing baby is first formed of hyaline cartilage and later replaced by bone. Hyaline cartilage supports the nose, larynx, and trachea, and it covers the ends of long bones at joints.

Elastic cartilage (Figure 11.7) is more flexible than hyaline cartilage because of the many elastic fibers in the matrix. It forms the supporting framework of the external ear.

Fibrocartilage (Figure 11.7) contains many nonelastic fibers that make it tough and strong. Cartilage cells in lacunae are arranged in short rows. Fibrocartilage forms the intervertebral discs, where it serves as a cushioning shock absorber.

Bone (Figure 11.8) is the hardest and most rigid connective tissue since its matrix is formed of calcium salts. The matrix is deposited in concentric rings (lamellae) around an **osteonic canal** that contains small blood vessels and nerves. Osteocytes (bone cells) are located in rings of **lacunae** located between the rings of matrix. **Canaliculi** (tiny canals) extending from each lacuna enable the diffusion of materials to and from the cells. The concentric rings of matrix and osteocytes around an osteonic canal form an **osteonic system,** and many osteonic systems are packed together to form solid bone.

Blood (Figure 11.9) is a special form of connective tissue since it consists of several types of blood cells—**red blood cells, white blood cells,** and **platelets**—in a liquid matrix, the **plasma.** Blood transports materials throughout the body and is involved in fighting disease organisms.

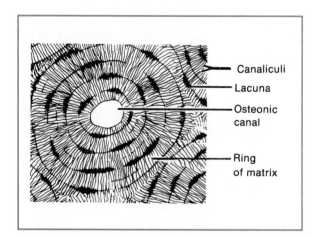

Figure 11.8. Compact bone, ground section

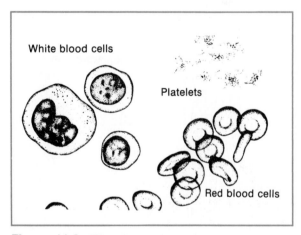

Figure 11.9. Blood

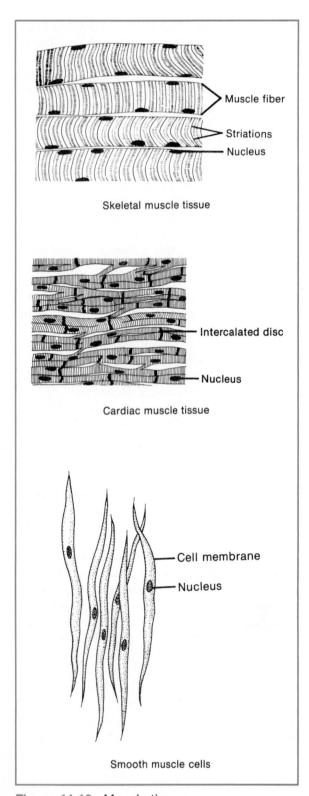

Skeletal muscle tissue

Cardiac muscle tissue

Smooth muscle cells

Figure 11.10. Muscle tissue

Materials

Per lab

Prepared slides of:
  loose fibrous connective tissue
  adipose tissue
  dense fibrous connective tissue
  hyaline cartilage
  fibrocartilage
  bone, ground, x.s.

Assignment 4

Examine the prepared slides of connective tissues, correlating their structure with their functions. ***Draw the tissues as they appear on your slide at 100× in item 4a on the laboratory report and complete item 4.***

Muscle Tissue

Muscle tissue is composed of muscle cells joined by fibrous connective tissue. Muscle cells are adapted for contraction (shortening), which results from the interaction of protein fibrils (small fibers) within muscle cells. There are three types of muscle tissue that are distinguished by (1) location, (2) structure, and (3) contraction characteristics. Your task is to learn these distinguishing characteristics.

Skeletal muscle tissue (Figure 11.10) occurs in organs known as skeletal muscles that are attached to bones. Individual cells are elongate and cylindrical in shape. Each cell (fiber) contains several peripherally located nuclei and exhibits **striations,** alternating light and dark bands extending across the width of the cells. Functionally, skeletal muscle is said to be **voluntary** because its contraction is under conscious control.

Cardiac muscle tissue (Figure 11.10) is the muscle tissue forming the walls of the heart. It consists of branching and interwoven muscle cells that are joined end-to-end by **intercalated discs.** Striations are present, and each cell contains a single, centrally located nucleus. Functionally, cardiac muscle is **involuntary** because its contractions are not consciously controlled.

Smooth muscle tissue (Figure 11.10) is the muscle tissue in the walls of hollow organs other than the heart, e.g., in intestines and blood vessels. Its spindle-shaped cells lack striations and have a single, centrally located nucleus. Functionally, smooth muscle tissue is involuntary because its contractions are not under conscious control.

Materials

Per lab
Prepared slides of:
    cardiac muscle tissue
    skeletal muscle tissue
    smooth muscle tissue, teased

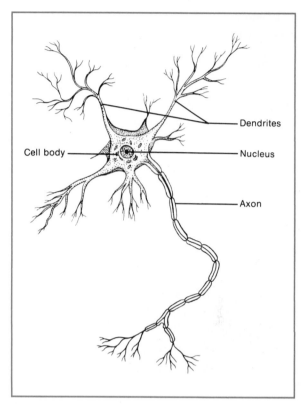

Figure 11.11. A neuron

Assignment 5

Examine the prepared slides of muscle tissues, noting their structural characteristics. ***Draw the muscle tissues as they appear on your slides in item 5a, and complete item 5 on the laboratory report.***

Nerve Tissue

Nerve tissue composes the brain, spinal cord, and nerves. It consists of **neurons** (nerve cells) (Figure 11.11), which form and transmit neural impulses, and **neuroglial cells,** which support the neurons. A neuron is the structural and functional unit of nerve tissue. A neuron consists of a cell body, which is where the nucleus is located, and two types of neuron processes: (1) an axon that carries impulses away from the cell body and (2) one or more dendrites that carry impulses toward the cell body. Neuron processes may be very short (less than 1 mm) or very long (over 1 m). It is the complex interconnections between neurons that enable the functions carried out by the nervous system.

Materials

Per lab
Prepared slides of:
    neurons, giant multipolar

Examine the prepared slides of multipolar neurons, noting their structural characteristics. *Draw a neuron as it appears on your slide in item 6a, and complete the laboratory report.*

12

DISSECTION OF THE FETAL PIG

In this exercise, you will study the basic body organization of mammals through the dissection of a fetal (unborn) pig. The dissection will focus on the internal organs located in the ventral body cavity. Your dissection will proceed easier if you understand the common directional terms used to describe the location of organs. These terms are noted in Table 12.1. Figure 12.1 shows the relationship of most of the directional terms, as well as the three common planes associated with a bilaterally symmetrical animal. The integumentary, skeletal, muscular, nervous, and endocrine systems will not be studied. Study the organ systems outlined in Table 12.2 so that you

know their functions and the major organs of each.

GENERAL DISSECTION GUIDELINES

The purpose of the dissection is to expose the various organs for study, and it should be done in a manner that causes minimal damage to the specimen. *Follow the directions carefully and in sequence.* Read the entire description of each incision and be sure that you understand it *before* you attempt it. Work in pairs, alternating the roles of reading the directions and performing

TABLE 12.1
Anatomical Terms of Direction

Anterior	Toward the head
Posterior	Toward the hind end
Cranial	Toward the head
Caudal	Toward the hind end
Dorsal	Toward the back
Ventral	Toward the belly
Superior	Toward the back
Inferior	Toward the belly
Lateral	Toward the sides
Medial	Toward the midline

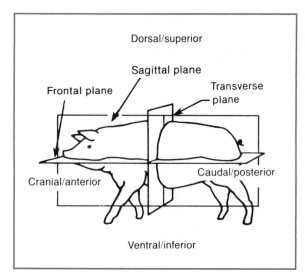

Figure 12.1. Directional terms and body planes

the dissection. Use scissors for most of the incisions. *Use a scalpel only when absolutely necessary.* Be careful not to damage organs that must be observed later.

Your instructor may have you complete the dissection over several sessions. If so, wrap the specimen in wet paper towels at the end of each session, and seal it in a plastic bag to prevent it from drying out.

Protect your hands from the preservative by applying a lanolin-based hand cream before starting the dissection or by wearing disposable plastic gloves. Wash your hands thoroughly at the end of each dissection session and reapply the hand cream.

Materials

Per student group
Dissecting instruments and pins
Dissecting microscope
Dissecting pan, wax-bottomed
Fetal pig (*Sus scrofa*), double-injected

TABLE 12.2
Major Organs and Functions of the Organ Systems

System	Major Organs	Major Functions
Integumentary	Skin, hair, nails	Protection, cooling
Skeletal	Bones	Support, protection
Muscle	Muscles	Movement
Nervous	Brain, spinal cord, sense organs, nerves	Rapid coordination via impulses
Endocrine	Pituitary, thyroid, parathyroid, testes, ovaries, pancreas, adrenal, pineal	Slower coordination via hormones
Digestive	Mouth, pharynx, esophagus, stomach, intestines, liver, pancreas	Digestion of food, absorption of nutrients
Respiratory	Nasal cavity, pharynx, larynx, trachea, bronchi, lungs	Exchange of oxygen and carbon dioxide
Cardiovascular	Heart, arteries, capillaries, veins, blood	Circulation of blood, transport of materials
Lymphatic	Lymphatic vessels, lymph, spleen, thymus, lymph nodes	Cleansing and return of extracellular fluid to bloodstream
Urinary	Kidneys, ureters, urinary bladder, urethra	Formation and removal of urine
Reproductive		
Male	Testes, epididymis, vas deferens, seminal vesicles, bulbourethral gland, prostate gland, urethra, penis	Formation and transport of sperm and semen
Female	Ovaries, oviducts, uterus, vagina, vulva	Formation of eggs, sperm reception, intrauterine development of offspring

Per lab
Hand cream, lanolin-based
String

EXTERNAL ANATOMY

Examine the fetal pig and locate the external features shown in Figure 12.2. Determine the sex of your specimen. The **urogenital opening** in the female is immediately ventral to the anus and has a small **genital papilla** marking its location. A male is identified by the **scrotal sac** ventral to the anus and a **urogenital opening** just posterior to the **umbilical cord.** Two rows of nipples of mammary glands are present on the ventral abdominal surface of both males and females, but the mammary glands later develop only in maturing females. Mammary glands and hair are two distinctive characteristics of mammals.

Make a transverse cut through the umbilical cord and examine the cut end. Locate the two **umbilical arteries** that carry blood from the fetal pig to the placenta, and the single **umbilical vein** that returns blood from the placenta to the fetal pig.

POSITIONING THE PIG FOR DISSECTION

Position your specimen in the dissection pan as follows:

1. Tie a piece of heavy string about 20 in. long around each of the left feet.
2. Place the pig on its back in the pan. Run the strings under the pan, and tie them to the corresponding right feet to spread the legs and expose the ventral surface of the body. This also holds the pig firmly in position.

HEAD AND NECK

This portion of the dissection will focus on structures of the mouth and pharynx. If your instructor wishes for you to observe the salivary glands, a demonstration dissection will be prepared for you to observe.

Dissection Procedure

To expose the organs of the mouth and pharynx, start by inserting a pair of scissors in the angle of the lips on one side of the head and cut posteriorly through the cheek. Open the mouth as you make your cut, and follow the curvature of the tongue to avoid cutting into the roof of the mouth. Continue through the angle of the mandible (lower jaw). Just posterior to the base of the tongue, you will see a small whitish projection, the **epiglottis,** extending toward the roof of the mouth. Hold down the epiglottis and surrounding tissue and continue your incision dorsal to it and on into the opening of the **esophagus.** Now, repeat the procedure on the other side

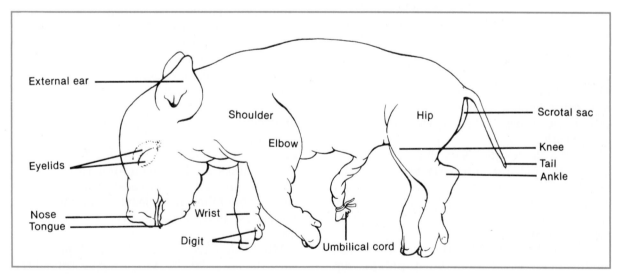

Figure 12.2. External features of a male fetal pig (*Sus scrofa*)

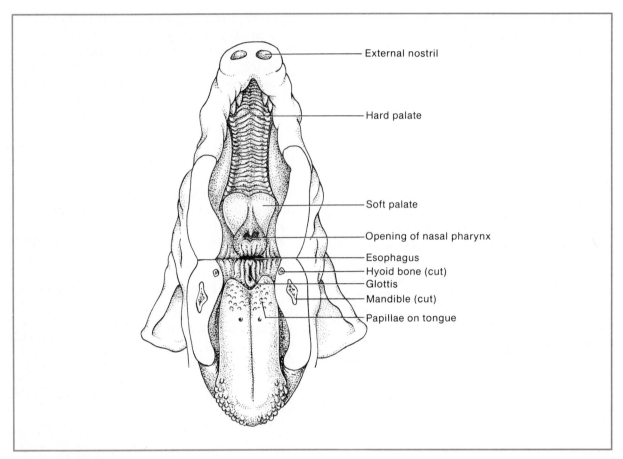

Figure 12.3. The oral cavity

Observations

so that the lower jaw can be pulled down to expose the structures as shown in Figure 12.3.

Only a few deciduous teeth will have erupted, usually the third pair of incisors and the canines. Other teeth are still being formed and may cause bulging of the gums. Make an incision in one of these bulges to observe the developing tooth.

Observe the tongue. Note that it is attached posteriorly and free anteriorly as in all mammals. Locate the numerous **papillae** on its surface, especially near the base of the tongue and along its anterior margins. Papillae contain numerous microscopic taste buds.

Observe that the roof of the mouth is formed by the anteriorly located **hard palate,** supported by bone and cartilage, and the posteriorly located **soft palate.** Paired nasal cavities lie dorsal to the roof of the mouth. The space posterior to the nasal cavities is called the **nasal pharynx,** and it is contiguous with the **oral pharynx,** the throat, located posterior to the mouth. Locate the opening of the nasal pharynx posterior to the soft palate. The oral pharynx may be difficult to visualize because your incision has cut through it on each side, but it is the region where the glottis, esophageal opening, and opening of the nasal pharynx are located.

Make a midline incision through the soft palate to expose the nasal pharynx. Try to locate the openings of the **eustachian tubes** in the dorsolateral walls of the nasal pharynx. Eustachian tubes allow air to move into or out of the middle ear to equalize the pressure on the eardrums.

When you have completed your observations, close the lower jaw by tying a string around the snout.

GENERAL INTERNAL ANATOMY

In this section, you will open the **ventral cavity** to expose the internal organs. The ventral cavity consists of the **thoracic cavity** that is located anterior to the diaphragm, and the **abdomino-pelvic cavity** that is located posterior to the diaphragm.

Dissection Procedure

Using a scalpel, carefully make an incision, through the skin only, from the base of the throat to the umbilical cord along the ventral midline. This is incision 1 in Figure 12.4. Separate the skin from the body wall along the incision just enough to expose the body wall. Do this by lifting the cut edge of the skin with forceps while separating the loose fibrous connective tissue that attaches the skin to the body wall with the handle of a scalpel or a blunt probe.

While lifting the body wall at the ventral midline with forceps, use scissors to make a small incision through the body wall about halfway between the **sternum** and the **umbilical cord.** (Always cut away from your body when using scissors. This gives you better control. Rotate the dissecting pan as necessary.) Once the incision is large enough, insert the forefinger and middle finger of your left hand and lift the body wall from below with your fingertips as you make the incision. This position gives you good control of the incision and prevents unnecessary damage.

Extend the incision along the ventral midline anteriorly to the tip of the sternum and posteriorly to the umbilical cord. Insert the blunt blade of the scissors under the body wall as you make your cuts. Be careful not to cut any underlying organs.

In this manner, continue cutting through both skin and body wall around the anterior margin of the umbilical cord and posteriorly along each side as shown for incision 2. Then, make the lateral incisions just in front of the hind legs (incision 3). Make incision 4 from the midline laterally following the lower margin of the ribs as shown. This will allow you to fold out the resulting lateral flaps.

The **umbilical vein,** which runs from the umbilical cord to the liver, must be cut to lay out the umbilical cord and attached structures posteriorly. Tie a string on each end of the umbilical vein so you will recognize the cut ends later. You will have an unobstructed view of the abdominal organs when this is done.

Extend incision 1 anteriorly through the sternum using heavy scissors. Locate the **diaphragm** and cut it free from the ventral body wall. Lift the left half of the ventral thoracic wall as you cut the connective tissue forming the mediastinal septum that extends from the ventral wall to and around the heart. While lifting the left ventral wall, cut through the thoracic wall (ribs and all) at its lateral margin (incision 5), and cut through the muscle tissue at the anterior end of the sternum so that the left thoracic ventral wall can be removed. Repeat the process for the right thoracic ventral wall. The heart and lungs are now exposed.

Continue incision 1 anteriorly to the chin, and separate the neck muscles to expose the thymus gland, thyroid gland, larynx, and trachea. Remove muscle tissue as necessary to expose these structures.

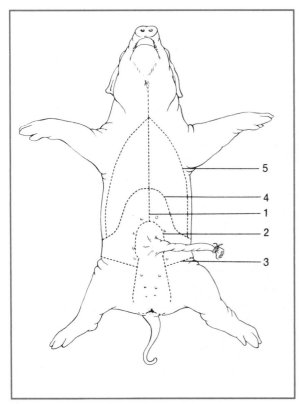

Figure 12.4. Sequence of the incisions (ventral view)

Wash out the cavities of the pig in a sink while being careful to keep the organs in place.

THE THORACIC ORGANS

The thoracic cavity is divided into left and right pleural cavities containing the **lungs.** See Figure 12.5. The left lung has three lobes, and the right lung has four lobes. The inner thoracic wall is lined by a membrane, the **parietal pleura,** and each lung is covered by a **visceral pleura.** In life, fluid secreted into the potential space between these membranes, the **pleural cavity,** reduces friction as the lungs expand and contract during breathing.

At the midline, the parietal pleurae join with fibrous membranes to form a connective tissue partition between the pleural cavities called the **mediastinum.** The **heart** is located within the mediastinum. The heart is enclosed in a pericardial sac formed of the **parietal pericardium** and an outer fibrous membrane that is attached to the diaphragm. Remove this sac to expose the heart, if you have not already done so. The **visceral pericardium** tightly adheres to the surface of the heart.

In the neck region, locate the **larynx** (voice box) that is composed of cartilage and contains the vocal folds (cords). The **trachea** (windpipe) extends posteriorly from the larynx and divides dorsal to the heart to form the **bronchi** that enter the lungs. You will see these structures more clearly later after the heart has been removed.

Locate the whitish **thymus gland** that lies near the anterior margin of the heart and extends into the neck on each side of the trachea.

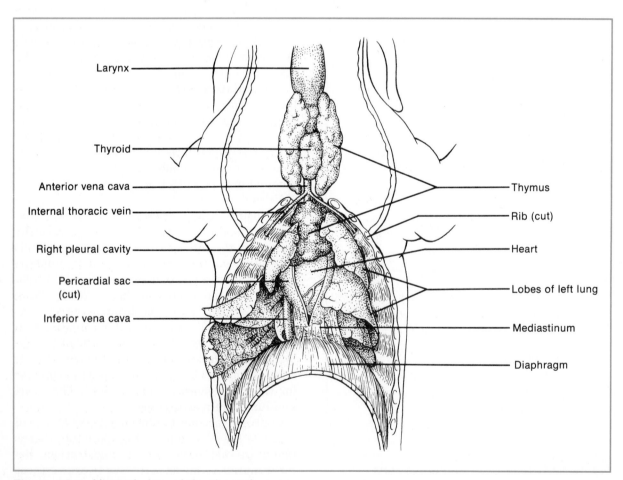

Figure 12.5. Ventral view of the thoracic organs

The thymus is an endocrine gland, and it is the maturation site for T-lymphocytes that are important components of the immune system.

Locate the small, dark-colored **thyroid gland** located on the anterior surface of the trachea at the base of the neck. The thyroid is an endocrine gland that controls the rate of metabolism in the body.

Remove the thymus and thyroid glands as necessary to get a better view of the trachea and larynx, but don't cut major blood vessels. Carefully remove the connective tissue supporting the trachea so that you can move it to one side to expose the **esophagus** located dorsal to it. The esophagus is a tube that carries food from the pharynx to the stomach. In the thorax, it descends through the mediastinum dorsal to the trachea and heart and then penetrates the diaphragm to open into the stomach. You will get a better view of it later on.

DIGESTIVE ORGANS IN THE ABDOMEN

The inner wall of the abdominal cavity is lined by the **parietal peritoneum,** while the internal organs are covered by the **visceral peritoneum.** The internal organs are supported by thin membranes, the **mesenteries,** that consist of two layers of peritoneum between which are located blood vessels and nerves that serve the internal organs.

The digestive organs are the most obvious organs within the abdominal cavity, especially the large liver that fits under the dome-shaped diaphragm. See Figure 12.6.

Stomach

Lift the liver anteriorly on the left side to expose the **stomach.** The stomach receives food from the esophagus and releases partially digested food into the small intestine. Food is temporarily stored in the stomach, and the digestion of proteins begins there. The saclike stomach is roughly J shaped. The longer curved margin on the left side is known as the **greater curvature.** The shorter margin between the openings of the esophagus and small intestine on the right side is the **lesser curvature.**

A saclike fold of mesentery, the **greater omentum,** extends from the greater curvature to the dorsal body wall. The elongate, dark organ supported by the greater omentum is the **spleen,** a lymphatic organ that stores and filters the blood in an adult. Locate the **lesser omentum,** a smaller fold of mesentery extending from the lesser curvature of the stomach and small intestine to the liver. The peritoneum also attaches the liver to the diaphragm.

Cut open the stomach along the greater curvature from the esophageal opening to the junction with the small intestine. The greenish material within it and throughout the digestive tract is the **meconium,** which is composed mostly of epithelial cells sloughed off from the lining of the digestive tract, mucus, and bile from the gallbladder. The meconium is passed in the first bowel movement of a newborn pig.

Wash out the stomach and observe the interior of the stomach. Note the numerous folds in the stomach lining that allow the stomach to expand when filled with food. Observe that the stomach wall is thicker just anterior to the junction with the small intestine. This thickening is the **pyloric sphincter** muscle that usually is closed, but it opens to release material from the stomach into the small intestine. Locate a similar, but less obvious, thickening at the esophagus—stomach junction. This is the **cardiac sphincter** muscle that opens to allow food to enter the stomach, but it usually remains closed to prevent regurgitation.

Small Intestine

The small intestine consists of two parts: duodenum and jejunum-ileum. The first part of the small intestine is the short **duodenum.** It extends from the stomach, curves posteriorly, then turns anteriorly toward the stomach. The duodenum ends and the **jejunum-ileum** begins where the small intestine again turns posteriorly. In life, most of the digestion of food and absorption of nutrients occurs within the small intestine.

Lift out the small intestine and observe how it is supported by mesenteries that contain blood vessels and nerves. Without damaging blood vessels, remove a 1-in. segment of the jejunum-ileum, cut through its wall lengthwise, open it up, and wash it out. The inner lining has enormous numbers of microscopic projections, the **villi,** that give it a velvety appearance. Place a small flattened section under water in a small

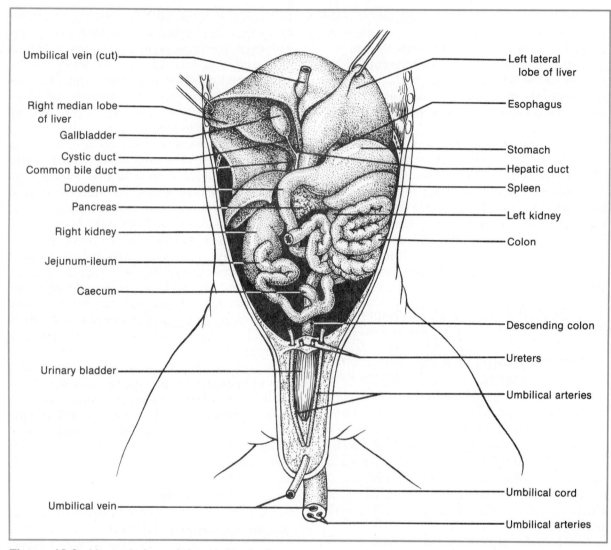

Figure 12.6. Ventral view of the abdominal organs

petri dish and examine the inner lining under a dissecting microscope for a better look. The absorption of nutrients occurs through the villi.

Pancreas and Liver

Locate the pennant-shaped **pancreas** located in the mesentery within the curve of the duodenum. It secretes the hormone insulin as well as digestive enzymes that are emptied into the duodenum via a tiny **pancreatic duct.** Try to find this small duct by carefully dissecting away the peritoneum from a little finger of pancreatic tissue that follows along the descending duodenum.

The **liver** is divided into five lobes. It carries out numerous vital metabolic functions and stores nutrients. Nutrients absorbed from the digestive tract are processed by the liver before being released into the general circulation. In addition, the liver produces **bile.** Bile consists of bile pigments, by-products of hemoglobin breakdown, and bile salts. Bile salts help emulsify fats, which facilitates their digestion in the small intestine.

Lift the right median lobe of the liver to locate the **gallbladder** on its undersurface just to the right of where the umbilical vein enters the liver. Careful dissection of the peritoneum will reveal the **hepatic duct** that carries bile from the liver and the **cystic duct** that carries bile to and from the gallbladder. These two ducts merge to form the **common bile duct** that carries bile into the anterior part of the duodenum. A sphincter muscle at the end of the common bile duct prevents bile from entering the duodenum except when food enters the duodenum from the stomach. When this sphincter is closed, bile backs up and enters the gallbladder via the cystic duct, where it is stored until needed. Food entering the duodenum triggers a hormonal control mechanism that causes contraction of the gallbladder forcing bile into the duodenum.

Large Intestine

Follow the small intestine to where it joins with the **large intestine** or **colon.** At this juncture, locate the small side pouch, the **caecum,** which is small and nonfunctional in pigs and humans. The **ileocaecal valve** is a sphincter muscle that prevents food material in the colon from reentering the small intestine. Make a longitudinal incision through this region to observe the ileocaecal opening and valve.

In life, the large intestine contains large quantities of intestinal bacteria that decompose the nondigested food material that enters it. Water is reabsorbed back into the blood, forming the feces that are expelled in defecation.

The pig's large intestine forms a unique, tightly coiled spiral mass. From the spiral mass, it extends anteriorly to loop over the duodenum before descending posteriorly against the dorsal wall into the pelvic region, where its terminal portion, the **rectum,** is located. The external opening of the rectum is the **anus.** The rectum will be observed in a later portion of the dissection.

CARDIOVASCULAR SYSTEM

The cardiovascular system consists of the heart, arteries, capillaries, veins, and blood. **Blood** is the carrier of materials that are transported by the circulatory system. It is pumped by the **heart** through **arteries** that carry it away from the heart and into the **capillaries** of body tissues. Blood is collected from the capillaries by **veins** that return the blood to the heart.

In this section, you are to (1) locate the major blood vessels, noting their location and function, and (2) identify the external features of the heart. In a double-injected fetal pig, the arteries are injected with red latex and the veins are injected with blue latex. However, a few veins in your pig may not be injected because valves in veins tend to restrict the flow of latex. Also, some blood vessels may not be in the exact location as shown in the figures. Sometimes it is necessary to trace a vessel to the organ it serves in order to positively identify it.

As you study the heart, major veins, and major arteries, carefully remove tissue as necessary to expose the vessels. This is best done by separating tissues with a blunt probe and by picking away connective tissue from the blood vessels with forceps. Make your observations in accordance with the descriptions and sequence that follow.

Before starting your dissection of the circulatory system, it is important to understand the basic pattern of circulation in an adult mammal, as shown in Figure 12.7, and contrast that with the circulation in a fetal mammal, as shown in Figure 12.8.

Adult Circulation

In the adult mammal, deoxygenated blood is returned from the body to the **right atrium** of the heart by two large veins. The **anterior vena cava** returns blood from regions anterior to the heart. The **posterior vena cava** returns blood from areas posterior to the heart. Blood from the digestive tract is carried by the **hepatic portal vein** to the liver, whose metabolic processes modify the nutrient content of the blood before it is carried by the **hepatic vein** into the posterior vena cava.

At the same time that deoxygenated blood enters the right atrium, oxygenated blood from the lungs is carried by the **pulmonary veins** into the **left atrium** of the heart.

When the atria contract, blood from each atrium is forced into its corresponding ventricle. Immediately thereafter, the ventricles contract. During ventricular contraction, blood pressure in the ventricles closes the **atrioventricular valves,** preventing a backflow of blood into the

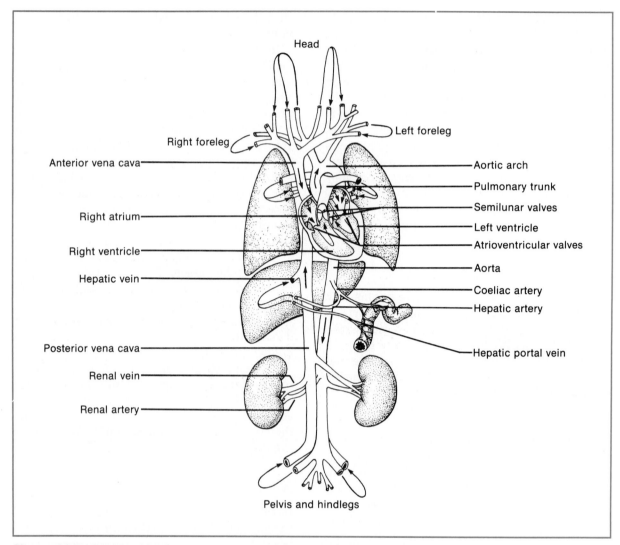

Figure 12.7. Diagrammatic representation of circulation in an adult mammal

atria, and opens the **semilunar valves** at the bases of the two large arteries that exit the heart. Deoxygenated blood in the **right ventricle** is pumped through the **pulmonary trunk** that branches into the **pulmonary arteries** carrying blood to the lungs. Oxygenated blood in the **left ventricle** is pumped into the **aorta** that divides into numerous branches to carry blood to all parts of the body except the lungs.

At the end of ventricular contraction, the semilunar valves close, preventing a backflow of blood into the ventricles, and the atrioventricular valves open, allowing blood to flow from the atria into the ventricles in preparation for another heart contraction.

Fetal Circulation

In a fetus, the placenta is the source of oxygen and nutrients, and it removes metabolic wastes from the blood. The lungs, digestive tract, and kidneys are nonfunctional. Circulatory adaptations to this condition make the circulation of blood in the fetus quite different from that in an adult.

Deoxygenated blood is returned from the body via the anterior and posterior vena cavae as in the adult. However, blood rich in oxygen and nutrients is carried from the placenta by the **umbilical vein** directly through the liver by a segment called the **ductus venosus** and on into

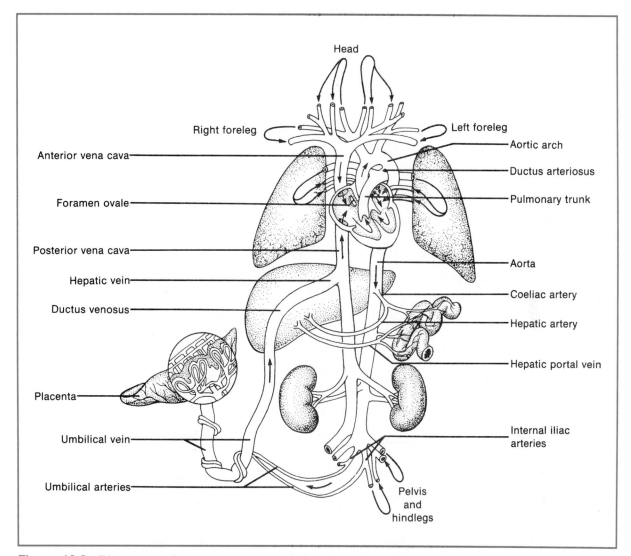

Figure 12.8. Diagrammatic representation of circulation in a fetal mammal

the posterior vena cava. Therefore, the posterior vena cava returns blood to the right atrium that has a high oxygen concentration, but less than that in the umbilical vein.

The fetal heart contains an opening between the right and left atria called the **foramen ovale** that allows much of the blood entering the right atrium from the posterior vena cava to pass directly into the left atrium. This enables blood high in oxygen to enter the left ventricle, which pumps it through the aorta to the body.

Much of the blood pumped from the right ventricle through the pulmonary trunk passes through a fetal vessel called the **ductus arteriosus** into the aorta to bypass the lungs and augment the blood supply to the body. Very little blood is carried by the pulmonary arteries to the nonfunctional lungs and returned to the heart by the pulmonary veins.

In the pelvic region, the **umbilical arteries** arise from the internal iliac arteries and carry blood to the placenta, where wastes are removed and oxygen and nutrients are picked up from the maternal blood.

At birth, the lungs are inflated by breathing, and the umbilical blood vessels, ductus venosus, and ductus arteriosus constrict, causing blood pressure changes that close the foramen ovale. These changes produce the adult pattern of circulation. Growth of fibrous connective tissue

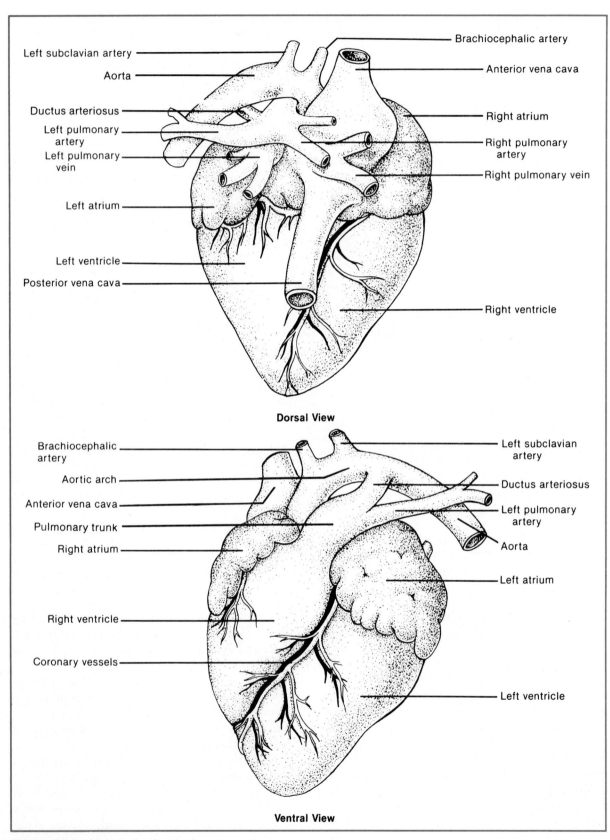

Figure 12.9. Heart of a fetal pig

subsequently seals the foramen ovale and converts the constricted vessels into ligamentous cords.

The Heart and Its Great Vessels

Study the structure of the heart and the great vessels shown in Figures 12.9, 12.10, and 12.11. Be sure that you know the relative positions of these vessels before trying to locate them in your specimen.

If you haven't done so, carefully cut away the pericardial sac from the heart and its attachment to the great vessels. Locate the **left** and **right atria,** blood receiving chambers, and the **left** and **right ventricles,** blood pumping chambers. Note the **coronary arteries** and **cardiac veins** that run diagonally across the heart at the location of the ventricular septum, a muscular partition separating the two ventricles. Coronary arteries supply blood to the heart muscle, and blockage of these arteries results in a heart attack.

While viewing the ventral surface of the heart, locate the large **anterior vena cava** that returns blood from the head, neck, and forelegs to the right atrium. By lifting up the apex of the heart you will see the large **posterior vena cava** that returns blood from regions posterior to the heart into the right atrium. Later, when you remove the heart, you will see where these veins enter the right atrium.

Now, locate the **pulmonary trunk** that exits the right ventricle and, just dorsal to it, the **aorta** that exits the left ventricle. These large arteries appear whitish due to their thick walls. Move the apex of the heart to your left and locate the whitish **ductus arteriosus** that carries blood from the pulmonary trunk into the aorta, bypassing the nonfunctional fetal lungs. You will see the **pulmonary arteries** and **veins** later when you remove the heart.

Major Vessels of the Head, Neck, and Thorax

As you read the description of each major vessel, locate it first in Figures 12.10 and 12.11; then locate it in your dissection specimen.

As noted earlier, the large **anterior vena cava** returns blood from the head, neck, and forelegs into the right atrium. It is formed a short distance anterior to the heart by the union of the left and right brachiocephalic veins. The **brachiocephalic veins** are large but very short. Each one is formed by the union of veins draining the head, neck, and forelegs.

There are two jugular veins on each side of the neck. The **external jugular** is more laterally located and drains superficial tissues. The **internal jugular** is more medially located and drains deep tissues including the brain. External and internal jugular veins may join just before their union with the subclavian vein. A **cephalic vein** drains part of the shoulder and joins with the external jugular near the union of the jugular veins.

The paired **subclavian veins** drain most of the shoulders and forelegs. Follow a subclavian vein laterally to locate the entrance of a **subscapular vein** that drains part of the shoulder. Distal to the subscapular vein the subclavian becomes the **axillary vein** in the axilla or armpit area and then becomes the **brachial vein** as it enters the foreleg. Closer to the heart, the anterior vena cava receives blood from the **internal thoracic veins,** draining the internal surface of the thoracic wall, and the **costocervical** veins, draining the upper back muscles and neck.

Once you have located these veins, cut the superior vena cava, leaving a stub at the heart, and lift it and the attached veins anteriorly to expose the underlying arteries. Locate the **aorta** and note that it forms the **aortic arch** as it curves posteriorly dorsal to the heart. Two major arteries branch from the aortic arch. The first and larger branch is the **brachiocephalic artery.** The second and smaller branch is the **left subclavian artery.**

Follow the brachiocephalic anteriorly, where it branches to form the **right subclavian artery** and a **carotid trunk** that divides to form a pair of **common carotid arteries** that parallel the internal jugular veins to carry blood to the head and neck. At the base of the head, each common carotid branches into internal and external carotid arteries.

The right subclavian artery gives off (1) an anterior branch, the **thyrocervical artery** that

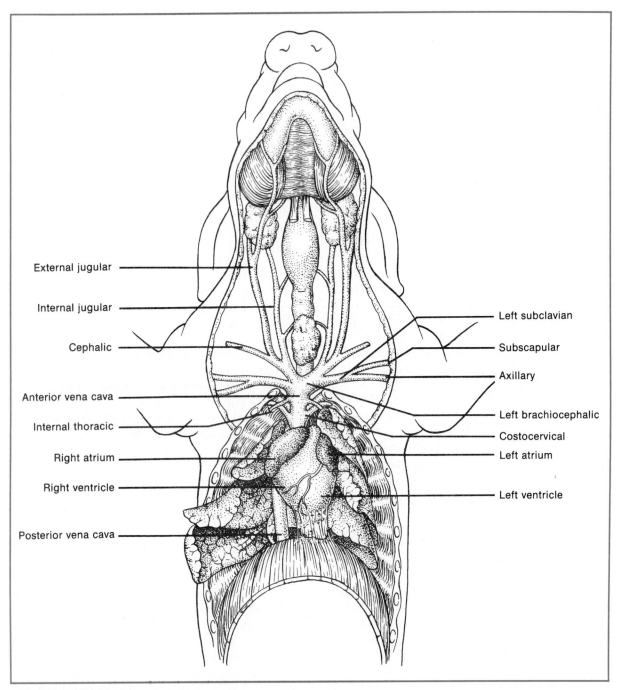

Figure 12.10. Ventral view of the heart and anterior veins

supplies organs of the neck, and (2) posterior branches, the **costocervical** and **internal thoracic arteries** that supply the back and neck muscles and the thorax wall. The continuation of the subclavian becomes the **right axillary artery** that enters the foreleg to become the **brachial artery.**

The **left subclavian artery** forms four major branches that match the branches of the right subclavian artery. From anterior to posterior they are (1) a thyrocervical artery, (2) a left axillary artery, (3) an external thoracic artery, and (4) an internal thoracic artery.

After you have located the major anterior ar-

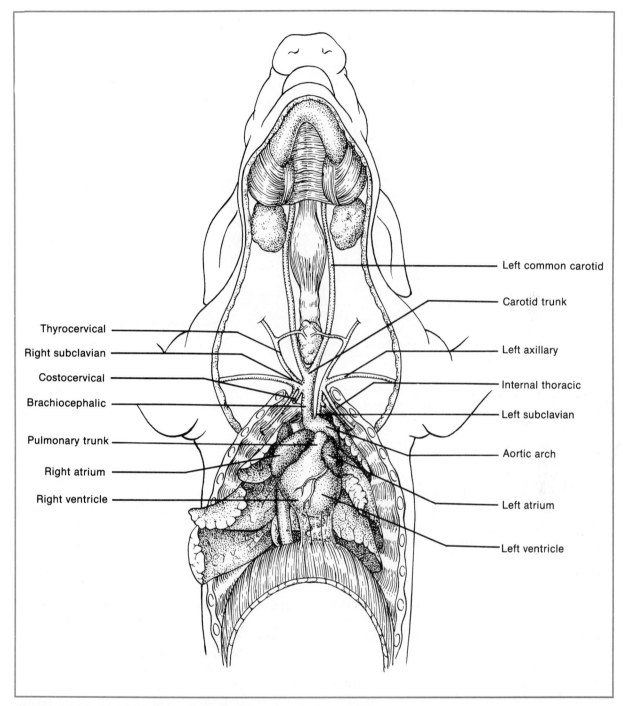

Figure 12.11. Ventral view of the heart and anterior arteries

teries and veins, carefully cut through the major vessels to remove the heart. Leave stubs of the vessels on the heart and identify them by comparing your preparation with Figure 12.9. Especially locate the **pulmonary arteries** and **veins.**

Major Vessels Posterior to the Diaphragm

Follow the aorta and posterior vena cava posteriorly through the diaphragm. Compare your observations with Figures 12.12 and 12.13.

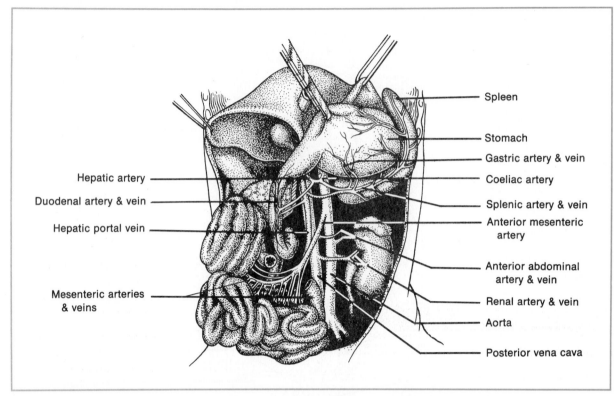

Figure 12.12. Ventral view of arteries and veins supplying the digestive organs

Just posterior to the diaphragm, locate the **hepatic vein** that emerges from the liver and enters the posterior vena cava. Immediately posterior to this union, the **ductus venosus** enters the posterior vena cava. Recall that the ductus venosus is an extension of the **umbilical vein** in the fetal pig. Dissect away liver tissue as necessary to expose the unions of the hepatic vein and ductus venosus with the posterior vena cava. Save the liver tissue adjacent to the hepatic vein and ductus venosus, but remove and discard the rest. Cut through the left side of the diaphragm, and push the abdominal organs to your left (specimen's right) to expose the posterior vena cava and aorta.

Lift the stomach anteriorly to expose the first branch from the aorta posterior to the diaphragm. This is the **coeliac artery.** It gives off branches to the stomach, spleen, and liver (**hepatic artery**). Posterior to the coeliac is a large branch, the **anterior mesenteric artery,** that gives off numerous branches to the intestine. Note that many small veins from the intestine join to form the **hepatic portal vein** (often not injected) that runs anteriorly to enter the liver. This vein carries nutrient-rich blood from the intestine to the liver for processing prior to being released into the general circulation.

Now remove most of the intestines, but save a small section associated with the hepatic portal vein. This will expose the lower portion of the aorta and posterior vena cava.

Locate the **renal arteries** and **renal veins** serving the kidneys. Since the kidneys are located dorsal to the parietal peritoneum, you will have to strip away some of this tissue to observe the vessels and kidneys clearly. The **adrenal glands** (endocrine glands), narrow strips of tissue, should be visible on the anterior margin of the kidneys.

Carefully dissect away the connective tissue to expose the posterior portions of the aorta, posterior vena cava, and the attached vessels. Locate the ureters that carry urine from the kidneys to the urinary bladder, which is located in the reflected tissue containing the umbilical arteries and umbilical cord. Free them from surrounding tissue so they are movable, but don't cut through them.

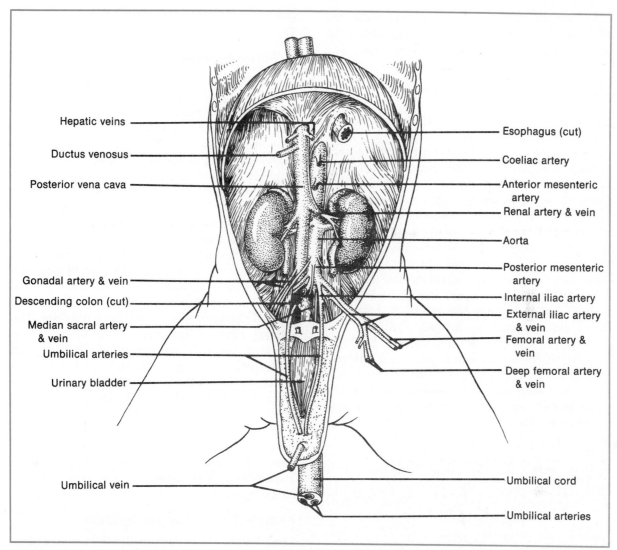

Figure 12.13. Ventral view of the arteries and veins in the abdominopelvic cavity

The small arteries and veins posterior to the renal vessels serve the gonads, ovaries or testes as the case may be. Just below the gonadal arteries, the **posterior mesenteric artery** arises from the ventral surface of the aorta and supplies most of the large intestine.

Locate the pair of **external iliac arteries** that branch laterally from the posterior end of the aorta. After giving off small branch arteries to the body wall, each external iliac artery extends into the thigh, where it divides into the medially located **deep femoral artery** and the laterally located **femoral artery.** Each of these arteries is associated with a corresponding vein, the **deep femoral vein** and the **femoral vein** that

drain the leg and merge to form the **external iliac vein.**

Now locate the medial terminal branches of the aorta, the **internal iliac arteries.** The major extensions of these arteries form the **umbilical arteries** that pass over the urinary bladder to enter the umbilical cord. The other branches of the internal iliac arteries supply the pelvic area and lie alongside the **internal iliac veins** that drain the pelvic area.

The external and internal iliac veins join to form a **common iliac vein** on each side of the body. The union of the common iliac veins forms the posterior beginning of the posterior vena cava.

Located between the common iliac veins you will see the **median sacral artery** and **vein** serving the dorsal wall of the pelvic area.

THE RESPIRATORY SYSTEM

The respiratory system functions to bring air into the lungs, where blood releases its load of carbon dioxide and receives oxygen for transport to body cells. Air enters the lungs via a series of air passageways. Air enters the **nostrils** and flows through the **nasal cavity,** where it is warmed and moistened. It passes into the **nasal pharynx** and on into the **oral pharynx,** where it enters the **larynx** via a slitlike opening called the **glottis.** You observed these components when you dissected the oral cavity and pharynx. From the larynx, air passes down the **trachea** that branches to form the **primary bronchi,** which carry air into the **lungs.**

Return now to the neck and thoracic cavity to observe the respiratory organs. Locate the larynx, trachea, and primary bronchi. Since the heart has been removed, the trachea and primary bronchi may be readily seen by dissecting away some of the surrounding connective tissue and blood vessels. See Figure 12.11.

Make a midventral incision in the larynx, open it, and identify the **vocal folds** located on each side. Note how the trachea branches to form the primary bronchi. Each primary bronchus divides within the lung to form smaller and smaller air passages until tiny microscopic **bronchioles** terminate in vast numbers of saclike **alveoli,** the sites of gas exchange. Dissect along a primary bronchus to locate the **secondary bronchi** that enter the lobes of a lung. Remove a section of trachea and observe the shape of the **cartilaginous rings** that hold it open. Make a section through a lung to observe its spongy nature and the air passages within it.

UROGENITAL SYSTEM

The urinary and reproductive organs compose the urogenital system. They are considered together because the organs are closely interrelated. Refer to Figures 12.14 or 12.15 depending on the sex of your specimen. You are responsible for knowing the urogenital system of each sex, so prepare your dissection carefully and ex-change it when finished for a specimen of the opposite sex.

Urinary System

If you haven't removed the intestines, do so now to expose the urinary organs. Leave a stub of the large intestine because you will want to locate the rectum later. Dissect away the parietal peritoneum to expose the **kidneys** located dorsal to it and against the dorsal body wall. The kidneys are held in place by connective tissue. An **adrenal gland** is located on the anterior surface of each kidney.

Locate the origin of a **ureter** on the medial surface of a kidney near where the renal vessels enter it. As urine is formed by the kidney, it passes into the ureter for transport by peristalsis (wavelike contractions) to the urinary bladder. Trace the ureter posteriorly to its dorsally located entrance into the **urinary bladder.** The urinary bladder lies between the umbilical arteries on the reflected portion of the body wall containing the umbilical cord. Note that the posterior portion of the urinary bladder narrows to form the **urethra** that enters the pelvic cavity. It will be observed momentarily.

If your instructor wants you to dissect a kidney, make a coronal section to expose its interior. Compare the internal structure with Figure 20.2 in Exercise 20 as you read the accompanying description.

Female Reproductive System

The **uterus** is located dorsal to the urinary bladder. It consists of two **uterine horns** that join posteriorly at the midline to form the **body of the uterus.** Locate the uterine horns and the body of the uterus.

Follow the uterine horns anteriolaterally to the small, almost nodulelike, **ovaries** just posterior to the kidneys, where they are supported by mesenteries. The uterine horns end at the posterior margin of the ovaries, where they are continuous with the tiny, convoluted **uterine tubes** that continue around the ovaries to their anterior margins. The anterior end of a uterine tube forms an expanded funnel-shaped opening, the **infundibulum,** that receives the immature ovum when it is released from an ovary. Fertilization occurs in the uterine tubes, and the fetal pigs develop in the uterine horns.

Now, remove or reflect the skin from the ven-

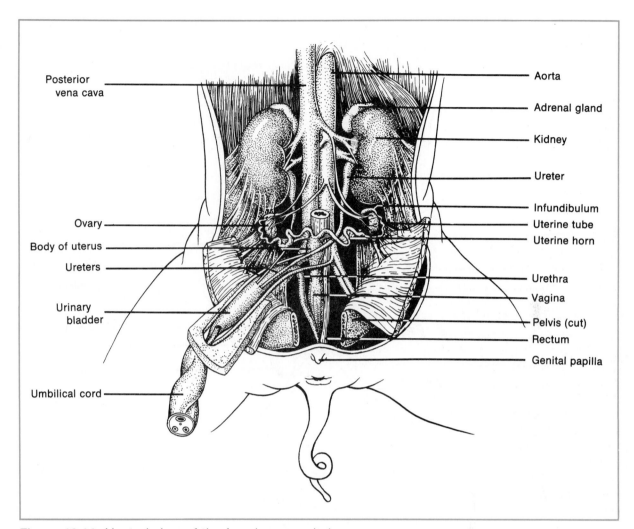

Figure 12.14. Ventral view of the female urogenital system

tral surface of the pelvis. Use your scalpel to cut carefully at the ventral midline through the muscles and bones of the pelvic girdle. Spread the legs and open the pelvic cavity to expose the **urethra** and, just dorsal to it, the **vagina** extending posteriorly from the body of the uterus. You can now see the **rectum,** the terminal portion of the large intestine, located dorsally to the vagina. Note that the vagina and urethra unite to form the **vaginal vestibule** a short distance from the external urogenital opening that is identified externally by the genital papilla.

Male Reproductive Organs

The **testes** develop within the body cavity just posterior to the kidneys. Later in fetal develop-

ment, the testes descend through the **inguinal canals** into the **scrotum,** an external pouch. The scrotum provides a temperature for the testes that is slightly less than body temperature. The lower temperature is necessary for the production of viable sperm. Follow a testicular artery and vein to locate the inguinal canal.

On one side, cut open the scrotum to expose a testis. Locate the **epididymis,** a tortuous mass of tiny tubule that begins on the anterior margin of the testis and extends along the lateral margin to join posteriorly with the **vas deferens** or sperm duct. Trace the vas deferens anteriorly from the scrotum, through the inguinal canal to where it loops over a ureter to enter the urethra.

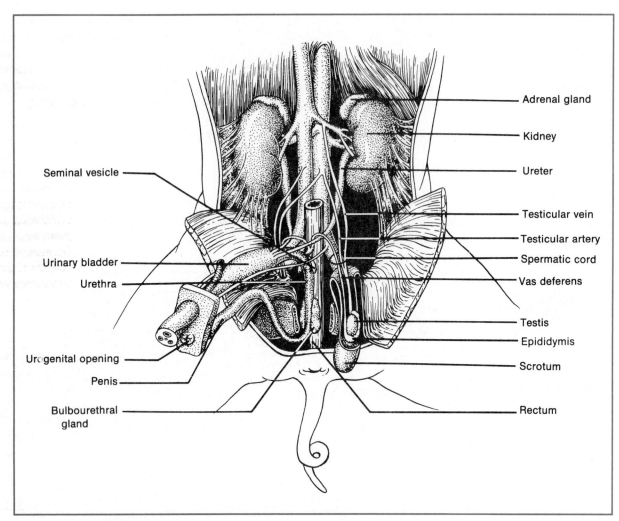

Figure 12.15. Ventral view of the male urogenital system

Locate the **urogenital opening** just posterior to the umbilical cord in the reflected flap of body wall containing the urinary bladder. The **penis** extends posteriorly from this point. Make an incision alongside the penis and free it from the body wall. Push it to one side, and use your scalpel to make a midline incision through the pelvic muscles and bones. Spread the legs and open the pelvic cavity so you can dissect out the pelvic organs. Locate the **urethra** at the base of the urinary bladder and carefully remove connective tissue around it, separating it from the **rectum** as it continues posteriorly into the pelvic cavity.

Where the vas deferentia enter the urethra, locate the small glands located on each side of the urethra. These are the **seminal vesicles.**

Between the seminal vesicles on the dorsal surface of the urethra is the small **prostate gland.** Follow the urethra posteriorly to locate the pair of **bulbourethral glands** on each side of the urethra where it enters the penis. At ejaculation, these accessory glands secrete the fluids that transport sperm.

CONCLUSION

This completes the dissection. If you have done it thoughtfully, you have gained a good deal of knowledge about the body organization of mammals, including humans. Dispose of your specimen as directed by your instructor. There is no laboratory report for this exercise.

13

CIRCULATION OF BLOOD

The **cardiovascular system** transports substances throughout the body by circulating the blood through a system of closed vessels. In this way, nutrients and oxygen are supplied to cells and organic wastes and carbon dioxide are removed from them. The components of the cardiovascular system and their functions are:

Heart   Pumps the blood through the blood vessels.
Arteries   Carry blood away from the heart.
Capillaries Microscopic vessels connecting the smallest arteries and smallest veins; sites of the exchange of materials between the blood and body cells.
Veins   Return blood to the heart.
Blood   Transporting medium.

In this exercise, you will investigate the heart and blood vessels. Blood will be considered in the next exercise.

THE HEART

As you read this section, refer to Figure 13.1. The human heart, like those of all mammals and birds, consists of four chambers: two atria and two ventricles. The two **atria** receive blood returning to the heart in veins, and the two **ventricles** pump blood away from the heart into arteries.

The right atrium receives blood from two large veins. The **superior vena cava** returns blood from the head, neck, shoulders, and arms, and the **inferior vena cava** returns blood from the rest of the body and the legs. The left atrium receives blood from the left and right **pulmonary veins.** The right ventricle pumps blood into the **pulmonary trunk,** which branches into the pulmonary arteries carrying blood to the lungs. The left ventricle pumps blood into the **aorta,** which divides to carry blood to all parts of the body except the lungs.

The heart actually is composed of two pumps. The right atrium and right ventricle form one pump, and the left atrium and left ventricle form the other pump. There are no connections between the two pumps. That is, there are no

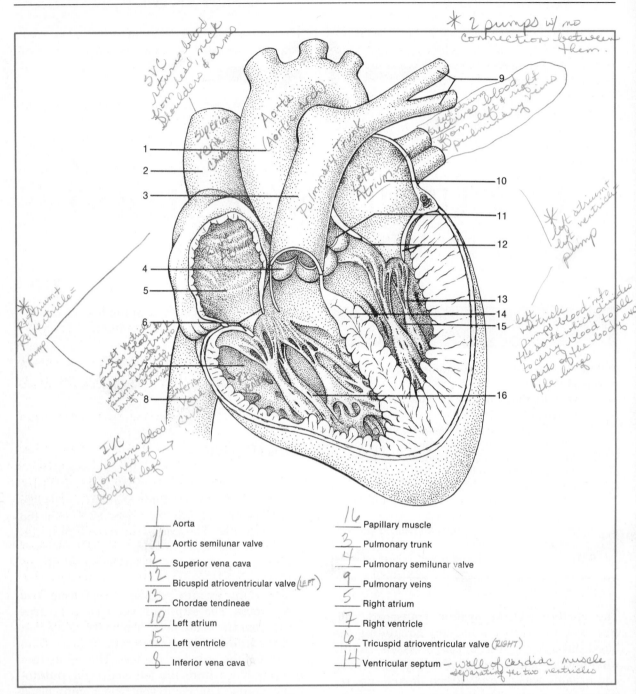

Figure 13.1. Basic structure of the heart

Handwritten annotations on figure:

* 2 pumps w/ no connection between them.

SVC returns blood from head, neck shoulders & arms

Superior vena cava

Aorta (Aortic arch)

receives blood from left & right pulmonary veins

Pulmonary Trunk

Left Atrium

* left atrium + left ventricle = pump

* Rt Atrium + Rt Ventricle = pump

right ventricle pumps blood into pulmonary trunk which divides into pulm arteries carrying blood to lungs

left ventricle pumps blood into the aorta which divides to carry blood to all parts of the body except the lungs

Inferior vena cava

IVC returns blood from rest of Body & legs

Labels (left column):

_1_ Aorta
_11_ Aortic semilunar valve
_2_ Superior vena cava
_12_ Bicuspid atrioventricular valve (LEFT)
_13_ Chordae tendineae
_10_ Left atrium
_15_ Left ventricle
_8_ Inferior vena cava

Labels (right column):

_16_ Papillary muscle
_3_ Pulmonary trunk
_4_ Pulmonary semilunar valve
_9_ Pulmonary veins
_5_ Right atrium
_7_ Right ventricle
_6_ Tricuspid atrioventricular valve (RIGHT)
_14_ Ventricular septum — wall of cardiac muscle separating the two ventricles

connections between the two atria or between the two ventricles. The **ventricular septum,** a wall of cardiac muscle, separates the two ventricles.

There is an opening between the atrium and ventricle on each side that is guarded by an atrioventricular (AV) valve. Each AV valve is named for the number of cusps or valve flaps that compose it. The **bicuspid atrioventricu-**

lar valve is located between the left atrium and left ventricle. The **tricuspid atrioventricular valve** lies between the right atrium and right ventricle. Heart valves keep the blood flowing in a single direction. Each AV valve permits blood to flow from atrium to ventricle but prevents reverse flow from ventricle to atrium.

The cusps of AV valves are anchored by **chordae tendineae,** thin, nonelastic cords of fibrous connective tissue, to the **papillary muscles,** small mounds of heart muscle on the inner surface of the ventricles. This arrangement holds the valve flaps so that they seal off each atrium during ventricular contraction and prevents the valve cusps from being forced into the atria. The valve cusps restrained by chordae tendineae resemble parachutes and their lines when the ventricles contract.

Another set of valves, the semilunar valves, are located at the base of the two large arteries carrying blood from the heart. The **pulmonary semilunar valve** and the **aortic semilunar valve** allow blood to enter their respective arteries during ventricular contraction but prevent the backflow of blood from the arteries into the ventricles.

The wall of the heart, best observed in the ventricles, consists mostly of **cardiac muscle tissue** sandwiched between two membranes. The interior of the heart is lined by the **endocardium,** a membrane of simple squamous epithelium, and the exterior surface is covered by the **epicardium,** which is also simple squamous epithelium. In the body, the heart is enveloped by a somewhat loose-fitting, double-layed membrane, the **pericardial sac.** Fluid secreted between the pericardial sac and the epicardium reduces friction, enabling the heart to move freely within the pericardial sac as it beats.

Flow of Blood Through the Heart

The flow of blood through the heart is related to the heart cycle of relaxation and contraction. **Diastole** is the relaxation phase, and **systole** is the contraction phase.

During atrial and ventricular diastole, blood flows into the right atrium from the superior and inferior venae cavae and into the left atrium from the pulmonary veins. The contraction of the atria in atrial systole forces blood into the relaxed ventricles until they are filled with blood. In ventricular systole, the ventricles contract and the atria relax. The sudden increase in blood pressure within the ventricles closes the AV valves and pumps blood through the semilunar valves into the arteries. The right ventricle pumps blood into the pulmonary trunk and on to the lungs, and the left ventricle pumps blood into the aorta and on to all other parts of the body. At the start of ventricular diastole, the semilunar valves close. Note that blood pressure

changes in the heart chambers cause the opening and closing of the heart valves.

Heart Sounds

The heart sounds are usually described as a *lub-dup* (pause) *lub-dup,* and so forth. These sounds are produced by the closing of the heart valves. The closure of the AV valves produces the first sound at the start of ventricular systole, and the closure of the semilunar valves at the start of ventricular diastole produces the second sound.

Materials

Per lab
Alcohol pads
Colored pencils
Heart model
Stethoscopes

Assignment 1

1. Label Figure 13.1 and color-code the major parts of the heart.
2. Identify the parts of the heart on a heart model.
3. Use a stethoscope to listen to your own heart sounds. A stethoscope amplifies the sounds so you can hear them more clearly. Clean the ear pieces with an alcohol pad. Fit them into your ears by directing them inward and forward. Place the stethoscope diaphragm over your heart just to the left of your sternum (breast bone) to hear the heart sounds. You don't need to place the stethoscope diaphragm against your skin—over a shirt or blouse usually is OK.
4. ***Complete item 1 on Laboratory Report 13 that begins on page 349.***

DISSECTION OF A SHEEP HEART

A study of external and internal structure of a sheep heart will extend your understanding of heart structure and function. Refer to Figures 13.1 and 13.2 as you proceed.

Materials

Per student group
Dissecting instruments
Dissecting pan
Sheep heart, fresh or preserved

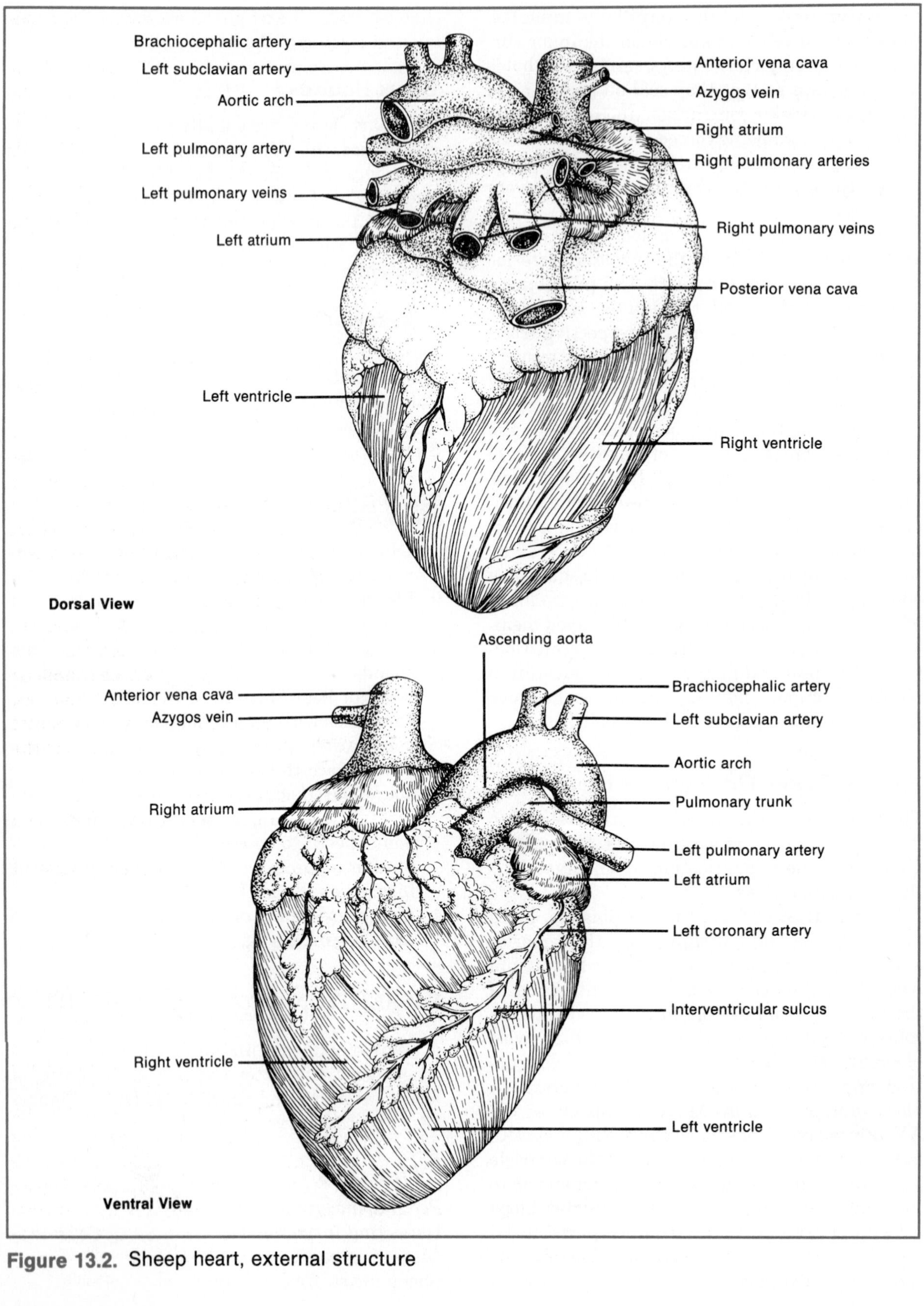

Brachiocephalic artery
Left subclavian artery
Aortic arch
Left pulmonary artery
Left pulmonary veins
Left atrium
Left ventricle

Anterior vena cava
Azygos vein
Right atrium
Right pulmonary arteries
Right pulmonary veins
Posterior vena cava
Right ventricle

Dorsal View

Ascending aorta

Anterior vena cava
Azygos vein
Right atrium
Right ventricle

Brachiocephalic artery
Left subclavian artery
Aortic arch
Pulmonary trunk
Left pulmonary artery
Left atrium
Left coronary artery
Interventricular sulcus
Left ventricle

Ventral View

Figure 13.2. Sheep heart, external structure

1. Study the sheep heart as described below. Because a sheep walks on four legs and humans walk on two legs, descriptive directional terms pertaining to the heart are different in the two organisms. A few comparative terms for each are:

Sheep	*Human*
Ventral	Anterior
Dorsal	Posterior
Anterior	Superior
Posterior	Inferior

External Features

a. Usually the pericardial sac has been removed, but look for remnants of it around the bases of the venae cavae, aorta, and pulmonary vessels.

b. Note that the epicardium is so tightly attached to the heart that it does not appear as a distinct membrane macroscopically.

c. The coronary arteries and veins that supply blood to the heart itself are obscured by fat deposits. Locate the fat deposits along the interventricular sulcus (groove) on the ventral surface of the heart. The right and left ventricles lie on either side of this sulcus. Pick away some of the fat to locate a coronary artery and cardiac vein.

d. Looking at the ventral surface of the heart and referring to Figure 13.2, locate the left and right atria, aorta, pulmonary trunk, and anterior vena cava. Compare the thickness of the vessel wall in the aorta with that of the anterior vena cava.

e. Looking at the dorsal view of the heart and referring to Figure 13.2, locate the aorta, left and right pulmonary arteries, left and right pulmonary veins (which are often embedded in fat), anterior vena cava, and posterior vena cava.

Internal Features

a. Hold the heart in your left hand, dorsal side up, with the anterior vena cava toward you. Insert a scissors blade into the anterior vena cava and cut through its wall into the right atrium. Locate the tricuspid AV valve. Holding the heart upright, pour water through the tricuspid valve into the right ventricle. Observe the action of the tricuspid valve as you gently squeeze the right ventricle.

b. Pour out the water from the right ventricle and continue your cut through the tricuspid valve and ventricle wall to the tip of the right ventricle. Spread the cut ventricular wall and locate the papillary muscles and chordae tendineae. Are the chordae tendineae fragile or tough, strong or weak, elastic or nonelastic?

c. Insert a probe into the cut end of the pulmonary trunk and into the right ventricle. Use scissors to cut from the right ventricle through the anterior ventricular wall along the probe and through the pulmonary semilunar valve. Note the arrangement and thickness of the cusps.

d. Insert a scissors blade into the top of the left atrium and cut through walls of the left atrium and ventricle to the tip of the left ventricle. Spread the ventricular walls and locate the bicuspid valve, chordae tendineae, and papillary muscles. Compare the thickness of the walls in the left and right ventricles.

e. Insert a probe into the cut end of the aorta and into the left ventricle. Cut from the left ventricle along the probe and through the aortic semilunar valve. Locate the two openings to coronary arteries just above (anterior to) the semilunar valve. Note the valve structure.

f. Dispose of the heart as directed by your instructor. Wash and dry your dissecting pan and instruments.

2. ***Complete item 2 on the laboratory report.***

THE PATTERN OF CIRCULATION

In vertebrates, blood flows through a system of closed vessels to transport materials to and from body cells. **Arteries** carry blood from the heart, and they divide into smaller and smaller arteries and ultimately lead to **arterioles,** which are nearly microscopic in size. Arterioles lead to the **capillaries,** the smallest blood vessels. The walls of the capillaries are composed of a single layer of squamous epithelial cells. See Figure 13.3. Exchange of materials occurs between blood in capillaries and the body cells. From capillaries, blood flows into tiny veins called **ven-**

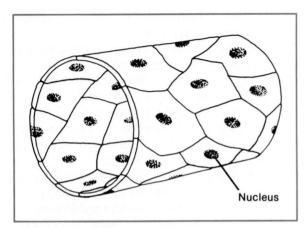

Figure 13.3. Capillary structure

ules that combine to form larger **veins** that carry blood back to the heart.

Figure 13.4 shows the relationship between an arteriole, capillaries, and a venule. The diameter of the arterioles affects the flow of blood and blood pressure and is controlled by circular smooth muscles that respond to impulses from the autonomic nervous system. The flow of blood into capillaries is controlled by a **precapillary sphincter muscle** that is governed by impulses from the autonomic nervous system and local chemical stimuli. An increase in the local CO_2 concentration opens the sphincter and increases capillary blood flow. Similarly, a decrease in CO_2 concentration reduces capillary blood flow.

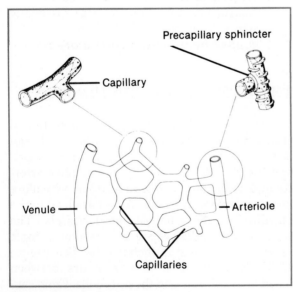

Figure 13.4. Capillary network

The exchange of materials between the body cells and capillary blood involves **tissue fluid** (interstitial fluid) as an intermediary. Tissue fluid is the thin layer of extracellular fluid that covers all cells and tissues. This is the pattern of the exchange:

$$\begin{array}{ccc} \text{Blood in} & \rightleftharpoons \text{Tissue} & \rightleftharpoons \text{Body} \\ \text{capillaries} & \text{fluid} & \text{cells} \end{array}$$

Since the heart is a double pump, there are two separate circuits (pathways) of circulation: the pulmonary circuit and the systemic circuit. Examine the diagrammatic scheme of circulation in Figure 13.5 as you read this section.

The **pulmonary circuit** carries deoxygenated (low in oxygen) blood to the lungs and returns oxygenated (high in oxygen) blood to the heart. The right ventricle pumps deoxygenated blood through the pulmonary trunk that branches forming the pulmonary arteries carrying blood to the lungs. In the lungs, oxygen is picked up by the blood as it releases carbon dioxide into air in the lungs. Then oxygenated blood is returned through the pulmonary veins to the left atrium.

The **systemic circuit** carries oxygenated blood to all parts of the body except the lungs and returns deoxygenated blood to the heart. The left ventricle pumps oxygenated blood through the aorta, and the aorta gives off smaller arteries that carry oxygenated blood throughout the body (except the lungs). After oxygen and carbon dioxide are exchanged with body cells, deoxygenated blood is carried in veins that merge to form the superior vena cava draining areas above the heart and the inferior vena cava draining areas below the heart. The venae cavae return deoxygenated blood into the right atrium.

Note that blood from the intestines is carried by the hepatic portal vein to the liver before it is returned to the inferior vena cava by the hepatic vein. This allows the liver to process nutrients absorbed from the intestine before they enter the systemic circulation.

Materials

Per lab

Demonstration setup of capillary flow in a frog's foot or a goldfish's caudal fin

Prepared slides of:

artery and vein, x.s.

atherosclerotic artery, x.s.

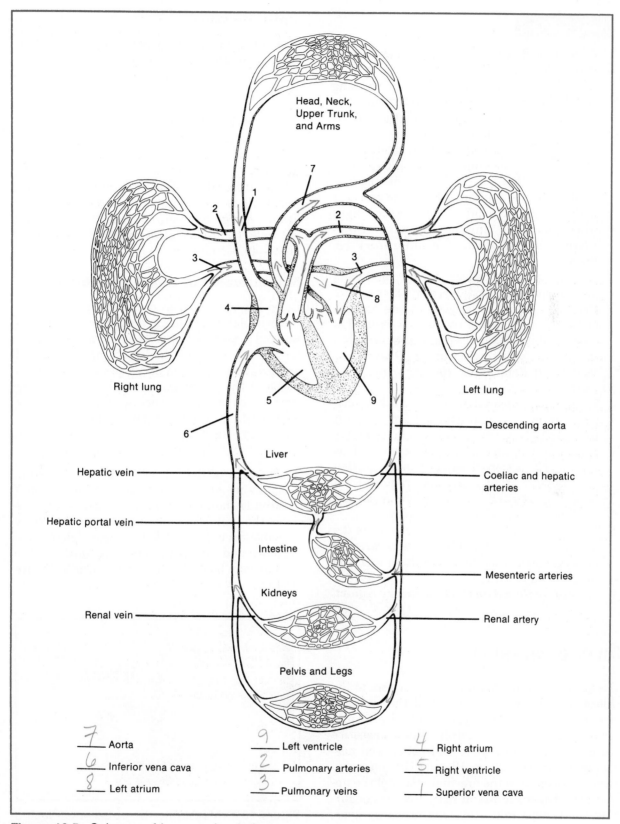

Head, Neck, Upper Trunk, and Arms

Right lung

Left lung

Descending aorta

Hepatic vein

Liver

Coeliac and hepatic arteries

Hepatic portal vein

Intestine

Mesenteric arteries

Kidneys

Renal vein

Renal artery

Pelvis and Legs

_7_ Aorta

_9_ Left ventricle

_4_ Right atrium

_6_ Inferior vena cava

_2_ Pulmonary arteries

_5_ Right ventricle

_8_ Left atrium

_3_ Pulmonary veins

_1_ Superior vena cava

Figure 13.5. Scheme of human circulation

1. *Complete items 3a–3c on the laboratory report.*
2. Examine the capillary blood in the webbing of a frog's foot that has been set up under a demonstration microscope. Note the pulsating flow in the feeding arteriole and the smooth flow in the collecting venule. Observe the blood cells moving in single file through a capillary. How would you describe the flow of blood in the capillary? Keep the foot wet with water. Turn off the microscope light when you have finished observing. *Complete items 3d–3f on the laboratory report.*
3. Examine prepared slides of artery and vein, x.s. Compare the difference in thickness of the walls due to differences in the amount of smooth muscle and connective tissue. Locate the single layer of squamous endothelial cells forming the interior lining. *Draw these vessels in item 3g on the laboratory report.*
4. Examine a prepared slide of an atherosclerotic artery, x.s., and note the fatty deposit that partially plugs the vessel. Cholesterol is primarily responsible for this deposit. This type of obstruction in coronary arteries often causes heart attacks and may require coronary bypass surgery. *Complete items 3g and 3h on the laboratory report.*
5. Label Figure 13.5. Add arrows to indicate the direction of blood flow. Color vessels carrying deoxygenated blood blue and those carrying oxygenated blood red.
6. *Complete item 3 on the laboratory report.*

BLOOD PRESSURE

Usually, the term "blood pressure" refers to the blood pressure within arteries of the systemic circuit. There are two types of blood pressure: systolic and diastolic. **Systolic blood pressure** occurs during ventricular contraction and normally averages 120 ± 10 mm Hg (mercury) when measured in the brachial artery of the upper arm. **Diastolic blood pressure** occurs during ventricular relaxation and normally averages 80 ± 10 mm Hg.

Pulse Pressure

The difference between systolic and diastolic blood pressures produces the **pulse pressure.** The alternating increase and decrease of arterial blood pressures causes a corresponding expansion and contraction of the elastic arterial walls. The pulsating expansion of arterial walls may be detected as the **pulse** by placing the fingers on the skin over a surface artery. See Figure 13.6.

The pulse rate indicates the number of heart contractions per minute. Normal pulse rates usually range between 65 and 80 per minute, but well-conditioned athletes may have rates as low as 40 per minute.

Measurement of Blood Pressure

Blood pressure is most commonly measured in the brachial artery using a **sphygmomanometer** and a **stethoscope.** The sphygmomanometer consists of an inflatable cuff that is wrapped around the upper arm and a pressure gauge to measure the air pressure within the cuff. The stethoscope is used to hear sounds produced by blood rushing through a partially closed brachial artery. Read through the following procedures completely before you begin the process.

1. The arm of the "patient" should be resting palm up on the table top. Wrap the cuff around the upper arm with the bottom edge of the cuff about an inch above the elbow joint. Secure the cuff with the Velcro® fastener and attach the pressure gauge so the dial may be easily read, as shown in Figure 13.7.

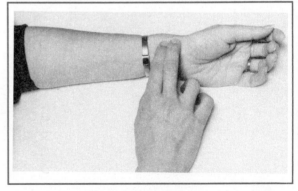

Figure 13.6. Measuring the pulse rate. Place the fingers over the radial artery at the wrist.

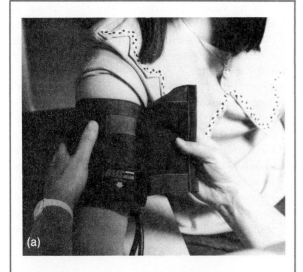

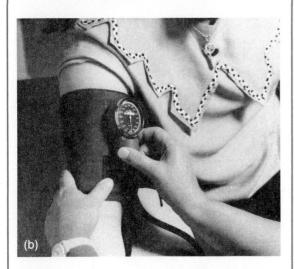

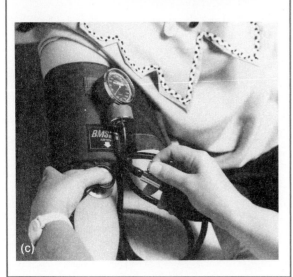

2. Insert the ear pieces of the stethoscope inward and forward into your ears. Hold the diaphragm of the stethoscope over the brachial artery with your left thumb, as shown in Figure 13.7. With your right hand, gently close the screw valve above the bulb of the sphygmomanometer and squeeze the bulb to inflate the cuff to about 150 mm Hg. This pressure closes the brachial artery.

3. Open the screw valve slightly to slowly release air from the cuff, decreasing the pressure of the cuff against the artery while listening with the stethoscope for a pulsating sound of blood squirting through the partially closed artery. As soon as you hear the first pulsating sound, read the pressure on the pressure gauge. This is the systolic blood pressure. (The pressure in the cuff equals the systolic pressure forcing blood through the partially closed artery.)

4. Continue to slowly release air from the cuff while listening to the pulsating sound as it deepens and then ceases. As soon as the sound ceases, read the pressure on the pressure gauge. This is the diastolic blood pressure.

5. Release the remaining air from the cuff and remove it from the "patient's" arm.

Materials

Per student pair
Sphygmomanometer
Stethoscope

Figure 13.7. Measuring blood pressure.
(a) The sphygmomanometer cuff is placed around the upper arm with its lower edge about an inch above the elbow. Attach it snugly with the Velcro® fastener and fold back the leftover flap. (b) Attach the pressure gauge to the holding strap so that it is easily readable. (c) Close the valve with the set screw and pump air into the cuff to 150 mm Hg. Place the diaphragm of the stethoscope over the brachial artery on the medial half of the anterior elbow joint. Slightly open the valve to slowly release air. When the first pulse sound is heard, read the systolic pressure. Continue releasing air and when the pulse sound suddenly disappears, read the diastolic pressure.

Assignment 4

1. *Complete items 4a and 4b on the laboratory report.*
2. Measure your pulse rate (beats/min) at rest by placing your fingers over the radial artery, as shown in Figure 13.6. Count the number of beats for 15 seconds and multiply by 4 to determine the beats per minute. Add your results to the class data being recorded on the board (by gender) by your instructor. *Record your resting pulse rate in item 4c on the laboratory report.*
3. If your have no physical disability, run in place for 3 min and measure your pulse rate again. Measure your pulse rate at 1-min intervals until your pulse rate returns to its resting rate. The shorter the time required for your pulse rate to return to its resting rate, the better is your physical condition. *Complete items 4c and 4d on the laboratory report.*
4. Working with your partner, take each other's blood pressure at rest following the procedures described above and shown in Figure 13.7. *Caution: Do not leave the brachial artery compressed for more than 30 seconds. Record your results in item 4e on the laboratory report.*
5. Measure the blood pressure after running in place for 3 min, and a second time 3 min after the exercise. *Complete the laboratory report.*

14

BLOOD

OBJECTIVES

After completing the laboratory session, you should be able to:
1. Describe the components and general functions of blood.
2. Describe the cellular components of blood and recognize them when viewed microscopically.
3. Perform these blood tests: differential white cell count, clotting time, volume of packed red cells, and blood typing.
4. Explain the basis of blood groups, blood typing, transfusions of compatible blood, and erythroblastosis fetalis.
5. Define all terms in bold print.

Human blood carries materials to and from body cells. It consists of **plasma,** a fluid carrier, and **blood cells,** also called formed elements, that are suspended in the plasma. Plasma forms 55% of the blood volume while blood cells compose the remaining 45%. About 92% of the plasma is water, and most of the remaining 8% consists of plasma proteins that give blood its viscous (sticky) character. Blood cells consist of three major types: **erythrocytes,** red blood cells that transport oxygen and carbon dioxide; **leukocytes,** white blood cells that fight infections; and **thrombocytes** or platelets that initiate blood clot formation.

BLOOD CELLS

In this section, you will study characteristics of blood cells and learn to recognize them when viewed microscopically. The cellular components of blood are illustrated in Figure 14.1.

Erythrocytes

Red blood cells (RBCs) are tiny cells with a diameter of only 7–8 μm. They have lost their nuclei during the maturation process so their shape is like biconcave discs. RBCs contain **hemoglobin,** a red pigment that gives the red color to blood. Hemoglobin combines loosely with oxygen and carbon dioxide, enabling these respiratory gases to be transported by RBCs. Erythrocytes are the most numerous blood cells. They average about 5.4 million per cubic millimeter (mm^3) of blood in males and about 4.8 million per mm^3 in females.

Leukocytes

White blood cells (WBCs) are much larger than RBCs, and they always have a nucleus. WBCs are divided into two groups, granulocytes or agranulocytes, depending upon the presence or absence of cytoplasmic granules. Whenever

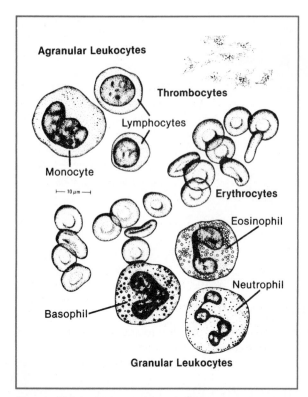

Figure 14.1. Human blood cells

blood cells are to be examined microscopically, they are stained with Wright's blood stain to make it easier to identify the different types of WBCs. The different types of cytoplasmic granules are stained different colors, and all nuclei are stained purple.

Leukocytes are much less numerous in blood than RBCs. WBC counts range from 5,000 to 8,000 per mm^3. Leukocytes play a vital role by combating infection and providing a major defense against disease organisms. WBCs move through capillary walls and wander among surrounding tissue cells by amoeboid movement. It is within tissues where most of their battles against disease organisms occur. Some engulf bacteria and damaged cells and destroy them by digestion in food vacuoles. Others secrete chemicals that destroy or immobilize disease organisms.

Granulocytes

Neutrophils have a nucleus with 2–5 lobes and tiny lavender-staining granules that cause the cytoplasm to appear pale lavender. They are the most numerous leukocytes, forming 60–70% of the total WBC count. Neutrophils fight disease by removing disease organisms and damaged cells by phagocytosis. Their numbers increase rapidly during acute bacterial infections.

Eosinophils are easily recognized by their bilobed or U-shaped nucleus and red-staining cytoplasmic granules. They form only 2–4% of the total WBC count. Eosinophils neutralize histamine, a chemical released during allergic reactions, and they destroy parasitic worms.

Basophils are unique in having either a lobed or U-shaped nucleus and large, blue-staining cytoplasmic granules. They are the least numerous of the leukocytes forming only 0.5–1% of the total WBC count. Basophils release **histamine** during allergic reactions. Histamine dilates blood vessels, increasing the blood flow to affected areas. They also release **heparin,** a chemical that inhibits the formation of blood clots.

Agranulocytes

Lymphocytes are the smallest white blood cells. They are slightly larger than red blood cells. They are easily recognized by the large spherical nucleus surrounded by a small amount of cytoplasm lacking granules. Lymphocytes form 20–25% of the total WBC count, and they play a vital role in immunity. There are two types of lymphocytes. **T-lymphocytes** directly attack and destroy virus-infected cells and tumor cells, while **B-lymphocytes** produce and release into the blood antibodies that destroy disease-causing organisms and provide immunity. The number of lymphocytes increases during viral infections and antigen-antibody reactions.

Monocytes are the largest leukocytes, and they are easily recognized by the presence of a large kidney-shaped nucleus and abundant cytoplasm lacking granules. Monocytes form 3–8% of the total WBCs in blood. They destroy disease-causing organisms by phagocytizing bacteria and cells infected with viruses. An increase in monocytes may indicate a chronic infection.

Thrombocytes

Thrombocytes or platelets are not cells but tiny, non-nucleated fragments of large cells. Thrombocytes are much smaller (2–4 μm in diameter) than RBCs, and their density ranges from

250,000 to 500,000 per mm^3 of blood. Thrombocytes play a vital role in the formation of blood clots.

Materials

Per student
Colored pencils
Compound microscope
Schilling blood chart

Per lab
Prepared slides of human blood:
  normal
  sickle-cell anemia

Assignment 1

1. Obtain a Schilling blood chart and study the appearance of blood cells in circulating blood as shown in the bottom row of the chart. Color the blood cells in Figure 14.1 as they appear on the blood chart.
2. Obtain a prepared slide of normal human blood and examine it using your oil immersion (100×) or high-dry (40×) objective. Locate each type of blood cell. It will take careful searching to locate a basophil and an eosinophil. **Draw each type of blood cell as it appears on your slide in item 1a on Laboratory Report 14 that begins on page 353.**
3. Examine a prepared slide of blood from a patient with **sickle-cell anemia,** an inherited blood disorder. In afflicted persons, RBCs contain abnormal hemoglobin and have a sickle-like shape. Sickled erythrocytes cannot effectively transport oxygen and carbon dioxide.
4. **Complete item 1 on the laboratory report.**

BLOOD TESTS

Blood tests are used by physicians in the diagnosis of certain illnesses. In our study, the following tests are used to increase your understanding of blood and are not to be considered diagnostic.

Your instructor may choose to have the tests requiring a blood sample performed using sheep blood, chemicals that simulate blood, or your own blood. If you use your own blood, your instructor will inform you of specific procedures to follow so that no student will be in contact with blood of another student as a precaution against the inadvertent transmission of viruses that cause AIDS or hepatitis. **Caution:** *Follow your instructor's directions carefully. Students who know that they are positive for HIV or hepatitis are* **not** *to perform these tests on their own blood.*

To obtain a drop of your own blood for test purposes, you will pierce a fingertip with a sterile, disposable lancet as described below.

1. Rub the fingertip to be pierced for a few seconds to increase the blood flow.
2. Use an alcohol pad to cleanse the fingertip and allow the fingertip to dry.
3. Remove a sterile lancet from its paper package so that the tip does not touch anything.
4. Pierce the fingertip with a quick, deep stab and *immediately place the used lancet in the biohazard sharps container.*
5. The resulting drop of blood is then used as described for one of the tests that follow.
6. Press the alcohol pad against the fingertip until the bleeding stops and *place the alcohol pad in the biohazard bag.*

Differential White Cell Count

A differential white cell count determines the percentage of each type of white blood cell in the total white cell count. It is used clinically in making a diagnosis of certain illnesses including acute and chronic infections.

Materials

Per lab
Prepared slides of human blood:
  normal
  infectious mononucleosis

Assignment 2

1. Examine a prepared slide of human blood with your microscope. Use the low power objective to locate an area where the blood cells are separated from each other. Then switch to the high-dry (40×) or oil immersion (100×) objective. Move the slide in the manner as shown in Figure 14.2 as you identify and **tabulate each WBC observed in item 2a on**

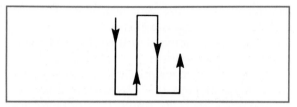

Figure 14.2. Pattern of slide examination when making a differential white cell count

the laboratory report until you have tabulated 100 white blood cells.

2. Obtain a prepared slide of blood from a patient with **infectious mononucleosis,** a contagious disease caused by the Epstein-Barr virus. It occurs mostly in children and young adults and is often transmitted by kissing. Perform a differential white cell count of this slide.

3. ***Complete item 2 on the laboratory report.***

Clotting Time

The clotting (coagulation) of blood is a protective mechanism to prevent the excessive loss of blood. Normal clotting time is 3–6 min. Figure 14.3 illustrates the chain reaction that leads to clot formation. Damaged tissue cells and ruptured thrombocytes release the enzyme **prothrombin activator** to start the chain reaction. This enzyme catalyzes the conversion of prothrombin (an inactive enzyme) into the active enzyme **thrombin** that promotes the conversion of **fibrinogen,** a soluble plasma protein, into **fi-**

brin, insoluble protein strands. The tangled threads of fibrin plus entangled blood cells compose a blood clot.

Materials

Per student
Alcohol pad
Capillary tube, nonheparinized
File, three-cornered
Lancet, sterile and disposable

Per lab
Biohazard bag
Biohazard sharps container
Household bleach, 10%

Assignment 3

1. Place a double layer of paper towels on the table top at your work station. All of the following procedures are to be done over the paper towels.

2. Pierce a fingertip to yield a drop of blood following the procedure described earlier. *Immediately place the used lancet in the biohazard sharps container.* Record the time.

3. Insert one end of a nonheparinized capillary tube into the drop of blood and lower the other end until the tube fills with blood. Lay the tube on the paper towel and press an alcohol pad against your fingertip until the bleeding stops. *Immediately place the alcohol pad in the biohazard bag.*

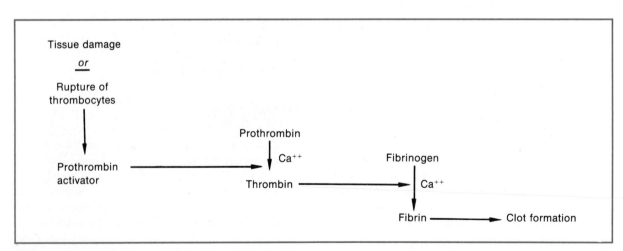

Figure 14.3. Reactions leading to clot formation

4. Exactly 2 min after the blood drop appeared, score the tube with the file and break off a small segment of the tube to see if clotting has occurred. When breaking off a segment, separate the ends carefully to see if a fibrin strand extends between them. Repeat at 1-min intervals until a fibrin strand appears. Record the time of the appearance of a fibrin strand. See Figures 14.4–14.6.

5. Place the file in the pan of 10% household bleach. Wrap the capillary tube and fragments in the paper towels and place both towels and tube fragments in the biohazard bag.

6. Wash your work station with 10% household bleach and place the paper towels in the biohazard bag.

7. Calculate the clotting time and ***complete item 3 on the laboratory report.***

Volume of Packed Red Cells

Anemia is a condition in which the oxygen-carrying capacity of the blood is decreased. It may result from a decreased concentration of hemoglobin in RBCs or from a decreased concentration of RBCs in the blood. One way to determine the concentration of erythrocytes in blood is to measure the **volume of packed red cells (VPRC)** or **hematocrit** as a percentage of the blood volume. In males, the normal VPRC range is 40–54%. In females, the normal VPRC range is 37–47%. The VPRC is determined using heparinized capillary tubes, a special micro-hematocrit centrifuge, and a micro-hematocrit tube reader.

Materials

Per student
Alcohol pad
Capillary tube, heparinized
Lancet, sterile and disposable

Per lab
Biohazard bag
Biohazard sharps container
Household bleach, 10%
Micro-hematocrit centrifuge (Adams)
Seal-Ease (Clay-Adams)
Micro-hematocrit tube reader (Adams)

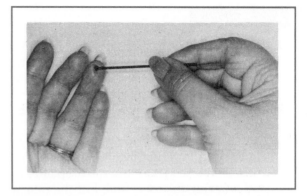

Figure 14.4. Fill a plain capillary tube by inserting one end into the drop of blood and lowering the other end.

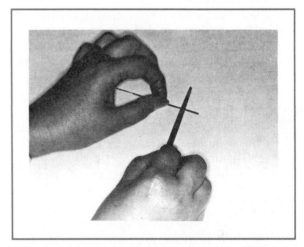

Figure 14.5. At 1 min intervals, use a three-cornered file to score the tube and break off a small section to see if the blood has clotted.

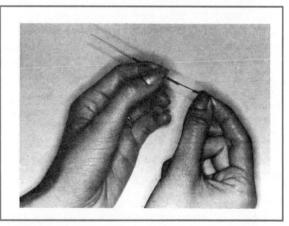

Figure 14.6. When the blood has clotted a strand of fibrin will be noted between the broken ends of the tube.

1. Place a double layer of paper towels on the table at your work station and perform the collection of blood over the paper towels as before.
2. Blot the first drop of blood with the alcohol pad.
3. Collect the second drop of blood in a heparinized capillary tube by inserting the heparinized end (red ring) into the drop of blood and lowering the other end. Fill the tube at least two-thirds full. See Figure 14.7.
4. Insert the blood end of the tube into Seal-Ease to seal that end of the tube with clay. See Figure 14.8.
5. Place the tube in a numbered slot of the centrifuge with the sealed end against the rubber ring around the perimeter of the centrifuge disc. Record your slot number. (The centrifuge must be loaded with an even number of tubes in opposite slots so that the centrifuge is balanced.) See Figure 14.9.
6. When loading is completed, screw on the inside cover and tighten it by hand. Then fasten the outer cover.
7. Start the centrifuge by setting the timer at 4 min.
8. After the centrifuge has stopped, open the outer cover and unscrew the inner cover using the centrifuge wrench if necessary.
9. Remove your tube and use the micro-hematocrit tube reader to determine your VPRC (hematocrit) following the procedures on the tube reader. See Figure 14.10.
10. Wrap your tube and anything else in contact with your blood in the paper towels and place them in the biohazard bag. Wash your work station with 10% household bleach and place the paper towels in the biohazard bag.
11. ***Complete item 4 on the laboratory report.***

Blood Typing

Human blood may be classified according to the presence or absence of various **antigens** (proteins) on the red blood cells. The presence of these antigens is genetically controlled, so an individual's **blood type** is the same from birth

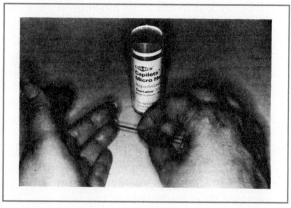

Figure 14.7. Fill a heparinized capillary tube by inserting the red-tipped end into the drop of blood and lowering the other end.

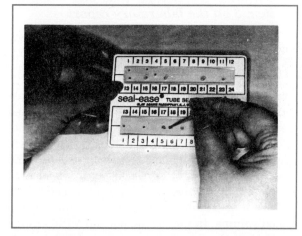

Figure 14.8. Insert the blood end of the tube into Seal-Ease to plug that end of the tube.

Figure 14.9. Place the capillary tube in a numbered slot of the centrifuge with the sealed end against the bumper at the perimeter of the head.

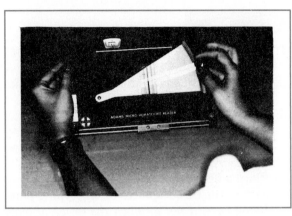

Figure 14.10. Use the tube reader to determine the percentage of the VPRC (hematocrit).

to death. See Table 14.1. The antigens most commonly tested for in blood typing tests are A, B, and D (Rh).

Testing for the presence of A and B antigens determines the ABO blood group. The presence or absence of the D (Rh) antigen determines the Rh blood type. The two factors are combined in designating an individual's blood type (e.g., A,Rh⁺; A,Rh⁻; AB,Rh⁺). Table 14.2 indicates the association of these **antigens** and their corresponding **antibodies** in blood.

All blood types are not compatible with each other. Therefore, knowing the blood types of both donor and recipient in blood transfusions is imperative. The *antigens of the donor* and the *antibodies of the recipient* must be considered in blood transfusions. Transfusion of incompatible blood results in clumping (agglutination) of erythrocytes in the transfused blood, which may plug capillaries and result in death. Clumping of erythrocytes occurs whenever the *antigens of the donor* and the *antibodies of the recipient,* that are designated by the same letter, are brought

TABLE 14.1
Percentage of ABO Blood Types

Blood Type	% U.S. Blacks	% U.S. Whites
A	25	41
B	20	7
AB	4	2
O	51	50

TABLE 14.2
Antigen-Antibody Associations

Blood Type	Antigen	Antibody
O	None	a,b
A	A	b
B	B	a
AB	AB	None
Rh⁺	D	None
Rh⁻	None	d*

* Antibodies are produced by an Rh⁻ person only after Rh⁺ red blood cells enter his or her blood.

together. See Table 14.2. The antibodies of the donor are so diluted in the recipient's blood that their effect is inconsequential.

The Rh Factor

About 85% of Caucasians and 99–100% of Chinese, Japanese, African blacks, and Native Americans possess the D (Rh) antigen and are therefore Rh⁺. If an Rh⁻ individual receives a single transfusion of Rh⁺ blood, antibodies are produced against the D (Rh) antigen. However, no clumping of cells occurs due to the gradual increase in antibodies and the loss of erythrocytes containing the D (Rh) antigen. If a second transfusion of Rh⁺ blood is received, clumping will occur, and death may result.

Antibodies against the D (Rh) antigen may be produced by an Rh⁻ woman carrying an Rh⁺ fetus due to "leakage" of fetal erythrocytes into the maternal blood. The accumulation of antibodies in maternal blood is gradual enough so that no complications result during the first pregnancy with an Rh⁺ fetus. In subsequent pregnancies with Rh⁺ fetuses, however, the maternal antibodies may diffuse into the fetal blood and destroy the fetal erythrocytes. This pathological condition, **erythroblastosis fetalis,** may be fatal to the fetus. The mother suffers no consequences unless she subsequently receives a transfusion of Rh⁺ blood.

Typing Procedure

Testing for the presence of A, B, and D antigens in blood is done by adding a drop of **antiserum** containing a specific antibody to a drop of blood. If the corresponding antigen is present, the antibody will cause the RBCs to agglutinate, forming clumps of RBCs that are visually detectable.

Since testing for the presence of the D antigen requires a temperature of 50 °C, you will use a slide warming box and a special typing plate to allow simultaneous testing for all three antigens: A, B, and D. See Figure 14.11. In addition, each typing station has the other materials needed for the typing procedure.

Materials

Per typing station
Alcohol pads
Biohazard bag
Biohazard sharps container
Blood typing box with typing plate
Blood typing antisera (anti-A, anti-B, anti-D)
Household bleach, 10%
Lancets, sterile and disposable
Microscope slides, new
Toothpicks, flat

Assignment 5

1. Turn on the slide warming box for at least 5 min prior to use.
2. Place a clean glass slide across the typing plate and allow it to warm for 3 min.
3. Cleanse your fingertip with an alcohol pad. After it has dried, pierce it with a sterile lancet. *Immediately, place the used lancet in the biohazard sharps container.*
4. Place a drop of blood on your glass slide over each of the three squares on the typing plate.
5. *Being careful not to touch the drops of blood with the droppers,* add a drop of anti-D antiserum to the drop of blood in the D square, a drop of anti-B serum to the drop of blood in the B square, and a drop of anti-A serum to the drop of blood in the A square.
6. Quickly mix the antiserum and blood in each square with a *different* toothpick. (Using the same toothpick for more than one square will invalidate the results.)

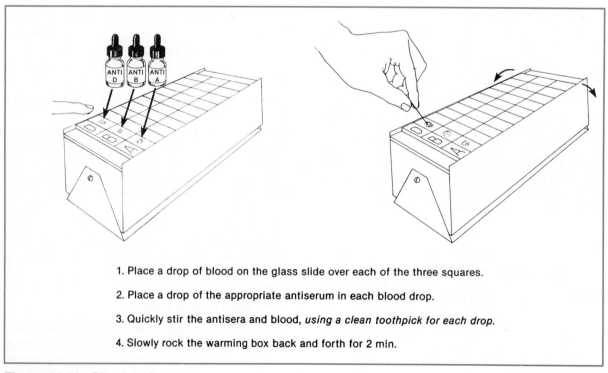

1. Place a drop of blood on the glass slide over each of the three squares.

2. Place a drop of the appropriate antiserum in each blood drop.

3. Quickly stir the antisera and blood, *using a clean toothpick for each drop.*

4. Slowly rock the warming box back and forth for 2 min.

Figure 14.11. Blood typing setup

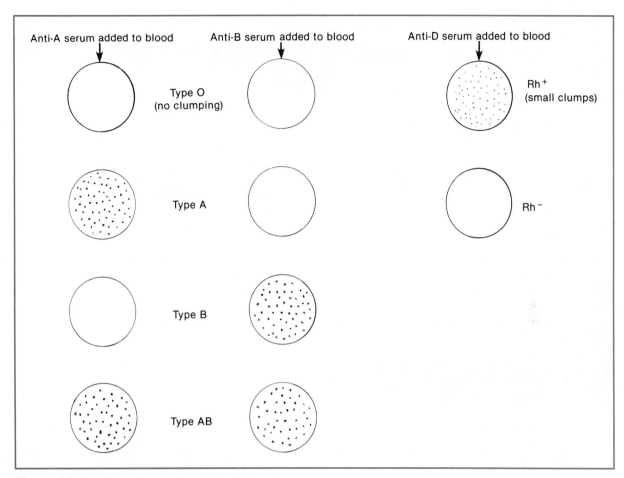

Figure 14.12. Blood type determination

7. Rock the warming box back and forth for exactly 2 min and then read the results. The presence of tiny red granules in the D square indicates presence of the D antigen. Clumping of RBCs in the B square indicates the presence of the B antigen, and clumping in the A square indicates presence of the A antigen. See Figure 14.12.

8. Place your slide, alcohol pad, and anything else in contact with your blood in the biohazard bag.

9. ***Complete item 5 on the laboratory report.***

GAS EXCHANGE

After completion of the laboratory session, you should be able to:
1. Identify the parts of the respiratory system on a human torso model or on charts and describe the function of each part.
2. Describe the mechanics of breathing.
3. Determine the capacities of your lungs.
4. Define all terms in bold print.

Body cells require a continuous supply of oxygen and the constant removal of carbon dioxide. These needs are met by the interaction of the respiratory and circulatory systems. The respiratory system brings air into the lungs, where oxygen diffuses into the blood and carbon dioxide from the blood diffuses into the air in the lungs. The air in the lungs is then exhaled, and the carbon dioxide is dispersed in the atmosphere. In this way, breathing enables the continuous exchange of oxygen and carbon dioxide between air and blood in the lungs. The circulatory system transports oxygen to the body cells and brings carbon dioxide to the lungs for exchange.

THE RESPIRATORY SYSTEM

Refer to Figure 15.1 as you read this section and label the parts indicated.

The nasal cavity is divided into left and right portions by the **nasal septum,** a midline partition composed of bone and cartilage. The **hard** and **soft palates** separate the nasal and oral cavities. The nasal cavity warms and filters the air as it enters the respiratory system. This is accomplished as the air passes over the warm, moist, mucus-producing membrane lining the nasal cavity. Foreign particles tend to be trapped in mucus that is carried to the pharynx by beating cilia and swallowed. The turbinates, or **nasal conchae,** which are shelflike protuberances from the lateral walls of the nasal cavity, play an important role by increasing the surface area in contact with the passing air.

The **pharynx** serves as a passageway for both air and food. From the pharynx, air enters the **larynx** through an opening called the **glottis.** The larynx is a cartilaginous box that contains the vocal cords. The **epiglottis,** a small cartilaginous projection from the upper part of the larynx, flops over to close the glottis when you swallow to prevent food and liquids from entering the larynx.

The larynx is attached to the upper end of the **trachea,** or windpipe, which is located just ante-

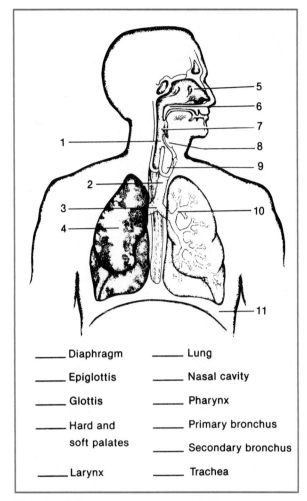

Figure 15.1. The human respiratory system

_____ Diaphragm   _____ Lung

_____ Epiglottis   _____ Nasal cavity

_____ Glottis   _____ Pharynx

_____ Hard and   _____ Primary bronchus
      soft palates
                         _____ Secondary bronchus

_____ Larynx   _____ Trachea

Figure 15.1. The human respiratory system

the **alveoli.** The alveoli are microscopic sacs arranged in clusters at the end of each tiny air duct, and they are formed of simple squamous epithelium. The exchange of oxygen and carbon dioxide occurs between the air in the alveoli and the blood in the capillaries that surround them. Each lung has about 300 million alveoli, which increase the respiratory surface of the lung to about 75 m^2. See Figure 15.2.

The outer surface of each lung is covered by a thin, tightly adhering membrane, the **visceral pleura,** while the inner surface of the thoracic cavity is lined by a similar membrane, the **parietal pleura.** Fluid secreted by these membranes into the **pleural cavity,** the very thin space between the membranes, reduces friction as the lungs expand and contract during breathing.

The **diaphragm** is the sheetlike muscle separating the thoracic and abdominopelvic cavities. It is the primary muscle involved in breathing.

Materials

Per lab
Anatomical charts
Corrosion preparation of mammalian lungs
Human torso model
Prepared slides of:
    trachea, x.s.
    lung tissue, normal
    lung tissue, emphysematous

rior to the esophagus. The trachea descends to about midway in the thorax, where it divides to form the **primary bronchi** that enter each lung.

Like the nasal cavity, the larynx, trachea, and bronchi are lined with **pseudostratified ciliated columnar epithelium** whose goblet cells produce a thin layer of mucus over the epithelium. Airborne particles (e.g., dust, pollen, and bacteria) are trapped in the mucus. Then the beating cilia of the epithelium move the mucus and entrapped particles upward to the pharynx, where it is swallowed. In this way foreign materials are removed from the air passages, which helps to prevent respiratory infections.

In the lungs, the primary bronchi divide to form the smaller **secondary bronchi** that lead to the still smaller **bronchioles** and finally to

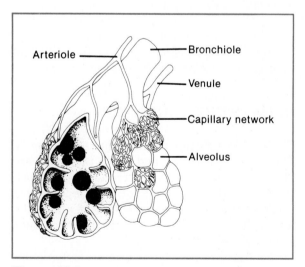

Figure 15.2. Alveoli in a lung

Assignment 1

1. Label Figure 15.1. Color-code the nasal cavity, pharynx, larynx, trachea, and primary bronchi.
2. Locate the parts of the respiratory system on the human torso model and anatomical charts.
3. Observe the corrosion preparation of mammalian lungs that shows how the bronchi divide into smaller and smaller air passages that ultimately terminate in alveoli.
4. ***Complete items 1a–1c on Laboratory Report 15 that begins on page 357.***
5. Examine a prepared slide of trachea, x.s. Using the 4× objective, observe a cartilaginous ring that supports the tracheal wall and holds the trachea open. Cartilaginous rings also support the bronchi. Now focus on the inner surface and switch to the 40× objective to observe the ciliated cells and goblet cells of the epithelial lining. ***Complete item 1d on the laboratory report.***
6. Place your fingers on your larynx (Adam's apple). Is its cartilaginous wall hard or soft? Can you feel the vibrations of your vocal cords when speaking? What happens to your larynx when you swallow? ***Complete items 1e–1h on the laboratory report.***
7. Examine prepared slides of normal and emphysematous lung tissue using the 4× objective. Locate the alveoli in the normal lung tissue and note their size and the thickness of their walls. In emphysematous lungs, the alveolar walls rupture, producing large air spaces in the lungs that reduce the respiratory surface area. Locate such areas on the slide of emphysematous lung tissue. ***Complete item 1 on the laboratory report.***

BREATHING MECHANICS

The air passages into the lungs are always open. Thus, atmospheric air and air in the lungs are always joined by air in the connecting air passages. Air moves into and out of the lungs because of changes in the pressure of air in the lungs, since atmospheric air pressure is constant at any given elevation.

The **diaphragm** is the primary muscle involved in breathing, but the **intercostal muscles** located between the ribs also participate. Recall that the diaphragm is a sheetlike muscle separating the thoracic and abdominal cavities.

Inspiration (inhalation) results from the contraction of the diaphragm, which increases the volume of the thoracic cavity and simultaneously decreases the air pressure in the lungs. Air flows into the lungs because atmospheric air pressure is greater than the air pressure in the lungs.

Expiration (exhalation) results from the relaxation of the diaphragm, which decreases the volume of the thoracic cavity and increases the air pressure in the lungs. Air flows out of the lungs because air pressure in the lungs is greater than the atmospheric air pressure.

Materials

Per lab
Breathing mechanics model

Assignment 2

1. Examine the breathing mechanics model. See Figure 15.3. The balloons represent the lungs, and the glass tubing represents the

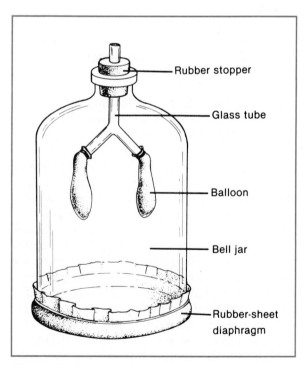

Rubber stopper

Glass tube

Balloon

Bell jar

Rubber-sheet diaphragm

Figure 15.3. Breathing mechanics model

trachea and primary bronchi. The glass jar corresponds to the thoracic wall, and the rubber sheet simulates the diaphragm. Note that the balloons are in an enclosed space, representing the thoracic cavity.

2. Observe what happens when the rubber sheet is pulled downward and pushed upward. Determine how this works.

3. ***Complete item 2 on the laboratory report.***

LUNG CAPACITY IN HUMANS

Lung capacities vary among males and females, primarily due to variations in the size of the thoracic cavity and the lungs. Capacities also vary with age. Average capacities are shown in Figure 15.4.

In this section, you will use a spirometer to determine your lung volumes. *Caution: Wipe the spirometer stem with an alcohol swab or a paper towel soaked in 10% household bleach before and after using the spirometer and use a fresh, sterile, disposable mouthpiece. Place all used mouthpieces in the biohazard bag after use.*

Materials

Per student group
Propper spirometer
Spirometer mouthpieces, sterile and disposable

Per lab
Biohazard bag
Household bleach, 10%

Assignment 3

Determine your lung volumes as described below and record your results in item 3 on the laboratory report.

1. The volume of air exchange during normal quiet breathing is the **tidal volume (TV)**. Determine it as described below.
   a. Rotate the dial of the spirometer so that the needle is at zero. See Figure 15.5.
   b. Disinfect the stem and add a sterile mouthpiece. Place the mouthpiece in your mouth, keeping the spirometer dial upward.

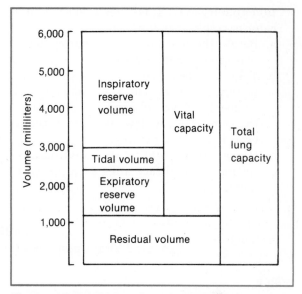

Figure 15.4. Human lung volumes

   c. Inhale through your nose and exhale through the spirometer for five normal quiet breathing cycles. Record the dial reading and divide by 5 to determine your tidal volume.

2. The amount of air that can be forcefully exhaled after a maximum inhalation is the **vi-**

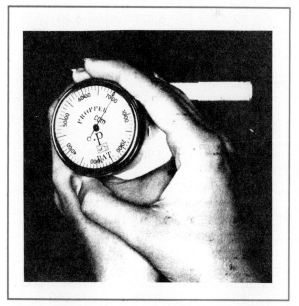

Figure 15.5. Propper spirometer. Slip on a sterile mouthpiece and rotate the dial face to zero before exhaling through the spirometer.

tal capacity (VC). It is often used as an indicator of respiratory function.

a. Rotate the dial face of the spirometer to place the needle at zero.
b. Take two deep breaths and exhale completely after each one. Then take a breath as deeply as possible and exhale through the spirometer. A slow, even expiration is best. Record the reading.
c. Repeat steps a and b two more times, resetting the dial face after each measurement.

d. Record the average of the three measurements as your vital capacity. Compare your vital capacity with those shown in Table 15.1.

3. The volume of air that can be exhaled *after* a normal tidal volume expiration is the **expiratory reserve volume (ERV).**

a. Rotate the dial face of the spirometer so that the needle is at 1,000. This is to compensate for the space on the dial between 0 and 1,000.

TABLE 15.1
Expected Vital Capacities (ml) for Adult Males and Females*

Males

Height (inches)	Age in Years					
	20	30	40	50	60	70
60	3,885	3,665	3,445	3,225	3,005	2,785
62	4,154	3,925	3,705	3,485	3,265	3,045
64	4,410	4,190	3,970	3,750	3,530	3,310
66	4,675	4,455	4,235	4,015	3,795	3,575
68	4,940	4,720	4,500	4,280	4,060	3,840
70	5,206	4,986	4,766	4,546	4,326	4,106
72	5,471	5,251	5,031	4,811	4,591	4,371
74	5,736	5,516	5,296	5,076	4,856	4,636

Females

Height (inches)	Age in Years					
	20	30	40	50	60	70
58	2,989	2,809	2,629	2,449	2,269	2,089
60	3,198	3,018	2,838	2,658	2,478	2,298
62	3,403	3,223	3,043	2,863	2,683	2,503
64	3,612	3,432	3,252	3,072	2,892	2,710
66	3,822	3,642	3,462	3,282	3,102	2,922
68	4,031	3,851	3,671	3,491	3,311	3,131
70	4,270	4,090	3,910	3,730	3,550	3,370
72	4,449	4,269	4,089	3,909	3,729	3,549

* Data from Propper Mfg. Co., Inc.

b. *After* a normal quiet expiration, forcefully exhale as much air as possible through the spirometer. Be sure not to take an extra breath at the start. Subtract 1,000 from the reading to determine your expiratory reserve volume. Repeat to obtain three replicates.

4. Based on your determinations above, calculate your inspiratory reserve volume. IRV = VC − (TV + ERV).

5. The volume of air exchanged during 1 min of quiet breathing is the **respiratory minute volume (RMV).**
   a. Count the number of breathing cycles during a 3-min period and divide by 3 to get the number of cycles per minute.
   b. Calculate the RMV as follows. RMV = TV × breathing cycles per minute.

6. ***Complete item 3 on the laboratory report.***

16

DIGESTION

The cells of the body require a continuous supply of nutrients derived from the food we eat. Most food molecules are too large to be absorbed into the blood, so they must be broken down into smaller, absorbable nutrient molecules by digestion. The major functions of the **digestive system** are:

1. Ingestion of food
2. Movement of food through the digestive tract
3. Digestion of food
4. Absorption of nutrients
5. Elimination of undigestable materials

THE DIGESTIVE SYSTEM

The digestive system consists of the digestive tract and associated organs that aid the digestive process. The digestive tract (alimentary canal), through which food materials pass from mouth to anus, is about 9 meters long. The functions of its major divisions are shown in Table 16.1. Refer to Figures 16.1, 16.2, and 16.3 as you read the following sections.

Oral Cavity

The mouth contains a number of organs that assist the digestive process. The roof of the mouth is formed of the anterior **hard palate** and the posterior **soft palate,** and it separates the oral and nasal cavities. This arrangement allows you to breathe while eating. The posterior fingerlike extension of the soft palate is the **uvula** that contracts upward when contacted by food during swallowing.

The **palatine tonsils** are located on each side of the base of the tongue. They have nothing to do with digestion but sometimes become enlarged and painful when infected and make swallowing difficult.

TABLE 16.1
Major Functions of the Digestive Tract Divisions

Structure	Major Function
Mouth	Eating, chewing, and swallowing food; digestion of starch begins here
Pharynx	Carries food to esophagus
Esophagus	Carries food to stomach by peristalsis
Stomach	Mixes gastric juice with food to form chyme; protein digestion begins here
Small intestine	Mixes chyme with bile and intestinal and pancreatic juices; digestion and absorption of nutrients completed here
Large intestine	Decomposition of undigested materials by bacteria; reabsorption of water to form feces
Anus	Defecation

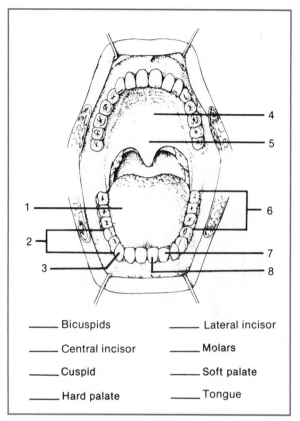

_____ Bicuspids   _____ Lateral incisor

_____ Central incisor   _____ Molars

_____ Cuspid   _____ Soft palate

_____ Hard palate   _____ Tongue

Figure 16.1. The oral cavity

During chewing, the **teeth** break the food into smaller pieces, and the **tongue** manipulates the food and mixes it with **saliva.** Saliva is produced by three pairs of **salivary glands** and discharged into the mouth through a number of salivary ducts. The **parotid glands** are located just in front of the ears and posterior to the angle of the jaw. If you have ever had the mumps, you are well acquainted with these glands since they swell and hurt when infected with the mumps virus. The **sublingual glands** are located in the anterior floor of the mouth under the tongue, and the **submandibular glands** are located posterior to them. Saliva cleans and lubricates the mouth and helps to hold the food together when swallowing. It also contains an enzyme that begins the digestion of starch.

Teeth

There are 32 teeth in a complete set of permanent teeth. See. Figure 16.1. From front to back on each side of each jaw, they are **central incisor, lateral incisor, cuspid,** first and second **bicuspids** (premolars), and first, second, and third **molars.** The third molars (wisdom teeth) often become impacted because of the evolutionary shortening of the jaws.

Figure 16.2 shows the basic structure of a tooth. A tooth consists of two major parts: a **crown** projecting above the bone, and a **root** embedded in bone. Most of a tooth is composed of **dentin,** but the dentin of the crown is covered by a layer of **enamel,** the hardest substance in the body. The **pulp cavity** is the hollow interior of a tooth, and it contains nerves and blood vessels that enter through **root canals.** The mucous membrane that covers the jaw bones and surrounds the bases of the crowns is the **gingiva** (gum).

Swallowing

When food is pushed posteriorly by the tongue into the **pharynx,** a swallowing reflex is set in motion that causes the larynx to move upward and the epiglottis to flop over the glottis so the food passes into the esophagus and not into the larynx.

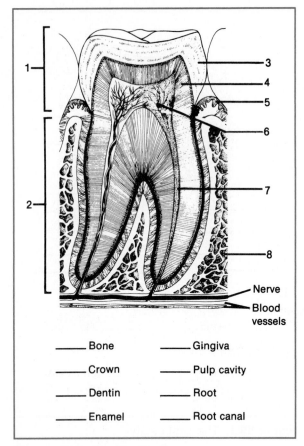

Figure 16.2. Tooth structure

_____ Bone   _____ Gingiva

_____ Crown   _____ Pulp cavity

_____ Dentin   _____ Root

_____ Enamel   _____ Root canal

Esophagus to Anus

Food is moved through the digestive tract (Figure 16.3) by **peristalsis,** the wavelike contraction of the muscles in its walls. The **esophagus** carries food from the pharynx to the **stomach.** The **cardiac sphincter,** a circular muscle located at the esophagus-stomach junction, opens to allow the passage of food and closes to prevent regurgitation. In the stomach, food is mixed with **gastric juice** and converted to a semiliquid mass called **chyme.** Enzymes in gastric juice begin the digestion of proteins and certain fats. Chyme is then released in small amounts into the small intestine. The **pyloric sphincter** controls the passage of chyme into the small intestine.

In the small intestine, chyme is mixed with bile, pancreatic juice, and intestinal juice. Bile and pancreatic juice enter the **duodenum,** the first portion of the small intestine. The bile and

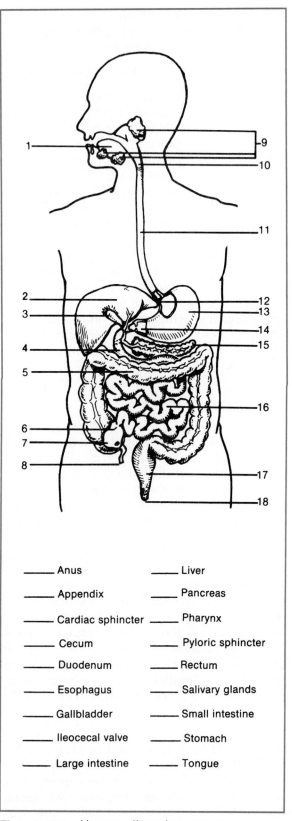

_____ Anus   _____ Liver

_____ Appendix   _____ Pancreas

_____ Cardiac sphincter _____ Pharynx

_____ Cecum   _____ Pyloric sphincter

_____ Duodenum   _____ Rectum

_____ Esophagus   _____ Salivary glands

_____ Gallbladder   _____ Small intestine

_____ Ileocecal valve   _____ Stomach

_____ Large intestine   _____ Tongue

Figure 16.3. Human digestive system

pancreatic ducts join to form a common opening into the duodenum.

Bile is secreted by the **liver,** the large gland in the upper right portion of the abdominal cavity. It is temporarily stored in the **gall bladder.** Bile emulsifies fats to facilitate fat digestion. **Pancreatic juice** is produced by the **pancreas,** a pennant-shaped gland located between the stomach and the duodenum. **Intestinal juice** is secreted by the inner lining (mucosa) of the small intestine. Enzymes in pancreatic juice and intestinal juice act sequentially to complete the digestion of food. Digestion of food and absorption of nutrients into the blood are completed in the small intestine.

The nondigestible material passes into the **large intestine** via the **ileocecal valve,** another sphincter muscle. The **appendix** is a small vestigial appendage attached to the pouchlike **cecum,** the first portion of the large intestine. The large intestine (colon) includes ascending, transverse, descending, and sigmoid regions. It ends with the **rectum,** a muscular portion that expels the feces through the **anus.** Decomposition of nondigestible materials by bacteria and the reabsorption of water are the major functions of the large intestine.

Materials

Per student
Colored pencils

Per lab
Anatomical charts
Human torso model
Human head model, midsagittal section
Tooth model

Assignment 1

1. Label Figures 16.1, 16.2, and 16.3. Color-code the organs to help you learn their locations.
2. Locate the parts of the digestive system on the models and anatomical charts.
3. ***Complete item 1 on Laboratory Report 16 that begins on page 361.***

Histology of the Small Intestine

Examine the structure of the small intestine in cross section in Figure 16.4. Note the four layers of the intestinal wall.

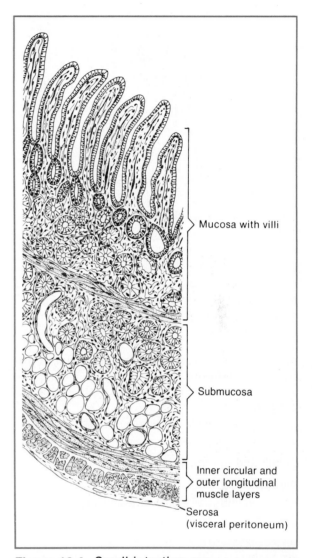

Mucosa with villi

Submucosa

Inner circular and outer longitudinal muscle layers

Serosa (visceral peritoneum)

Figure 16.4. Small intestine, x.s.

Peritoneum. The outer protective membrane that lines the coelom and covers the digestive organs.

Muscle layers. Outer longitudinal and inner circular layers whose contractions mix food with digestive secretions and move the food mass by peristalsis.

Submucosa. Connective tissue containing blood vessels and nerves serving the digestive tract.

Mucosa. The inner epithelial lining that secretes intestinal juice and absorbs nutrients.

Locate the **villi,** fingerlike extensions of the mucosa that project into the lumen of the small intestine. Villi increase the surface area of the

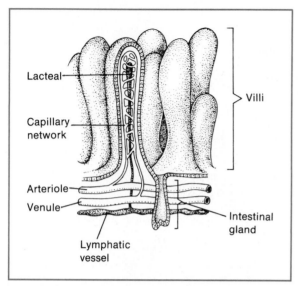

Figure 16.5. A diagrammatic representation of villi in the small intestine

mucosa, which facilitates the absorption of nutrients. See Figure 16.5.

Materials

Per student
Colored pencils
Compound microscope

Per lab
Prepared slides of small intestine, x.s.

Assignment 2

1. Color-code the four layers of the small intestine in Figure 16.4.
2. Examine a prepared slide of small intestine, x.s., at 40×. Locate the four layers of the small intestine as shown in Figure 16.4. Note the villi that greatly increase the surface area of the intestinal mucosa. Focus on the columnar epithelial lining. Are goblet cells present?
3. *Complete item 2 on the laboratory report.*

DIGESTION AND ENZYMES

The conversion of large, nonabsorbable food molecules into small, absorbable nutrient molecules occurs through the action of digestive enzymes. The role of digestive enzymes is to speed up the hydrolysis of food molecules. In **enzymatic hydrolysis,** digestive enzymes catalyze (speed up) reactions where the addition of water molecules breaks bonds of food molecules, forming smaller nutrient molecules.

$$\text{Nonabsorbable food molecules} \xrightarrow[\text{H}_2\text{O}]{\text{Digestive enzymes}} \text{Absorbable nutrient molecules}$$

Many different digestive enzymes are required to complete the digestion of food since a particular enzyme acts on only a single type of food molecule. Saliva, gastric juice, pancreatic juice, and intestinal juice contain specific digestive enzymes that act on specific food molecules. Table 16.2 shows the absorbable end products of enzymatic hydrolysis.

Digestion of Starch

In this section, you will assess the effects of temperature and pH on the action of **pancreatic amylase.** This enzyme accelerates the hydrolysis of **starch,** a polysaccharide, to **maltose,** a disaccharide and reducing sugar. See Figure 16.6. The experiments are best done by groups of four students, with each group assigned a different temperature and with all groups sharing their results.

You will dispense solutions from dropping bottles and use a number of test tubes. To avoid contamination, the droppers must not touch other solutions. All glassware must be clean and rinsed with distilled water.

You will dispense "droppers" and "drops" of solutions. As noted earlier, a "dropper" means *one dropper full* of solution (about 1 ml). When dispensing solutions by the "drop," hold all droppers at the same angle to dispense equal-sized drops. All drops should land squarely in the bot-

TABLE 16.2
End Products of Digestion

Food	End Products
Carbohydrates	Monosaccharides
Proteins	Amino acids
Fats	Fatty acids and monoglycerides

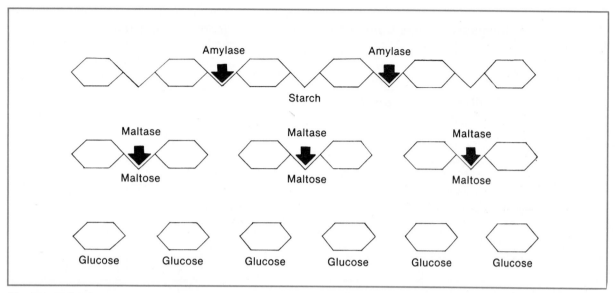

Figure 16.6. Digestion of starch

tom on the test tube. If they run down the side of the tube, your results may be affected.

Thoroughly mix the solutions placed in a test tube. If a vortex mixer is not available, shake the tube vigorously from side to side to mix the contents.

You will use the iodine test for starch and Benedict's test for reducing sugars to determine if digestion has occurred. Examine the demonstration of positive and negative tests prepared by your instructor and use these as standards for comparison.

Iodine Test

1. Add 2 drops of iodine solution to 1–2 droppers of solution to be tested.
2. A blue-black color indicates the presence of starch.

Benedict's Test

1. Add 5 drops of Benedict's solution to 1 dropper of solution to be tested, and heat the mixture to near boiling in a water bath for 3 min.
2. A light green, yellow, orange, or red color indicates the presence of a reducing sugar, in that order of increasing concentration.

Materials

Per student group
Beaker, 400 ml
Boiling chips

Glass marking pen
Hot plate
Test tubes, 18
Test-tube racks, submersible type
Dropping bottles of:
　Benedict's solution
　buffers, pH 5, 8, 11
　iodine solution (IKI)
　pancreatic amylase, 0.1%
　soluble starch, 0.1%

Per lab
Water baths, 5, 37, 70 °C
Celsius thermometers, 1 per water bath

Assignment 3

1. Label nine test tubes 1–9.
2. Add the pH buffer solutions to the test tubes as follows.
　Tube 1: 2 droppers pH 5
　Tube 2: 2 droppers pH 5
　Tube 3: 3 droppers pH 5

　Tube 4: 2 droppers pH 8
　Tube 5: 2 droppers pH 8
　Tube 6: 3 droppers pH 8

　Tube 7: 2 droppers pH 11
　Tube 8: 2 droppers pH 11
　Tube 9: 3 droppers pH 11
3. Add 5 drops of pancreatic amylase to tubes

TABLE 16.3
Summary of Tube Contents for the Starch Digestion Experiment

| Tube | Droppers of Solution | | | | Drops of Amylase |
| | Buffers | | | Starch | |
	pH 5	pH 8	pH 11		
1	2			1	5
2	2			1	
3	3				5
4		2		1	5
5		2		1	
6		3			5
7			2	1	5
8			2	1	
9			3		5

1, 3, 4, 6, 7, and 9. Shake the tubes to mix well.

4. Place the test tubes in a test-tube rack and place the rack in the water bath at your assigned temperature for 10 min.

5. Add 1 dropper of starch solution to tubes 1, 2, 4, 5, 7, and 8. Shake the tubes to mix well. Leave the tubes in the water bath for 15 min. Table 16.3 summarizes the contents of each tube.

6. After 15 min, remove the test-tube rack and tubes and return to your workstation.

7. Number another set of nine tubes 1B–9B.

8. Pour half the liquid in tube 1 into tube 1B, pour half the liquid in tube 2 into tube 2B, and so on until you have divided the liquid equally between the paired tubes. You now have nine pairs of tubes with the members of each pair containing liquid of identical composition.

9. Test all the original nine tubes (1–9) for the presence of starch by adding 2 drops of iodine solution to each tube. **Record your results in the table in item 3a on the laboratory report.**

10. Test all of the B tubes (1B–9B) for the presence of maltose, a reducing sugar, as follows. Add 5 drops of Benedict's solution to each tube, and place the tubes in a water bath (400-ml beaker half full of water) on your hot plate. Place some boiling chips in the beaker and heat the tubes to near boiling for 3–4 min. Watch for any color change in the solution. **Record your results in item 3a on the laboratory report.**

11. Exchange results with other groups using different temperatures. **Record their results in item 3a on the laboratory report.**

12. **Complete item 3 on the laboratory report.**

17

NEURAL CONTROL

Humans possess the most highly developed nervous system, and it may be subdivided into two major components. The **central nervous system (CNS)** is composed of the **spinal cord** and **brain.** The **peripheral nervous system (PNS)** is composed of **cranial nerves,** which emanate from the brain and extend to the head and certain internal organs, and the **spinal nerves,** which extend from the spinal cord to all parts of the body except the head. Of course, well-developed **sensory receptors** are evident.

In this exercise, you will investigate the structure and function of the nervous system. The nervous system provides rapid coordination and control of body functions by transmitting **impulses** over neuron processes. The impulses may originate in either the brain or receptors, and they are carried to **effectors** (muscles and glands) where the action occurs.

NEURONS

Neural tissue is composed of two basic types of cells. **Neurons,** or nerve cells, transmit impulses. Neuroglial cells provide structural support and prevent contact of neurons except at certain sites. Although neurons may be specialized in structure and function in various parts of the nervous system, they have many features in common. Figure 17.1 shows the basic structure of neurons, and Table 17.1 identifies the characteristics of the three functional types of neurons.

The **cell body** is an enlarged portion of the neuron that contains the nucleus. Two types of neuron processes or fibers extend from the cell body. **Dendrites** receive impulses from receptors or other neurons and carry impulses *toward* the cell body. **Axons** carry impulses *away from* the cell body. A neuron may have many dendrites, but only one axon is present.

Neuron processes of the peripheral nervous system are enclosed by a covering of **Schwann**

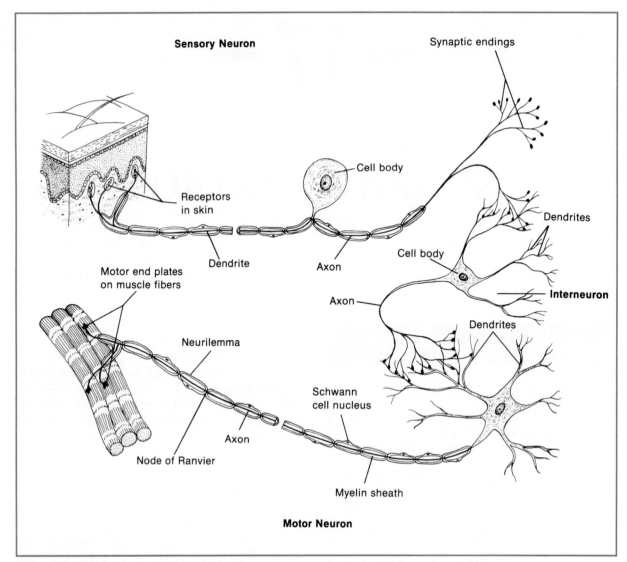

Figure 17.1. Neuron structure

cells. In larger processes, the multiple wrappings of Schwann cells form an inner **myelin sheath,** a fatty, insulating material, while the outer layer constitutes the **neurilemma.** The minute spaces between Schwann cells, where the neuron process is exposed, are called **nodes of Ranvier.** Impulses are transmitted more rapidly by myelinated fibers than by unmyelinated fibers. Schwann cells provide a pathway for the regeneration of neuron processes and are essen-

TABLE 17.1
Functional Types of Neurons

Neuron Type	Structure	Function
Sensory	Long dendrite, short axon	Carry impulses from receptors to the CNS
Interneuron	Short dendrites, short or long axon	Carry impulses within the CNS
Motor	Short dendrites, long axon	Carry impulses from the CNS to effectors

tial for regrowth. Schwann cells are absent in the central nervous system, but another type of neuroglial cell forms the myelin sheath of CNS neurons.

The junction of an axon tip of one neuron and a dendrite or cell body of another neuron is called a **synapse.** See Figure 17.2. Impulses passing along the axon cause the release of a **neurotransmitter** from the axon tip, or **synaptic knob,** into the **synaptic cleft,** the minute space between the neurons. The neurotransmitter binds with receptor sites on the postsynaptic membrane, producing either stimulation or inhibition of impulse formation, depending on the type of neurotransmitter involved. Immediately thereafter, an enzyme breaks down or inactivates the neurotransmitter, which prevents continuous stimulation or inhibition of the postsynaptic membrane. Transmission of neural impulses across synapses is always in one direction, axon to dendrite, because only axon tips can release neurotransmitters to activate the adjacent neuron.

Numerous substances are either known or suspected to be neurotransmitters. Among known neurotransmitters, **acetylcholine** and **norepinephrine** are stimulatory neurotrans-

mitters, while **glycine** and **gamma aminobutyric acid (GABA)** are inhibitory neurotransmitters.

THE BRAIN

The **brain** is the control center of the nervous system. It is enclosed within the cranium and covered by the **meninges,** three layers of protective membranes. **Cerebrospinal fluid** within the meninges provides an additional cushion to absorb shocks. Twelve pairs of cranial nerves are attached to the brain. All but one pair innervate structures of the head and neck; vagus nerves innervate the internal organs. Refer to Figures 17.3–17.7 as you study the major parts of the brain.

Cerebrum

The **cerebrum** is the largest part of the brain. It consists of right and left **cerebral hemispheres** that are separated by a median **longitudinal fissure.** A mass of neuron fibers, the **corpus callosum,** enables impulses to pass between the two hemispheres. The outer portion of the cere-

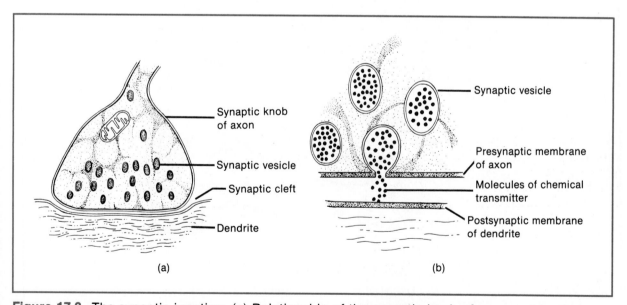

(a)

(b)

Figure 17.2. The synaptic junction. (a) Relationship of the synaptic knob of an axon to a dendrite. Synaptic vesicles migrate from the cell body to the synaptic knob. (b) Synaptic transmission. An impulse moving down the axon causes a synaptic vesicle to release a neurotransmitter that either stimulates or inhibits the formation of an impulse in the dendrite.

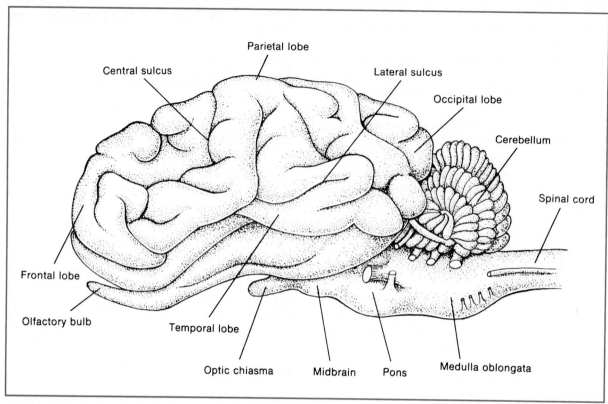

Figure 17.3. Lateral view of sheep brain

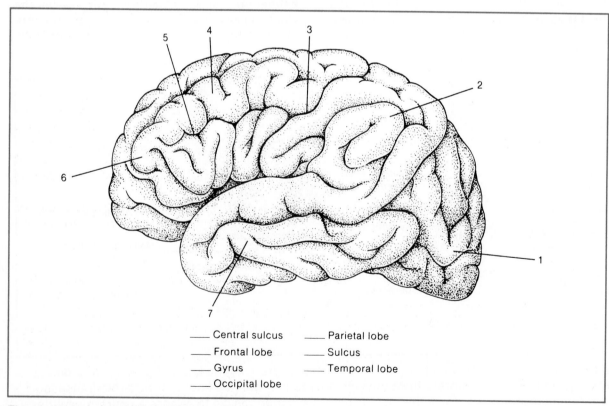

___ Central sulcus	___ Parietal lobe
___ Frontal lobe	___ Sulcus
___ Gyrus	___ Temporal lobe
___ Occipital lobe	

Figure 17.4. Lateral view of human cerebrum

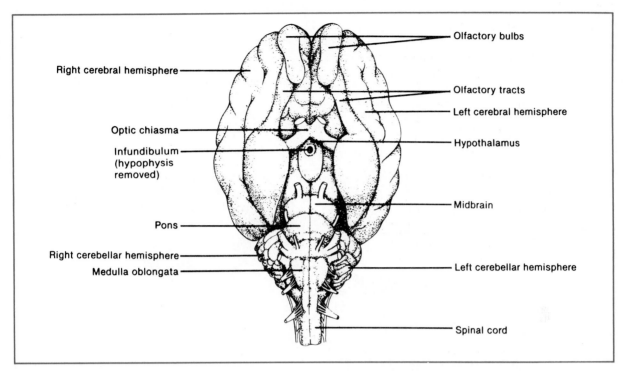

Figure 17.5. Ventral view of sheep brain

brum, the **cerebral cortex,** is composed of neuron cell bodies and unmyelinated neuron processes. Its surface area is increased by numerous **gyri** (ridges) and **sulci** (grooves). The cerebrum initiates voluntary actions and interprets sensations. In humans, it is the seat of will, memory, and intelligence.

Each hemisphere is divided into four lobes by fissures (deep grooves). Table 17.2 indicates the location and major functions of these lobes.

Cerebellum

The **cerebellum** lies just below and posterior to the occipital lobe of the cerebrum. It is divided into left and right hemispheres by a shallow fissure. Muscle tone and muscular coordination are subconsciously controlled by the cerebellum.

Brain Stem

The **brain stem** is composed of a number of structures located between the cerebrum and the spinal cord. A midsagittal section is required to locate the various parts. A major function is the linking of higher and lower brain areas, but the individual portions also have some specific functions.

The **thalamus,** which is located at the upper end of the brain stem, consists of two lateral

TABLE 17.2
Location and Function of the Cerebral Lobes

Lobe	Location	Function
Frontal	Anterior to the central sulcus	Voluntary muscular movements; intellectual processes
Parietal	Between frontal and occipital lobes	Interprets sensations from skin; speech interpretation
Temporal	Inferior to frontal and parietal lobes	Hearing; interprets auditory sensations
Occipital	Posterior part of cerebrum	Vision; interprets visual sensations

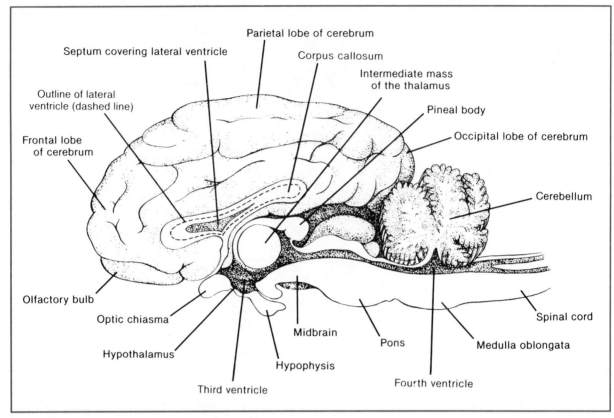

Figure 17.6. Sheep brain, midsagittal section

globular masses joined by an isthmus of tissue called the **intermediate mass.** It provides an uncritical awareness of sensations such as pain and pleasure and is a relay station between lower brain centers and the cerebrum.

The **hypothalamus** is located just below the thalamus. It plays a major role in homeostasis of the body by controlling things such as appetite, sleep, body temperature, and water balance. A major endocrine gland, the **hypophysis** or **pituitary gland,** is attached to the ventral wall of the hypothalamus by a short stalk, the **infundibulum.**

The **midbrain** is a small area between the thalamus and pons that is associated with certain visual reflexes. It contains the **pineal body.**

The **pons** is a rounded bulge on the ventral side of the brain stem. It is an important connecting pathway for higher and lower brain centers.

The **medulla oblongata** lies between the pons and the spinal cord. It is the lowest part of the brain, and all impulses passing between the brain and spinal cord must pass through it. In addition, it controls heart rate, blood pressure, and breathing.

Materials

Per student group
Colored pencils
Dissecting instruments and pans

Per lab
Model of human brain
Sheep brains, whole and sectioned
Prepared slides of:
   giant multipolar neurons
   nerve, x.s.

 Assignment 1

1. Add arrows to Figure 17.1 to indicate the direction of impulse transmission in motor and sensory neurons. Color-code the three types of neurons.
2. ***Complete items 1a and 1b on Laboratory Report 17 that begins on page 363.***

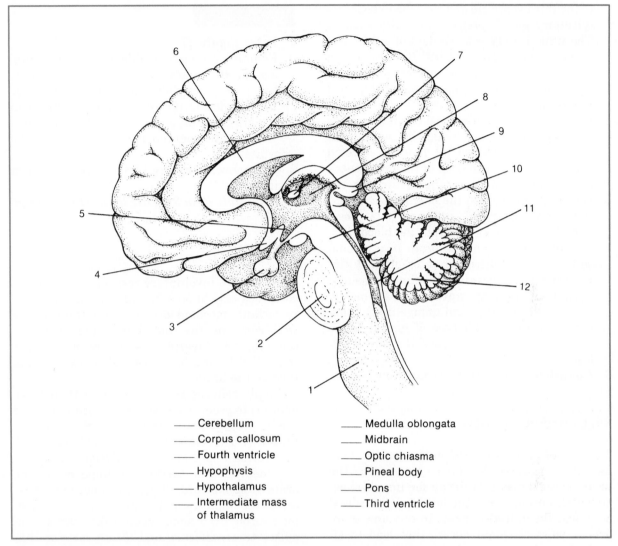

Cerebellum

_____ Corpus callosum

_____ Fourth ventricle

_____ Hypophysis

_____ Hypothalamus

_____ Intermediate mass
of thalamus

_____ Medulla oblongata

_____ Midbrain

_____ Optic chiasma

_____ Pineal body

_____ Pons

_____ Third ventricle

Figure 17.7. Midsagittal section of human brain

3. Examine a slide of giant multipolar neurons. Draw a neuron and label the cell body, nucleus, and neuron processes.

4. Examine a slide of nerve, x.s., and make a drawing of your observations. Note how the axons are arranged in bundles separated by connective tissue. Label an axon, myelin sheath, and connective tissue.

5. ***Complete item 1 on the laboratory report.***

Assignment 2

1. Study and color-code Figures 17.3, 17.5, and 17.6 until you are familiar with the structures of the sheep brain.

2. Label Figures 17.4 and 17.7 and locate these parts on a model of a human brain. Color-code the cerebral lobes in Figure 17.4 and the parts of the brain in Figure 17.7.

3. Obtain an entire sheep brain for study. Note the meninges, if present, and remove them with scissors. Locate the major parts of the brain shown in Figures 17.3 and 17.5. Observe the ridges and furrows that increase the surface area of the cerebrum. Is there an advantage in this?

4. Obtain half of a sheep brain that has been sectioned along the longitudinal fissure. Locate the structures shown in Figure 17.6. Note that the hypothalamus forms the floor

of the third ventricle and that the hypophysis (pituitary gland) projects ventrally from it. The **pineal body** is a remnant of a third eye found in primitive reptiles. Its function in some mammals may be to control seasonal reproductive activity based on photoperiod variation.

5. Examine the demonstration coronal sections of a sheep brain and locate the two **lateral ventricles.** Cerebrospinal fluid is secreted from blood vessels in each of the four ventricles of the brain. It circulates through the ventricles and within the meninges covering the brain and spinal cord, and it is subsequently reabsorbed back into the blood.

   Note that parts of the brain in a coronal section appear either white or gray. **White matter** consists mostly of myelinated neuron processes, and **gray matter** consists mostly of neuron cell bodies and unmyelinated neuron processes. Which type of neural tissue composes the cerebral cortex? the corpus callosum?

6. *Complete item 2 on the laboratory report.*

THE SPINAL CORD

The **spinal cord** is located in the vertebral canal and is covered by the meninges like the brain. It serves as a pathway for impulses between the brain, different levels of the spinal cord, and the **spinal nerves.** In humans, there are 31 pairs of spinal nerves that lead to all parts of the body except the head. Each spinal nerve joins the spinal cord to form ventral and dorsal roots. The **ventral root** contains fibers of motor nerves only. The **dorsal root** contains fibers of sensory nerves only, and the cell bodies of the sensory neurons are located in the **dorsal root ganglion.** See Figure 17.8. Unlike the cerebrum, the gray matter of the spinal cord is located interiorly, while the white matter is found exteriorly. What causes the difference in appearance of gray and white matter?

Materials

Per lab
Chart or model of the spinal cord
Prepared slides of cat spinal cord, x.s.

1. Label Figure 17.8.
2. Examine a prepared slide of cat spinal cord, x.s., at 40× and observe the gross features. Compare your slide with Figure 17.8. Use 100× to locate the cell bodies of neurons in the gray matter, and examine them at 400×. What do the neuron fibers look like in the white matter?
3. Add arrows to Figure 17.8 to show the path of impulses.
4. *Complete item 3 on the laboratory report.*

REFLEXES

Reflexes are involuntary responses (require no conscious act) to specific stimuli, and they are important mechanisms for maintaining the well-being of the individual. For example, coughing and sneezing are respiratory reflexes controlled by the brain. Spinal reflexes do not involve the brain.

Simple reflexes require no more than three neurons to produce a reaction to a stimulus, and this is accomplished through a **reflex arc.** See Figure 17.8. A reflex arc consists of (1) a receptor that forms impulses on stimulation, (2) a sensory neuron that carries the impulses to the brain or spinal cord, (3) an interneuron that receives the impulses and transmits them to a motor neuron, (4) a motor neuron that carries impulses to an effector, and (5) an effector that performs the action.

Materials

Per student pair
Reflex hammer
Laboratory lamp
Penlight

Patellar Reflex

Physicians use reflex tests to assess the condition of the nervous system. The **patellar reflex** is one commonly used. When the patellar tendon is struck just below the kneecap with a reflex hammer, the reflex action is a slight, instantaneous contraction of the large muscle (quadriceps femoris) on the front of the thigh that extends the lower leg. Striking the patellar tendon

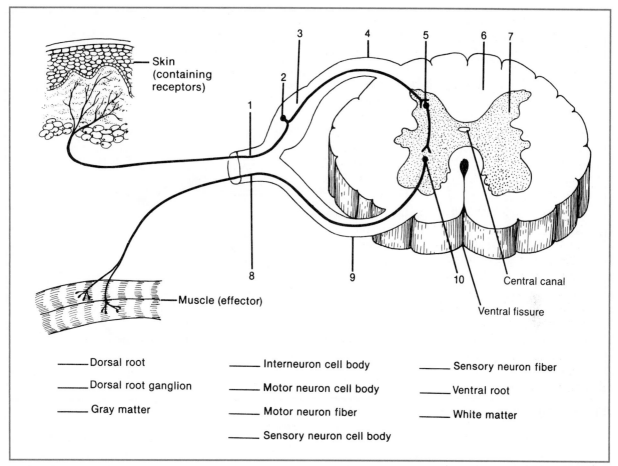

Figure 17.8. Spinal reflex arc

causes the muscle to be slightly stretched for an instant. This stretching causes impulses to be formed and carried along a sensory neuron to the spinal cord, where they are passed to a motor neuron that carries the impulses to the muscle, thereby causing a weak, brief contraction. Work in pairs to perform the reflex test.

Photopupil Reflex

This reflex enables a rapid adjustment of the size of the pupil to the existing light intensity, and it is coordinated by the brain. It is most easily observed in persons with light-colored eyes. When a bright light stimulates the retina of the eye, impulses are carried to the brain by sensory neurons. In the brain, the impulses are transmitted to interneurons and on to motor neurons that carry impulses to the muscles of the iris, causing them to contract. Contraction of

the iris muscles decreases the size of the pupil and controls the amount of light entering the eye.

Assignment 4

1. Perform the patellar reflex as follows.
   a. The subject should sit on the edge of a table with the legs hanging over the edge but not touching the floor.
   b. Strike the patellar tendon (see Figure 17.9) with the small end of the reflex hammer and observe the response.
   c. Divert the subject's attention by having the subject interlock the fingers of both hands and pull the hands against each other while you strike the patellar tendon again. Is the response the same as before? If not, how do you explain it?

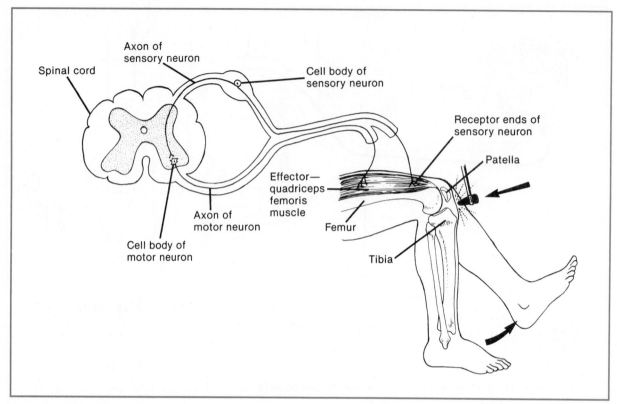

Figure 17.9. The patellar reflex

d. Test both legs and record the results.

e. Add arrows to Figure 17.9 to show the path of impulses.

f. ***Complete items 4a and 4b on the laboratory report.***

2. Perform the photopupil reflex as follows.

a. Have the subject sit with eyes closed facing a darkened part of the room for 1–2 min. When the subject opens his or her eyes, note any change in pupil size.

b. While the subject is looking into the darkened area, shine a desk lamp in his or her eyes (from about 3 ft away). Note any change in pupil size.

c. To observe the effect of unilateral stimulation, have the subject look into the darkened area and, while observing the right eye, shine a penlight in the left eye. Note any response. Then observe the left eye while stimulating the right eye.

d. ***Complete item 4 on the laboratory report.***

REACTION TIME

The nervous system controls and coordinates our reactions to thousands of stimuli each day. Some of these reactions are reflexes, but many are **voluntary reactions,** responses that are consciously initiated. **Reaction time** is the time interval from the instant of stimulation to the instant of a voluntary response. All responses result from the formation of impulses by stimulation and the transmission of these impulses along neurons to effectors that bring about the response.

In reflexes, impulses flow over predetermined "automated" neural pathways involving very few neurons, and they do not require processing by the cerebral cortex. In contrast, voluntary reactions involve a greater number of neurons and synapses and require processing of impulses by the cerebral cortex. Therefore, reflexes have much shorter response times than voluntary reactions.

The reaction time for a voluntary response is the sum of the times required for:

1. A receptor to form impulses in response to a stimulus.
2. Transmission of impulses to an integration center of the cerebral cortex.
3. Processing the impulses in the integration center.
4. Transmission of impulses to effectors.
5. Response of the effectors.

Do you think people differ in their reaction times to the same stimulus? In this section, you will test the null hypothesis that there is no difference in the reaction times of different persons in responding to the same stimulus with the same predetermined response.

Measuring Reaction Time

You will measure reaction time using a reaction-time ruler and the following procedure.

1. The subject sits on a chair or stool with the experimenter standing facing the subject.
2. The experimenter holds the *release end* of the reaction-time ruler between thumb and forefinger at about eye level or higher. See Figure 17.10.

(a) Subject holds thumb and forefinger 1 inch apart with the upper margin of the thumb at the thumb line of the reaction time ruler and, when ruler is released, closes them as quickly as possible to grasp ruler.

(b) Read the time in milliseconds at the upper edge of the thumb. In this photo, upper edge of thumb is at 235 msec.

Figure 17.10. Measuring visual reaction time

3. The subject places the thumb and forefinger of his or her dominant hand about an inch apart and on each side of the *thumb line* at the lower end of the ruler. The subject's attention is focused on the ruler at the thumb line.
4. When the subject says he or she is ready, the experimenter, within 10 sec, releases the ruler. The subject seeing the falling ruler catches it between thumb and forefinger as quickly as possible.
5. The reaction time is read in milliseconds at the upper edge of the thumb and recorded.
6. The test is repeated 5 times and the average reaction time is calculated. If any reaction time is grossly different, discard it and repeat the test to obtain 5 results that are fairly consistent.

Materials

Per student pair
Reaction-time ruler

Assignment 5

1. Have your partner measure your reaction time 5 times and ***record them in item 5a on the laboratory report.***
2. Calculate your average reaction time. Write it on the board for the class tabulation. ***Complete items 5b–5f on the laboratory report.***
3. Do you think practice and learning will decrease your reaction time? Repeat the reaction time test 20 times without recording the reaction time. Then repeat the test 5 times and record your reaction times.
4. ***Complete item 5 on the laboratory report.***

18

SENSORY PERCEPTION

Sensations result from the interaction of three components of the nervous system.

1. **Sensory receptors** generate impulses upon stimulation.
2. **Sensory neurons** carry the impulses to the brain or spinal cord, and **interneurons** carry the impulses to the sensory interpretive centers in the brain.
3. The **cerebral cortex** interprets the impulses as sensations.

The type of sensation is determined by the part of the brain receiving the impulses rather than by the type of receptors being stimulated. For example, the auditory center interprets all impulses it receives as sound stimuli regardless of their origin.

In this exercise, you will study the structure and function of the eye and the ear and perform tests that will demonstrate certain characteristics of sensory perception.

THE EYE

The eyes contain the receptors for light stimuli, and they are well protected by the surrounding skull bones and the eyelids. The eyelids and the anterior surface of the eye are covered by a mucous membrane, the **conjunctiva**, which contains many blood vessels and pain receptors except where it covers the cornea. The **lacrimal gland**, located over the upper, lateral portion of the eye, produces tears that cleanse the surface of the conjunctiva and keep it moist.

Structure of the Eye

Refer to Figure 18.1 as you study this section.

The eye is a hollow ball, roughly spherical in shape. Its wall is composed of three distinct layers.

The outer layer is composed of (1) the fibrous **sclera**, which forms the white portion of the eye, and (2) the transparent **cornea**, which forms the anterior bulge where light enters the eye.

The middle layer consists of three parts. The **choroid** coat, a black layer, absorbs excess light that passes through the retina and contains the blood vessels that nourish the eye. Anteriorly,

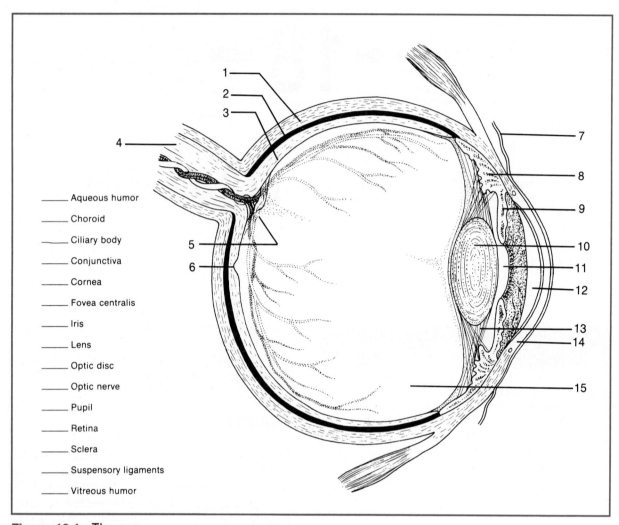

Aqueous humor

Choroid

Ciliary body

Conjunctiva

Cornea

Fovea centralis

Iris

Lens

Optic disc

Optic nerve

Pupil

Retina

Sclera

Suspensory ligaments

Vitreous humor

Figure 18.1. The eye

the **ciliary body** forms a ring of muscle that controls the shape of the **lens,** which is suspended from the ciliary body by the **suspensory ligaments.** The **iris** controls the amount of light entering the eye by controlling the size of the **pupil,** the opening in the center of the iris.

The **retina** composes the inner layer of the wall of the eye. It contains the photoreceptors: **rods** for black and white vision and **cones** for color vision. The neuron fibers coalesce at the **optic disc** where they enter the **optic nerve,** which carries impulses to the brain. The optic disc is known as the blind spot since it has no receptors. A tiny depression just lateral to the optic disc, the **fovea centralis,** contains densely packed cones for sharp, direct vision.

The interior of the eye behind the lens is filled with **vitreous humor,** a transparent, jellylike substance that helps hold the retina in place and gives shape to the eye. A watery fluid, the **aqueous humor,** fills the space between the lens and cornea.

Materials

Per student group
Colored pencils
Dissecting instruments and pan
Beef eye, preferably fresh

Per lab
Eye model

Assignment 1

1. Label Figure 18.1 and color-code the sclera, cornea, choroid, ciliary body, iris, retina, and lens.
2. Locate the parts of the eye on an eye model.
3. **Complete items 1a and 1b on Laboratory Report 18 that begins on page 367.**
4. Dissect a beef eye following these procedures.
   a. Examine the external surface of the eye. Locate the optic nerve and trim away any remnants of the extrinsic eye muscles and conjunctiva. Is the curvature of the cornea greater than the rest of the eyeball? What is the shape of the pupil?
   b. Use a sharp scalpel to make a small incision in the sclera about 0.5 cm from the edge of the cornea. See Figure 18.2. Holding the eye firmly but gently, insert scissors into this incision and cut through the sclera around the eyeball while holding the cornea upward. The fluid that exudes when making this cut is aqueous humor.
   c. Now gently lift off the anterior part of the eye and place it on the dissecting pan with its inner surface upward. In preserved eyes, the lens often remains with the cornea, but in fresh eyes it usually remains with the vitreous humor.
   d. Examine the anterior portion. Locate the ciliary body, a thickened black ring, and the iris. There may be a little aqueous humor remaining next to the inner surface of the cornea.
   e. Observe the vitreous humor and the attached lens in the posterior part of the eye. Pour the vitreous humor onto the dissecting pan and note its jellylike consistency.
   f. Use a dissecting needle to separate the periphery of the lens from the ciliary body on the surface of the vitreous humor. Hold the lens to your eye and look through it across the room. What is unusual about the image? Place it on this printed page. Does it magnify the print?
   g. Look at the interior of the posterior portion of the eye to see the thin, beige retina that now is wrinkled since the vitreous humor has been removed. Note that the retina easily separates from the underlying choroid but is attached at the blind

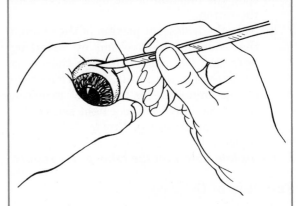

(a) Holding the eye as shown, make a small cut through the sclera with a sharp scalpel about 0.5 cm from the edge of the cornea.

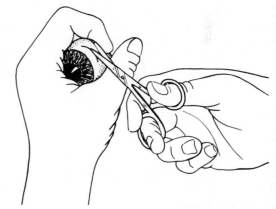

(b) Insert the point of a scissors into the cut and continue the cut through the wall around the eye

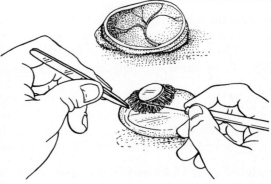

(c) After pouring the vitreous body onto the dissecting pan, carefully separate the lens from the vitreous body with a dissecting needle.

Figure 18.2. Three steps in dissecting a beef eye

spot, the junction of the optic nerve with the retina.

h. Note the iridescent portion of the choroid that aids dim-light vision by reflecting light that has passed through the retina back through the retina. This iridescence causes animals' eyes to "shine" at night, reflecting light back to a light source.

i. Dispose of the eye as directed by your instructor.

5. ***Complete item 1 on the laboratory report.***

Function of the Eye

Light waves reflecting from objects are bent as they pass through the cornea. They continue through the pupil and lens, which focuses the light rays on the retina and accommodates for near and distance vision as its shape is changed by muscles in the ciliary body. Photoreceptors in the retina form impulses that are transmitted via the optic nerve to the visual center in the brain, where they are interpreted as visual images.

Materials

Per lab
Meter sticks
Color-blindness test plates (Ishihara)
Astigmatism charts
Snellen eye charts

Assignment 2

Perform the following visual tests to learn more about visual sensations. Work in pairs. If you wear corrective lenses, perform the tests with and without them and note any differences in your results.

Blind Spot. No rods or cones are located at the junction of the retina and optic nerve. This site is known as the optic disc or blind spot. It can be located by using the following procedure.

1. Hold Figure 18.3 about 50 cm (20 in.) in front of your eyes.
2. Cover your left eye and focus with the right eye on the cross. You will be able to see the dot as well.
3. Slowly move the figure toward your eyes while focusing on the cross, until the dot disappears.

Figure 18.3. Blind spot test

4. Have your partner measure and record the distance from your eye to the figure at the point where the dot disappears.
5. Test the left eye in a similar manner but focus on the dot and watch for the cross to disappear.
6. ***Complete item 2a on the laboratory report.***

Near Point. The shortest distance from your eye that an object is in sharp focus is called the near point. The closer the distance, the more elastic the lens and the greater the eye's ability to accommodate for changes in distance. Elasticity of the lens is greatest in infants, and it gradually decreases with age. See Table 18.1. Accommodation is minimal after 60 yr of age, a condition called presbyopia. How does this relate to the common usage of bifocal lenses by older persons?

1. Hold this page in front of you at arm's length. Close one eye, focus on a word in this sentence, and slowly move the page toward your face until the image is blurred. Then move the page away until the image is sharp. Have

TABLE 18.1
Age and Accommodation

	Near Point	
Age	*Inches*	*Centimeters*
10	3.0	7.5
20	3.5	8.8
30	4.5	11.3
40	6.8	17.0
50	20.7	51.8
60	33.0	82.5

your partner measure the distance between your eye and the page.

2. Test the other eye in the same manner.
3. ***Complete item 2b on the laboratory report.***

Astigmatism. This condition results from an unequal curvature of either the cornea or the lens, which prevents the light rays from being focused sharply on the retina.

1. Cover one eye and focus on the circle in the center of Figure 18.4. If the radiating lines appear equally dark and in sharp focus, no astigmatism exists. If astigmatism exists, record the number of the lines that appear lighter in color or blurred.
2. Test the other eye in the same manner.
3. ***Complete item 2c on the laboratory report.***

Acuity. Visual acuity refers to the ability to distinguish objects in accordance to a standardized scale. It may be measured using a Snellen eye chart. If you can read the letters that are designated to be read at 20 ft at a distance of 20 ft, you have 20/20 vision. If the smallest letters that can be read at 20 ft are those designated to be read at 30 ft, you have 20/30 vision.

Figure 18.4. Astigmatism test

1. Stand 20 ft from the Snellen eye chart on the wall while your partner stands next to the chart and points out the lines to read.
2. Cover one eye and read the lines as requested. Record the rating of the smallest letters read correctly.
3. Test the other eye in the same manner.
4. ***Complete item 2d on the laboratory report.***

Color Blindness. The sensation of color vision depends on the degree to which impulses are formed by the three types of cones (receptors for red, green, and blue light) in the retina. The most common type of color blindness is red-green color blindness, which is caused by a deficit in cones stimulated by either red or green light. People with such a deficit have difficulty distinguishing reds and greens. A totally color-blind person sees everything as shades of gray.

1. Your partner holds the color-blindness test plates about 30 in. from your eyes in good light and allows you 5 sec to view each plate and give your response. Your partner records your responses in item 2e on your laboratory report.
2. ***Complete item 2e on the laboratory report.***

THE EAR

The ear contains not only receptors for sound stimuli but also receptors involved in maintaining equilibrium. For ease of study, the ear may be subdivided into the external ear, middle ear, and internal ear. Refer to Figure 18.5 as you study this section.

External Ear

The **external ear** includes the **auricle** or **pinna,** the flap of cartilage and skin commonly called the "ear," and the **auditory canal** that leads inward through the temporal bone to the **tympanic membrane,** or eardrum.

Middle Ear

The **middle ear,** a small cavity in the temporal bone, is connected to the pharynx by the **eustachian tube.** It is filled with air that enters or leaves via the eustachian tube depending on the

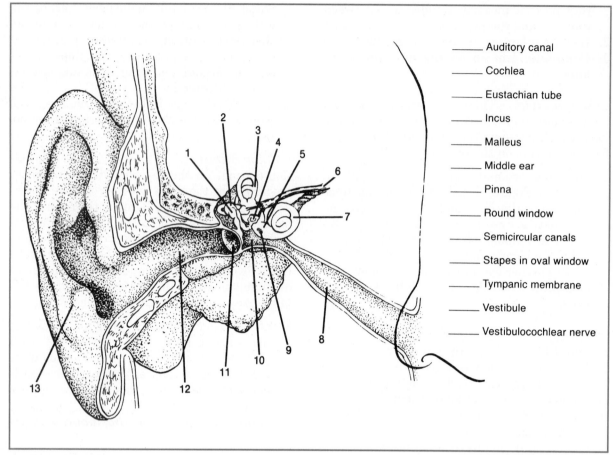

Auditory canal

Cochlea

Eustachian tube

Incus

Malleus

Middle ear

Pinna

Round window

Semicircular canals

Stapes in oval window

Tympanic membrane

Vestibule

Vestibulocochlear nerve

Figure 18.5. The ear

air pressure at each end of the tube. Three small bones, the **ear ossicles,** form a lever system that extends from the eardrum to the inner ear. In sequence, they are (1) the **malleus** (hammer), which is attached to the ear drum, (2) the **incus** (anvil), and (3) the **stapes** (stirrup), which is inserted into the **oval window** of the inner ear.

Inner Ear

The **inner ear** consists of a complex of interconnecting tubes and chambers that are embedded in the temporal bone and that are filled with fluid. It is subdivided into three major parts.

1. The **cochlea** is coiled like a snail shell and contains the receptors for sound stimuli.
2. The **vestibule** is the enlarged portion at the base of the cochlea. The stirrup is inserted into the oval window of the vestibule, and the **round window** is covered by a thin mem-

brane. These two windows are involved in the transmission of sound stimuli. In addition, the vestibule contains receptors for static equilibrium that inform the brain of the position of the head.
3. The three **semicircular canals** contain receptors for dynamic equilibrium that inform the brain when the head is turned or when the entire body is rotated.

The Hearing Process

The auricle channels sound waves into the auditory canal, and as they strike the eardrum, it vibrates at their frequency. This vibration causes a comparable movement of the ear ossicles (hammer, anvil, and stirrup), which, in turn, transfer the movements to the fluid in the vestibule and cochlea of the inner ear. The movement of the fluid is enabled by the thin membrane of the round window, which moves *out and in* in synchrony with the *in and out* movement of the stirrup in the oval window since the fluid cannot be compressed. The movement of the fluid in the cochlea stimulates the sound receptors, which send impulses to the auditory center in the brain, where they are interpreted as sound sensations.

Tone or pitch is determined by which receptors are stimulated and which part of the brain receives the impulses. Loudness is determined by the frequency of impulses formed and transmitted, which, in turn, depends on the intensity of the vibrations.

There are two kinds of hearing loss. **Conduction deafness** results from damage that prevents sound vibrations from reaching the inner ear. It is usually correctable by surgery or hearing aids. **Nerve deafness** is caused by damage to the sound receptors or neurons that transmit impulses to the brain. Nerve deafness usually results from exposure to loud sounds and is not correctable.

Materials

Per student group
Colored pencils
Cotton for ear plugs
Meter sticks
Pocket watches, spring wound
Tuning forks

Per lab
Model of the ear

Assignment 3

1. Label and color-code Figure 18.5.
2. Locate the structures shown in Figure 18.5 on the ear model or chart.
3. ***Complete item 3 on the laboratory report.***

Assignment 4

Watch-Tick Test. This is a simple test to detect hearing loss at a single sound frequency. It requires a quiet area. Work in teams of three students. One student is the subject, another moves the watch, and the third measures and records distances.

1. Have the subject sit in a chair and plug one ear with cotton. The subject is to look straight ahead and indicate by hand signals when the first and last ticks are heard.
2. Start with the watch about 3 ft laterally from the ear being tested and move it *slowly* toward the ear until the subject indicates the first tick is heard. Measure and record the distance from ear to watch.
3. Start with the watch close to the ear and *slowly* move it away from the ear until the last tick is heard. Measure and record the distance from ear to watch.
4. Calculate the average of the two measurements.
5. Test the hearing of the other ear in the same manner.
6. ***Complete item 4a on the laboratory report.***

Rinne Test. A tuning fork is used in this test, which distinguishes between nerve and conduction deafness *where some hearing loss exists*. Work in pairs.

1. Have the subject sit in a chair and plug one ear with cotton. The subject is to indicate by hand signals when the sound is heard or not heard.
2. Strike the tuning fork against the heel of your hand to set it in motion. *Never strike it against a hard object.*
3. Hold the tuning fork 6–9 in. away from the ear being tested with the edge of the tuning fork toward the ear as shown in Figure 18.6.
4. The sound will be heard initially by persons with normal hearing and those with minimal hearing loss. As the sound fades, a point will be reached at which it will no longer be heard. (The subject is to indicate this point by a hand signal.) When this occurs, place the end of the tuning fork against the temporal bone behind the ear. See Figure 18.6. *Where a slight hearing loss exists* and the sound reappears, some conduction deafness is present.
5. Persons with a severe hearing loss will not hear the sound in #4 or will hear it only briefly. If the sound reappears when the end of the tuning fork is placed against the temporal bone, conduction deafness is evident. If it does not reappear, nerve deafness exists.
6. ***Complete item 4b on the laboratory report.***

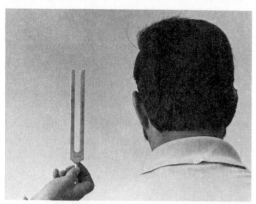

(a) Hold a vibrating tuning fork 6–9 in. from the ear with the edge of the tuning fork toward the ear.

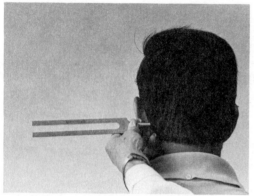

(b) When the sound is no longer heard, place the end of the tuning fork against the temporal bone behind the ear.

Figure 18.6. Rinne test

Static Balance. The inner ear is not the only receptor involved in the maintenance of equilibrium. Pressure and touch receptors in the skin, stretch receptors in the muscles, and light receptors in the eyes also are involved. The brain constantly receives impulses from these receptors and subconsciously initiates any necessary corrective motor actions. The following test is a simple way to observe how this interaction functions. Work in groups of two to four students.

1. Use a meter stick to draw a series of vertical lines on the chalkboard about 5 cm (2 in.)

apart. Cover an area about 1 m wide. This will help you detect body movements.
2. Have the subject remove his or her shoes and stand in front of the lined area facing you.
3. With feet together and arms at the side, the subject is to try to stand perfectly still for 30 sec while you watch for any swaying motion. Record the degree of swaying motion as slight, moderate, or great.
4. Repeat the test with the subject's eyes closed. Record the degree of movement. Try it again to see if extending the arms laterally helps the subject to maintain balance. What happens when the subject stands on only one foot with the hands at the side and with the eyes closed?
5. ***Complete item 4 on the laboratory report.***

SKIN RECEPTORS

Human skin contains receptors for touch, pain, pressure, hot, and cold stimuli. The following tests will enable you to detect certain characteristics of sensations involving some of these receptors. Work in pairs to perform these experiments.

Materials

Per student group
Beakers, 400 ml, 3
Celsius thermometer
Clock or watch with second hand
Coins
Dividers
Hot plate
Metric ruler

Per lab
Crushed ice

Assignment 5

Distribution of Touch Receptors. For you to perceive two simultaneous stimuli as two touch sensations, the stimuli must be far enough apart to stimulate two touch receptors that are separated by at least one unstimulated touch receptor. This characteristic can be used to determined the density of touch receptors in the skin.

1. With the subject's eyes closed, touch his or her skin with one or two points of the dividers. The subject reports the sensation as either "one" or "two." Start with the points of the dividers close together and gradually increase the distance between the points until the subject reports a two-point stimulus as a two-touch sensation about 75% of the time. Measure and record the distance between the tips of the dividers as the minimum distance evoking a two-point sensation.
2. Use the procedure in step 1 to determine the minimum distance giving a two-point sensation on the (1) inside of the forearm, (2) back of the neck, (3) palm of the hand, and (4) tip of the index finger.
3. **Complete item 5a on the laboratory report.**

Adaptation to Stimuli. Your nervous system has the ability to ignore stimuli or impulses so that you are not constantly bombarded with insignificant sensations.

1. Have the subject rest a forearm on the top of the table with the palm of the hand up.
2. With the subject's eyes closed, place a coin on the inner surface of the forearm. The subject is to indicate awareness of the presence of the coin and also the instant the sensation disappears. Record the time between these two events as the adaptation time.
3. Repeat the test using several coins stacked up to make a heavier object that increases the intensity of the stimulus. Determine the adaptation time.
4. **Complete items 5b and 5c on the laboratory report.**

Intensity of Sensations. The intensity of a sensation usually is proportional to the intensity of the stimulus. This occurs because the receptors form more impulses when the strength of the stimulus is increased, and the brain interprets the arrival of more impulses as a greater sensation.

1. Fill three 400-ml beakers with ice water, tap water, and warm water (about 50 °C), respectively).
2. Place your index finger in the water in each beaker, in sequence, and note the sensations.

Can you recognize the temperature differences?
3. Use some small beakers to prepare water with slight differences in temperature to determine the smallest difference that is detectable. Record this differential.
4. Now return to the original three beakers. Place your index finger in the ice water and note the sensation. Then immerse your whole hand and note the sensation. Repeat this procedure with the warm water.
5. **Complete items 5d–5f on the laboratory report.**
6. Use the three beakers of ice water, tap water, and warm water as in the previous experiment. Be sure that the warm water has not cooled.
7. Place one hand in ice water and the other in warm water. Note the sensation. Does it change with time? Explain.
8. After 2 min, place both hands in the beaker of tap water. Note the sensation. Does it change with time?
9. **Complete item 5 on the laboratory report.**

TASTE

The taste of food not only increases the appeal of food but it also increases the flow of saliva. In fact, just the thought of delicious food will stimulate salivation. Try it and see. What we usually refer to as taste is a combination of both taste and smell. This is why food "loses its taste" when you have a bad head cold that prevents airborne molecules of food from reaching your odor receptors in roof of the nasal cavity.

Taste buds, receptors for taste, are located on the sides of very small projections, **papillae,** on the upper surface of the tongue. There are only four basic tastes—sweet, sour, bitter, and salt—and the distribution of taste buds sensitive to these tastes is not uniform, as shown in Figure 18.7.

As shown in Figure 18.8, the sensory **taste cells** of a taste bud are embedded in epithelial tissue with only their taste hairs protruding through a tiny taste pore. When the substances capable of activating the taste cells contact the taste hairs, the taste cells are stimulated, form-

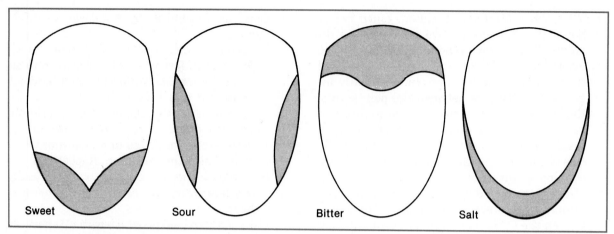

Figure 18.7. Taste zones of the tongue

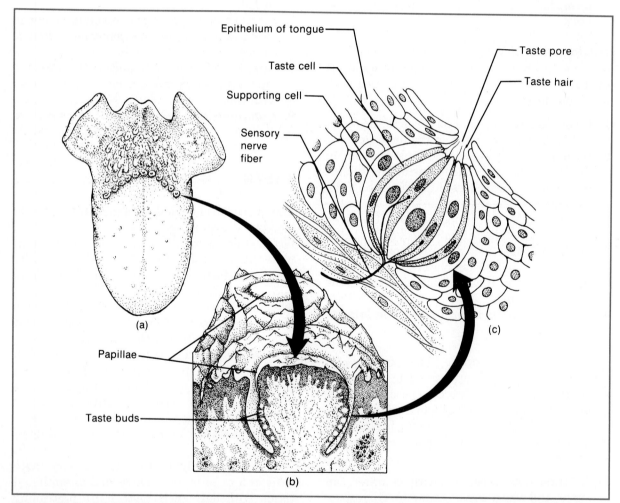

Figure 18.8. (a) Papillae containing taste buds are located on the upper surface of the tongue. (b) Taste buds are located along the outer margins of papillae on the tongue. (c) A taste bud consists of a bulb-shaped cluster of taste cells and supporting cells embedded in epithelial tissue. The hairlike tips of taste cells protrude slightly through a taste pore.

ing impulses that are carried by neurons to the brain.

Work in pairs to perform the following experiments and alternate the roles of subject and experimenter.

Materials

Per student pair
Cotton swabs, 6
Facial tissues, 6
Paper cups, 2
Sucrose granules
Dropping bottles of:
  10% sodium chloride (salt)
  10% sucrose (sweet)
  10% vinegar (sour)

Assignment 6

1. Verify the locations of the sweet, sour, and salt taste receptors as shown in Figure 18.7 as follows.
   a. Obtain 2 paper cups, several facial tissues, and 6 cotton swabs.
   b. Have the subject blot the upper surface of the tongue with a facial tissue and stick it out.
   c. Place several drops of sucrose solution on a cotton swab. Swab the tip of the tongue. Can the subject detect a sweet taste without withdrawing the tongue into his/her mouth? Is there any difference in the taste sensation after withdrawing the tongue?
   d. Have the subject rinse his/her mouth with water and blot the tongue and stick it out as before. Now swab an area where taste buds for sweet should be absent. Can the subject detect a sweet taste without withdrawing the tongue? Is there any difference in the taste sensation after withdrawing the tongue?
   e. Use the above procedure to verify the distribution of taste receptors for sour and salt.
   f. ***Complete item 6a on the laboratory report.***

2. Determine if taste buds can detect molecules that are not in solution.
   a. Obtain several facial tissues and about half of a teaspoon of sugar granules on a folded paper towel.
   b. Have the subject blot and stick out his/her tongue as before.
   c. Sprinkle a number of sugar granules on the "sweet zone" of the tongue. Can the subject detect a sweet taste without withdrawing the tongue into his/her mouth? After 15 sec, have the subject withdraw the tongue. Is there any difference in the sensation?
   d. ***Complete item 6 on the laboratory report.***

19

SUPPORT AND MOVEMENT

OBJECTIVES

After completion of the laboratory session, you should be able to:
1. Identify the major bones and types of articulations in a human skeleton.
2. Identify the parts of a typical long bone.
3. Describe the sexual differences in male and female pelvic girdles.
4. Describe and identify the types of levers formed by skeletal muscles and bones.
5. Describe the action of selected skeletal muscles.
6. Contrast the microscopic appearance of striations in relaxed and contracted skeletal muscle tissue.
7. Describe the roles of actin and myosin in muscle contraction.
8. Define all terms in bold print.

Support and movement of the body results from the interaction of the skeletal and muscle systems. These two body systems are so closely associated both structurally and functionally that we will consider them together in this exercise.

The skeletal system consists mainly of **bones** that join together, forming articulations (joints). At movable articulations, the bones are held together by **ligaments,** bands of dense connective tissue that are slightly elastic. In addition, there are many associated **cartilages.** The skeletal system provides: (1) protection for vital organs, (2) support for the body, (3) sites for muscle attachment, (4) a storehouse for calcium salts, and (5) formation of blood cells.

The muscle system consists of **skeletal muscles** that are attached to bones by **tendons,** bands or cords of inelastic dense fibrous connective tissue. Skeletal muscles are specialized for contraction (shortening), and they always span across a joint formed by bones. Thus, contraction of a muscle causes movement of one of the bones at the joint. In this way, the contraction of skeletal muscles enables body movements.

THE SKELETON

The skeleton consists of two major subdivisions. The **axial skeleton** is composed of the skull, vertebral column, ribs, and sternum. The **appendicular skeleton** consists of the bones of the upper extremities and the pectoral girdle and the lower extremities and the pelvic girdle. See Figure 19.1.

Axial Skeleton

The **skull** is composed of the **cranium** (8 fused bones encasing the brain), 13 fused **facial bones,** and the movable **mandible** (lower jaw).

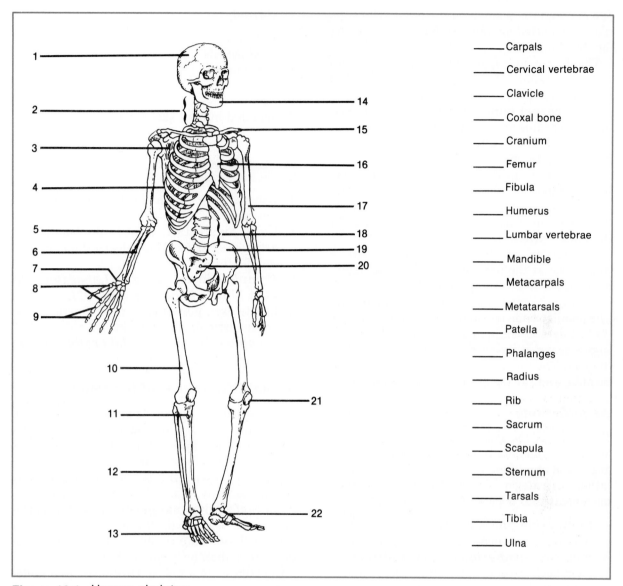

_____	Carpals
_____	Cervical vertebrae
_____	Clavicle
_____	Coxal bone
_____	Cranium
_____	Femur
_____	Fibula
_____	Humerus
_____	Lumbar vertebrae
_____	Mandible
_____	Metacarpals
_____	Metatarsals
_____	Patella
_____	Phalanges
_____	Radius
_____	Rib
_____	Sacrum
_____	Scapula
_____	Sternum
_____	Tarsals
_____	Tibia
_____	Ulna

Figure 19.1. Human skeleton

The **vertebral column** consists of vertebrae separated by **intervertebral discs** composed of fibrocartilage. Vertebrae are subdivided as follows:

Cervical vertebrae: 7 vertebrae of the neck
Thoracic vertebrae: 12 vertebrae of the thorax to which ribs are attached
Lumbar vertebrae: 5 large vertebrae of the lower back
Sacrum: bone formed of 5 fused vertebrae
Coccyx: tailbone formed of 3–5 fused, rudimentary vertebrae

There are 12 pairs of **ribs.** The first 10 pairs are joined to the **sternum** (breastbone) by costal cartilages to form the thoracic cage. The last 2 pairs are short and unattached anteriorly. They are called floating ribs.

Appendicular Skeleton

The **pectoral girdle** supports the upper extremities. It consists of a **clavicle** (collarbone) and **scapula** (shoulder blade) on each side of the body. The clavicle is attached to the sternum on one end and the scapula on the other. The scapula is supported by muscles that allow mobility for the shoulder.

The **humerus** (upper arm bone) articulates with the scapula at the shoulder and the **ulna** and **radius** at the elbow. The wrist is composed of eight **carpal bones** that lie between the (1) ulna and radius and (2) **metacarpals,** the bones of the hand. The **phalanges** are the bones of the fingers and thumb.

The **pelvic girdle** consists of two **coxal bones** (hipbones) that join together anteriorly at the pubic symphysis and are fused posteriorly to the sacrum. This provides a sturdy support for the lower extremities.

The **femur** (thighbone) articulates with a coxal bone at the hip and with the **tibia** (shinbone) at the knee. The **patella** (kneecap) is embedded in the tendon anterior to the knee joint. The smaller bone of the lower leg is the **fibula.** Both tibia and fibula articulate with the **tarsal bones** forming the ankle and posterior part of the foot. The anterior foot bones are five **metatarsals,** and the toe bones are the **phalanges.**

Articulations

The bones of the skeleton are attached to each other by ligaments in such a way that varying degrees of movement occur at the joints. Articulations are categorized according to the degree of movement that is possible.

1. **Immovable joints** are rigid, such as those that occur between the skull bones.
2. **Slightly movable joints** allow a little movement such as those between vertebrae and at the pubic symphysis.
3. **Freely movable joints** are the most common and allow the broad range of movement noted in the appendages. There are several types:
   a. **Hinge joints** allow movement in one direction only.
   b. **Ball-and-socket joints** allow angular movement in all directions plus rotation.
   c. **Gliding joints** occur where bones slide over each other, e.g., wrist and ankle bones.
   d. **Pivot joints** enable rotation in only one axis, e.g., between the first and second cervical vertebrae.

Materials

Per student
Colored pencils
Pipe cleaners

Per lab
Articulated human skeletons

Assignment 1

1. Label Figure 19.1 and color the axial skeleton blue.
2. Using Figure 19.1, identify the bones of an articulated skeleton and learn their recognition features.
3. ***Complete items 1a–1d on Laboratory Report 19 that begins on page 371.***
4. Locate examples of each type of articulation on the articulated skeleton.
5. ***Complete item 1 on the laboratory report.***

Sexual Differences of the Pelvis

The structure of the pelvic girdle is different in males and females, primarily because the female pelvis is adapted for childbirth. Table 19.1 lists some of the major differences between male and female pelvic girdles. Compare these characteristics with Figure 19.2. Since the characteristics of a pelvis may have considerable variations from the ideal, sex determination of a pelvis is based on a combination of characteristics rather than on a single characteristic.

Calculation of a **pelvic ratio** is helpful in determining the sex of a pelvic girdle. Two measurements are required. The ratio is calculated by dividing the first measurement by the second.

1. The distance between the tips of the ischial spines
2. The distances between the inner surface of the pubic symphysis and the upper, inner surface of the sacrum

$$\text{Pelvic ratio} = \frac{\text{first measurement}}{\text{second measurement}}$$

Females usually have a ratio of 1.0 or more, while males usually have a ratio of 0.8 or less.

TABLE 19.1
Comparison of the Male and Female Pelvis

Characteristic	Male	Female
General structure	Not tilted forward; narrower and longer; heavier bones	Tilted forward; broader and shorter; lighter bones
Acetabula	Larger and closer together	Smaller and farther apart
Pubic angle	Less than 90°	Greater than 90°
Sacrum	Longer and narrower	Shorter and wider
Coccyx	More curved; less movable	Straighter; more movable
Pelvic brim	Narrower and heart shaped	Wider and oval shaped
Ischial spines	Longer, sharper, closer together, and project more medially	Shorter, blunt, farther apart, and project more posteriorly

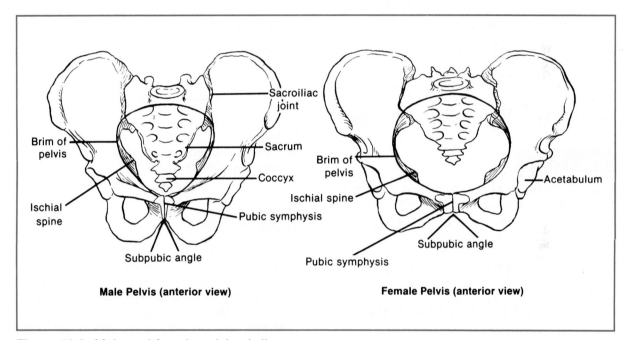

Figure 19.2. Male and female pelvic girdles

Fetal Skeleton

During development, bones (except skull bones) of a fetus are first formed of hyaline cartilage. Then ossification centers appear within the cartilage, and the formation of bone gradually replaces the cartilage. At the time of birth, the bones still have a large cartilage content. Skull bones form within membranes rather than in cartilage, but they also are incompletely formed at birth. The incompletely formed fetal skeleton provides greater flexibility during birth and the ability to grow after birth. Skeletal growth is complete around 25 years of age.

Materials

Per student group
Dividers or calipers
Measuring tape
Metric ruler
Pipe cleaners

Per lab
Articulated skeleton
Fetal skeleton
Pelvic girdles, male and female

Assignment 2

1. Study the differences of the male and female pelvic girdles noted in Table 19.1 and Figure 19.2.
2. Examine the male and female pelvic girdles provided and compare their distinguishing characteristics with Table 19.1 and Figure 19.2.
3. Use dividers or calipers to make the two measurements needed to calculate the pelvic ratio. Calculate the pelvic ratio for both male and female pelvic girdles.
4. Determine the pelvic ratio for the pelvic girdle of the articulated skeleton and examine its other features to determine the gender of the skeleton.
5. *Complete item 2 on the laboratory report.*

Assignment 3

1. Examine the fetal skeleton. Note the incomplete ossification of the bones. Observe how the brain of a baby is incompletely protected by cranial bones because of the spaces (fontanels) between the skull bones.
2. *Complete item 3 on the laboratory report.*

MACROSCOPIC BONE STRUCTURE

Figure 19.3 depicts the structure of a typical long bone, a human humerus. A long bone is characterized by a shaft of bone, the **diaphysis,** which extends between two enlarged portions forming the ends of the bone, the **epiphyses.** The articular surface of each epiphysis is covered by an **articular cartilage** (hyaline cartilage), which reduces friction in the joints and protects the ends of the bone. The rest of the bone is covered by the **periosteum,** a tough, tightly adhering membrane containing tiny blood vessels that penetrate into the bone. Bone deposition by the periosteum contributes to the growth in diameter of the bone. Larger blood vessels and nerves enter the bone through a channel called a **foramen.**

A longitudinal section of the bone reveals that the epiphyses consist of **cancellous** (spongy) **bone** covered by a thin layer of **compact bone,** while the diaphysis is formed of heavy, compact bone. **Red marrow** fills the spaces in the spongy bone. It forms red and white blood cells and platelets. The **medullary cavity** is filled with fatty **yellow marrow.** In immature bones, an **epiphyseal disc** of cartilage is located between the diaphysis and each epiphysis; this is the site of linear growth. A mature bone lacks this cartilage since it has been replaced by bone, and only an **epiphyseal line** of fusion remains.

Materials

Per student group
Beef femur or tibia, fresh and split lengthwise
Colored pencils
Dissecting instruments and pan

Per lab
Beef knee joint split lengthwise
Human femur, split

Assignment 4

1. Label Figure 19.3. Color-code spongy bone red, medullary cavity yellow, articular cartilage blue, and epiphyseal disc green. *Complete item 4a on the laboratory report.*
2. Examine the split human femur. Locate the spongy and compact bone, epiphyseal line, and medullary cavity.
3. Obtain a split beef femur. Locate the medullary cavity, compact bone, spongy bone, epiphyseal disc, and articular cartilage.
4. Remove some of the yellow marrow to observe the size of the medullary cavity. Does yellow marrow contain fat? Is red marrow well supplied with blood?
5. Feel the articular cartilage. How is it suited for its function?
6. Use a scalpel to cut off a small piece of periosteum. Is the periosteum loosely attached? Is it weak or strong? *Complete items 4b–4f on the laboratory report.*

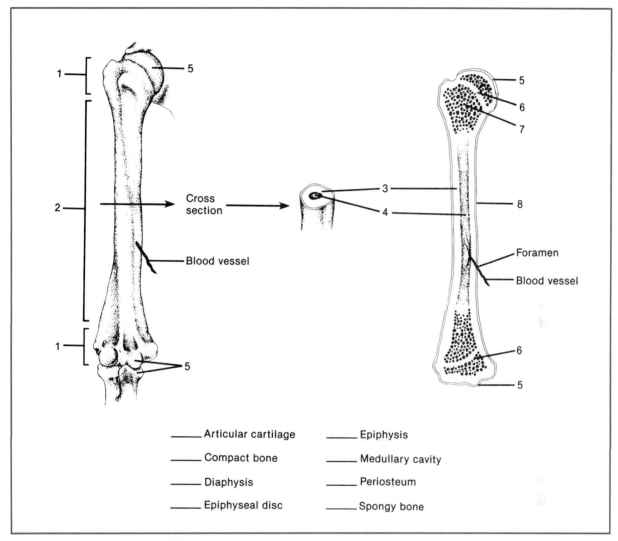

Figure 19.3. Structure of a long bone

_____ Articular cartilage   _____ Epiphysis

_____ Compact bone   _____ Medullary cavity

_____ Diaphysis   _____ Periosteum

_____ Epiphyseal disc   _____ Spongy bone

7. Examine the split beef knee joint. Note how the ligaments hold the bones together. Examine the ligaments. ***Complete item 4 on the laboratory report.***

SKELETAL MUSCLES

Skeletal muscles span a joint and are attached at each end to different bones by **tendons.** The end of a muscle that moves during contraction is the **insertion** while the stationary (nonmovable) end is the **origin.**

Skeletal muscles and bones of the skeleton are arranged to form a system of levers that work together to produce a variety of movements. A lever consists of a rigid rod that moves about a fixed point called a **fulcrum.** Two opposing forces act on a lever. The **resistance** in the weight to be moved by the **force** applied at a specific point on the lever. In the arrangement of muscles and bones, the rigid rod of the lever is formed by a *bone* and the fulcrum is a *joint*. The resistance may be the weight of an arm, leg, or body part, and it may also include an additional weight, such as a weight held in a hand. The

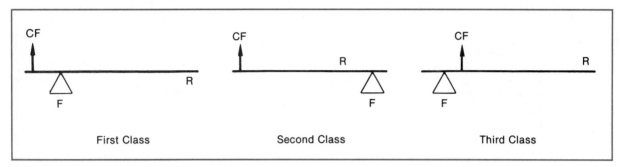

Figure 19.4. The three classes of levers

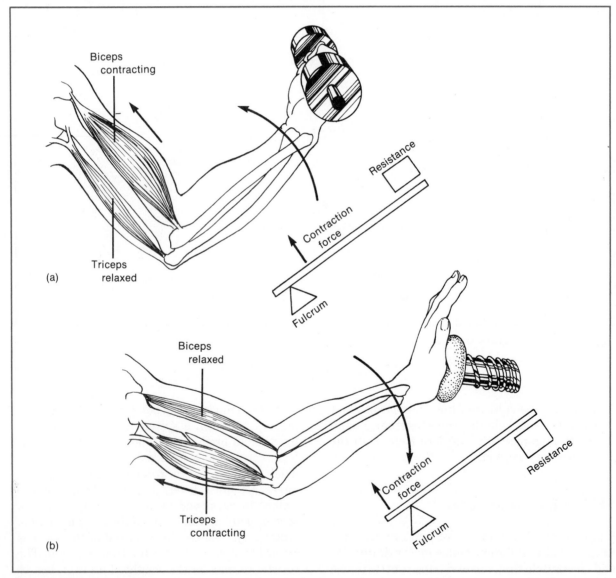

Figure 19.5. Antagonistic functions of biceps and triceps. (a) Contraction of the biceps flexes the forearm via a third-class lever. (b) Contraction of the triceps extends the forearm via a first-class lever.

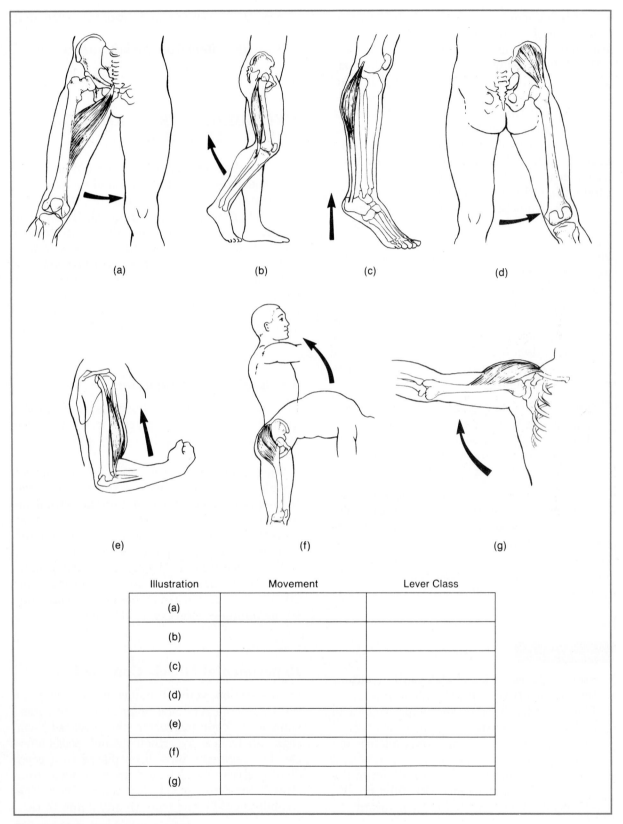

Figure 19.6. Body movements

force causing movement is always the **contraction force** of a muscle applied at the point of insertion on a bone.

The three types of levers formed in the body are diagrammed in Figure 19.4. Each type is characterized by the relative positions of the resistance (R), fulcrum (F), and contraction force (CF). In a **first-class lever,** the fulcrum is located between the contraction force and resistance. In a **second-class lever,** the resistance is located between the contraction force and the fulcrum. In a **third-class lever,** the contraction force is located between the fulcrum and the resistance. The nearer the contraction force is located to the fulcrum, the lower is the mechanical advantage and the greater is the speed of movement of the body part.

Skeletal muscles are arranged in **antagonistic groups,** meaning that opposing muscles move the same body part in opposite directions. This arrangement is necessary because muscles can only contract with force. They cannot extend with force. Here are two examples of antagonistic muscle actions.

> **Flexors** decrease the angle of a joint, while **extensors** increase the angle of a joint.
> **Abductors** move a body part away from the midline of the body, while **adductors** move a body part toward the midline of the body.

The biceps and triceps muscles of the upper arm are examples of antagonistic muscles. The biceps muscle flexes the forearm while the triceps is relaxed, and the triceps extends the forearm while the biceps is relaxed. See Figure 19.5 and note the action of these muscles and the types of levers involved.

Assignment 5

1. Study Figures 19.4 and 19.5 until you know the characteristics of each type of lever.
2. Place your left hand on your upper right arm and note the contraction and relaxation of the biceps and triceps muscles as you flex and extend your right forearm.
3. On Figure 19.6, record on each diagram the location of the resistance (R), fulcrum (F), and contraction force (CF). Then record the

type of movement and lever for each diagram.
4. *Complete item 5 on the laboratory report.*

MYOFIBRIL ULTRASTRUCTURE AND CONTRACTION

The striations of a skeletal muscle fiber result from the arrangement of actin and myosin myofilaments within each myofibril. See Figure 19.7. When a myofibril is relaxed, the thicker **myosin myofilaments** are the main components of, and only occur within, the dark-colored A bands. The thinner **actin myofilaments** are the only myofilaments in the light-colored I bands, but they extend into the A bands from the **Z lines.** Z lines form the boundaries of a **sarcomere,** the contractile unit of a myofibril.

Mechanics of Contraction

Contraction of a muscle fiber results from the interaction of actin and myosin in the presence of calcium and magnesium ions. ATP supplies the required energy.

When a muscle fiber is activated by a neural impulse, the cross-bridges of the myosin myofilaments attach to active sites on the actin myofilament and bend to exert a power stroke that pulls the actin filaments toward the center of the A band. After the power stroke, the cross-bridges separate from the first actin active sites, attach to the next active sites, and produce another power stroke. This process is repeated until maximum contraction is attained.

Experimental Muscle Contraction

In this section, you will induce muscle fibers to contract using ATP and magnesium and potassium ions. Your instructor has prepared 2-cm segments from a glycerinated rabbit psoas muscle. The segments have been placed in a petri dish of glycerol and teased apart to yield thin strands consisting of very few muscle fibers. The strands should be no more than 0.2 mm thick.

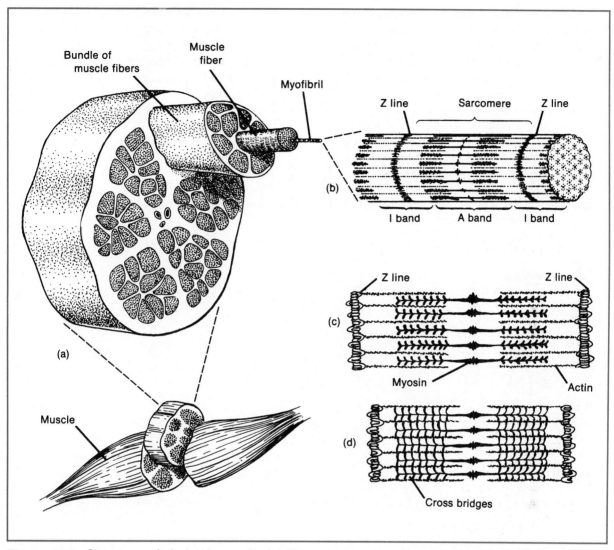

Figure 19.7. Structure of skeletal muscle. (a) The arrangement of muscle fibers within a muscle. (b) The ultrastructure of a myofibril showing the relationship of actin and myosin filaments. Note how the arrangement of actin and myosin differs in (c) a relaxed state and (d) a partially contracted state.

Materials

Per student

Microscopes, compound and dissecting
Dissecting instruments
Microscope slides and cover glasses
Plastic ruler, clear and flat
Glycerinated rabbit psoas muscle, 3–6 strands
Dropping bottles of:
    ATP, 0.25%, in triple-distilled water
    glycerol

magnesium chloride, 0.001 M
potassium chloride, 0.05 M

Assignment 6

1. Place a few muscle strands in a small drop of glycerol on a microscope slide and add a cover glass. Observe them with the high-dry or oil-immersion objective. ***Draw the pattern of***

striations of the relaxed muscle fibers in item 6a of the laboratory report.

2. Place three to five of the thinnest strands on another slide in just enough glycerol to moisten them. Arrange the strands straight, parallel, and close together.

3. Use a dissecting microscope to measure the length of the relaxed strands by placing the slide on a clear plastic ruler. *Record their lengths in item 6b on the laboratory report.* Note the width of the strands.

4. While observing through the dissecting microscope, add one drop from each of the solutions: ATP, magnesium chloride, and potassium chloride. Note any changes in length or width of the strands.

5. Remeasure and *record the length of the contracted strands in item 6b on the laboratory report.*

6. Place a few contracted strands in a small drop of glycerol on another slide and add a cover glass. Observe the strands with the high-power or oil-immersion objective. *Draw the pattern of striations in item 6a on the laboratory report.*

7. *Complete item 6 on the laboratory report.*

EXCRETION

Metabolic processes produce waste products that are harmful to body cells, so the wastes must be removed from body fluids. The principal metabolic wastes are carbon dioxide and nitrogenous wastes that accumulate in the blood. They must be removed as fast as they accumulate, so their concentrations in the blood remain within normal limits. Carbon dioxide is removed by the respiratory system. **Nitrogenous wastes** are removed by the **urinary system** during the formation of **urine** as the blood is cleansed by the kidneys.

The primary nitrogenous wastes are ammonia, urea, and uric acid. When amino acids are deaminated by the liver, the amine groups tend to form **ammonia (NH$_3$),** a substance toxic to body cells. To get rid of the ammonia, the liver converts it into **urea.** Urea is the most abundant nitrogenous waste in urine. A relatively small amount of **uric acid** is present as a result of the breakdown of nucleic acids.

The process of urinary excretion not only removes nitrogenous wastes, but it also regulates the concentration of ions, water, and other substances in body fluids.

THE URINARY SYSTEM

The urinary system consists of (1) a pair of **kidneys** that remove nitrogenous wastes, excess ions, and water from the blood to form **urine;** (2) a pair of **ureters** that carry urine by peristalsis from the kidney to the urinary bladder; (3) a **urinary bladder** that serves as a temporary storage container; and (4) a **urethra** that carries urine from the bladder during urination. See Figure 20.1.

The Kidney

The basic structure of a kidney is shown in coronal section in Figure 20.2. The outer portion is the **cortex,** which contains vast numbers of capillaries and **nephrons.** The inner **medulla** contains the **renal pyramids** that are composed mainly of **collecting tubules.** The tip (papilla)

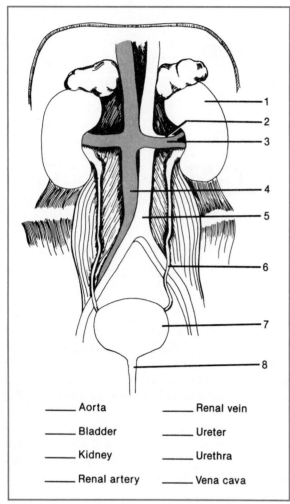

Aorta _____   Renal vein _____

Bladder _____   Ureter _____

Kidney _____   Urethra _____

Renal artery _____   Vena cava _____

Figure 20.1. Human urinary system

of each pyramid is inserted into a funnellike **calyx** that unites with the **renal pelvis.**

The Nephron

The nephron is the functional unit of the kidney. Each human kidney contains about 1 million nephrons. A nephron consists of (1) a **Bowman's** (nephron) **capsule** that surrounds an arteriole capillary tuft, a **glomerulus,** and (2) a tortuous, thin-walled **tubule** that leads to a collecting tubule. The nephron tubule consists of three parts. The **proximal convoluted tubule** leads from the nephron capsule to a U-shaped **loop of Henle** that is contiguous with a **distal convoluted tubule.** Many renal tubules are joined to a single collecting tubule.

Materials

Per student group
Colored pencils
Dissecting kit and pan
Microscope, compound
Sheep kidney, preferably fresh, mid-coronally
  sectioned

Per lab
Model of human urinary system
Model of human kidney, mid-coronal section
Prepared slides of kidney cortex
Sheep kidneys, triple injected and preserved

Assignment 1

Complete item 1 on Laboratory Report 20 that begins on page 375.

Assignment 2

1. Label and color-code parts of the urinary system in Figure 20.1 and parts of the kidney and nephron in Figure 20.2.
2. Study the models of the urinary system and kidney. Locate the parts shown in Figures 20.1 and 20.2.
3. Examine the demonstration whole kidney. Note its distinctive shape and locate the ureter and renal blood vessels.
4. Obtain a sheep kidney that has been coronally sectioned and locate the parts shown in Figure 20.2a. Correlate the structure of the parts with their functions that were previously described.
5. Examine the sectioned triple-injected kidney set up under a dissection microscope. Locate the glomeruli and nephron capsules.
6. Examine a prepared slide of kidney cortex. Locate a glomerulus and nephron capsule. Note the thin wall of the tubules. How many cells thick is the tubule wall?
7. ***Complete item 2 on the laboratory report.***

Urine Formation

Blood is carried from the aorta to the kidneys by renal arteries and returned from the kidneys by renal veins into the vena cava. While passing through the kidneys, the concentration of nitrogenous wastes and other substances in the

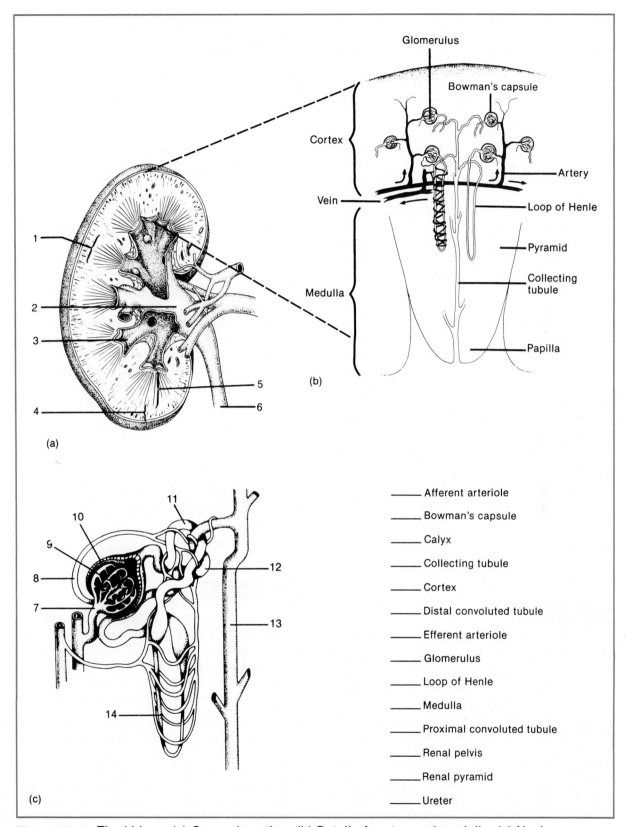

Figure 20.2. The kidney. (a) Coronal section. (b) Detail of cortex and medulla. (c) Nephron.

blood is controlled by (1) the partial removal of diffusable materials in the glomerular blood by **filtration** into Bowman's capsules, (2) **selective reabsorption** of useful substances from the tubules back into the blood, and (3) **secretion** of certain ions from the capillary blood into the tubules.

The renal artery branches into progressively smaller arteries leading ultimately to the afferent arterioles supplying blood to the glomeruli. The **afferent arteriole** leading to a glomerulus is larger in diameter than the **efferent arteriole** carrying blood from the glomerulus. This causes an increase in blood pressure in the glomerulus and accelerates the filtration of materials from the glomerular blood into the capsule. Thus, some of all diffusable materials pass from the glomerulus into Bowman's capsule.

The fluid in the capsule, the **filtrate,** flows through the nephron tubule, which is enveloped by a capillary network. Selective tubule reabsorption of needed materials—especially water, nutrients, and mineral ions—from the filtrate into the capillary blood occurs as the filtrate moves through the tubule. At the same time, certain excess ions (especially H^+ and K^+) may be secreted from the capillary blood into the filtrate. Both **tubule reabsorption** and **tubule secretion** involve active and passive transport mechanisms. The remaining fluid flows from the nephron tubule into a collecting tubule of a renal pyramid, continues into a calyx, and moves on into the renal pelvis. This fluid, now called **urine,** leaves the kidney via the ureter and is carried to the urinary bladder by peristalsis.

<hr>

Assignment 3

1. Study Table 20.1. Note the relationships among the daily volumes of blood circulated through human kidneys, filtrate removed, and urine formed. *Complete items 3a–3d on the laboratory report.*
2. Study Table 20.2. Note how the concentration of selected substances varies in different parts of the nephron and blood vessels. Use Figure 20.2 to orient yourself to the antomical association of the structures involved.

    Note that the concentration of urea is constant in columns 1, 2, and 3 of Table 20.2. This indicates that urea is a small molecule

TABLE 20.1

Volume (liters) of Blood, Filtrate, and Urine Formed in Humans in a 24-Hr Period

Blood flow through the kidneys	1,700
Filtrate removed from blood	180
Urine formed	1.5

that easily passes from the glomerulus into Bowman's capsule. How do you explain the change in urea concentration in columns 4 and 5?

3. *Complete item 3 on the laboratory report.*

URINALYSIS

In humans, the composition of urine is commonly used to assess the general functioning of the body. Diet, exercise, and stress may cause variations in composition and concentration, but significant deviations usually result from malfunctions of the body. See Table 20.3.

Materials

Per student
Collecting cup, plastic
Multistix 9 SG reagent strips, 6
Set of test tubes of simulated urine samples

Per lab
Biohazard bag

<hr>

Assignment 4

In this section, you will use "dip sticks" to analyze several simulated urine samples to determine if they are normal or if their characteristics suggest possible disease. Table 20.3 indicates normal values and some abnormal characteristics. Study it before proceeding.

1. Obtain six Multistix 9 SG reagent strips (dip sticks) and a color chart that indicates how to read the results. Study the color chart carefully to be sure that you understand the time requirements for reading the dip sticks and the location and interpretation of each reagent band. Note: The sequence of the reagent bands from the tip to the handle of a

TABLE 20.2

Concentration (mg/100 ml) of Dissolved Substances in Various Regions of the Nephron

Materials in Water Solution	1 Afferent Arteriole	2 Efferent Arteriole	3 Bowman's Capsule (Filtrate)	4 Collecting Tubule (Urine)*	5 Renal Vein
Urea	30	30	30	2,000	25
Uric acid	4	4	4	50	3.3
Inorganic salts	720	720	720	1,500	719
Protein	7,000	8,000	0	0	7,050
Amino acids	50	50	50	0	48
Glucose	100	100	100	0	98

* Trace amounts are not included.

TABLE 20.3

Urine Components Evaluated by Urinalysis

Component	Normal	Abnormal*
Color	Straw to amber	Pink, red-brown, or smoky urine may indicate blood in the urine. The higher the specific gravity, the darker is the color. Nearly colorless urine may result from excessive fluid intake, alcohol ingestion, diabetes, insipidis, or chronic nephritis.
Turbidity	Clear to slightly turbid	Excess and persistent turbidity may indicate pus or blood in the urine.
pH	4.8–8.0; Ave. = 6.0	Acid urine may result from a diet high in protein (meats and cereals) or a high fever; alkaline urine results from a vegetarian diet or bacterial infections of the urinary tract.
Specific gravity	1.003–1.035	Low values result from a deficiency of antidiuretic hormone or kidney damage that impairs water reabsorption. High values result from diabetes mellitis or kidney disease, allowing proteins to enter filtrate.
Blood or hemoglobin	Absent	Presence of intact RBCs may result from lower urinary tract infections, kidney disease allowing RBCs to enter filtrate, lupus, or severe hypertension. Presence of hemoglobin occurs in extensive burns and trauma, hemolytic anemia, malaria, and incompatible transfusions.
Protein	Absent or trace	Proteins are present in kidney diseases allowing proteins to enter the filtrate, and they may be present in fever, trauma, anemia, leukemia, hypertension, and other nonrenal disorders. They may occur due to excessive exercise and high-protein diets.
Glucose	Absent or trace	Presence usually indicates diabetes mellitis.
Ketones	Absent or trace	Presence results from excessive fat metabolism as in diabetes mellitis and starvation.
Bilirubin	Small amounts	Excess may indicate liver disease (hepatitis or cirrhosis) or blockage of bile ducts.
Nitrite (bacteria)	Absent	Presence indicates a bacterial urinary tract infection.
Leukocytes (pus)	Absent	Presence indicates a urinary tract infection.

* Only a few causes of abnormal values are noted.

dip stick matches the sequence of specific tests listed from top to bottom on the color chart.

2. Obtain six numbered test tubes containing simulated urine samples. Place them in a test-tube rack at your workstation.

3. Analyze the urine samples one at a time by dipping a dip stick completely into the sample so that all reagent bands are immersed. Remove the dip stick and place it on a paper towel with the reagent bands facing upward. Read your results after 1 min.

4. If you wish to test your own urine, obtain a plastic collecting cup and a Multistix 9 SG reagent strip from your instructor and perform the test in the restroom. Dispose of the urine in a toilet, place the collecting cup in the biohazard bag, and read your results. Then place the dip stick in the biohazard bag.

5. ***Record your results in item 4a on the laboratory report. Compare your results with Table 20.3 and complete item 4 on the laboratory report.***

21

REPRODUCTION

OBJECTIVES

After completion of the laboratory session, you should be able to:
1. Describe the structure and function of the male and female reproductive systems.
2. Identify parts of the reproductive systems on models and charts.
3. Describe the process of gametogenesis.
4. Identify the stages of sperm and egg development when viewed microscopically.
5. Describe the mode of action and relative effectiveness of birth control methods.
6. Define all terms in bold print.

The human male and female reproductive systems are specially adapted for their roles in gametic sexual reproduction. **Gametes** are sex cells, sperm and eggs, and they are produced by **gonads,** the sex glands. Sperm are produced by **testes,** the male sex glands, and eggs are produced by **ovaries,** the female sex glands. During copulation (sexual intercourse), sperm are deposited in the female reproductive tract, and they quickly move into the uterus and oviducts in search of an egg. As in all mammals, **internal fertilization,** the fusion of sperm and egg, usually occurs in an oviduct. Fertilization results in the formation of a **zygote,** the first cell of a potential child. As the embryo develops, it is carried into the uterus, where **internal development** takes place, leading to the birth of a baby in about 280 days.

MALE REPRODUCTIVE SYSTEM

Locate the organs described here in Figure 21.1. The male gonads are paired testes that are held in the saclike **scrotum.** This arrangement holds the testes outside the body cavity and at a temperature of 94–95 °F, which is necessary for the production of viable **spermatozoa.** Muscles in the wall of the scrotum relax or contract to position the testes farther from or closer to the body, and in this way regulate the temperature of the testes.

A testis contains numerous **seminiferous tubules** that produce the spermatozoa. **Interstitial cells** are located between the tubules, and they secrete **testosterone,** the male hormone responsible for the sex drive and the development of the sex organs and secondary sexual characteristics. The secretion of the **interstitial cell stimulating hormone (ICSH)** by the hypophysis (pituitary gland) activates the interstitial cells.

The **penis,** the male copulatory organ, contains three cylinders of spongy **erectile tissue** that fill with blood during sexual excitement to produce an erection. A circular fold of tissue, the **prepuce** (foreskin) covers the **glans penis.** For

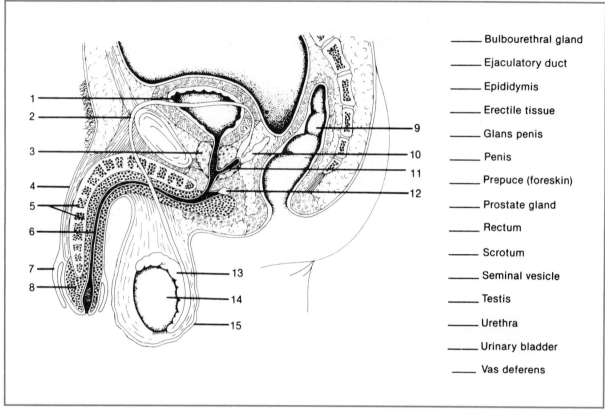

_____ Bulbourethral gland

_____ Ejaculatory duct

_____ Epididymis

_____ Erectile tissue

_____ Glans penis

_____ Penis

_____ Prepuce (foreskin)

_____ Prostate gland

_____ Rectum

_____ Scrotum

_____ Seminal vesicle

_____ Testis

_____ Urethra

_____ Urinary bladder

_____ Vas deferens

Figure 21.1. Male reproductive system

hygienic reasons, the prepuce of male babies is often removed by a surgical procedure called **circumcision.**

Mature, but inactive, sperm are carried down the seminiferous tubules to the **epididymis,** a long, coiled tube on the surface of the testis. Sperm are stored here until they are propelled through the reproductive tract by wavelike contractions during **ejaculation.**

The **bulbourethral glands** open into the urethra below the prostate gland and secrete an alkaline liquid that neutralizes the acidity of the urethra prior to ejaculation. At the male orgasm, sperm pass from the epididymis into the **vas deferens,** a duct that exits the scrotum and enters the body cavity via the inguinal canal. It continues across the surface of the urinary bladder to join with the **ejaculatory duct** within the **prostate gland** that is located around the urethra just below the bladder. Alkaline secretions from the **seminal vesicles** are mixed with the sperm just before the vas deferentia enter the prostate gland, where sperm-activating pros-

tatic secretions are added. Muscular contractions force **semen,** the mixture of sperm and glandular secretions, out through the urethra.

FEMALE REPRODUCTIVE SYSTEM

Locate the organs described here in Figure 21.2. The external female genitalia consist of (1) two folds of skin surrounding the vaginal and urethral openings, the **labia majora** (outer folds) and **labia minora** (inner folds), and (2) the **clitoris,** a nodule of erectile tissue homologous to the penis in the male. Collectively, these structures are called the **vulva.**

The **vagina** is a collapsible tube extending 4–6 in. from the external opening to the uterus. It serves as both the female copulatory organ and the birth canal. The **uterus** is a pear-shaped organ located over and posterior to the urinary bladder. The **cervix** of the uterus extends a short distance into the upper end of the vagina.

A pair of **ovaries,** the female gonads, are lo-

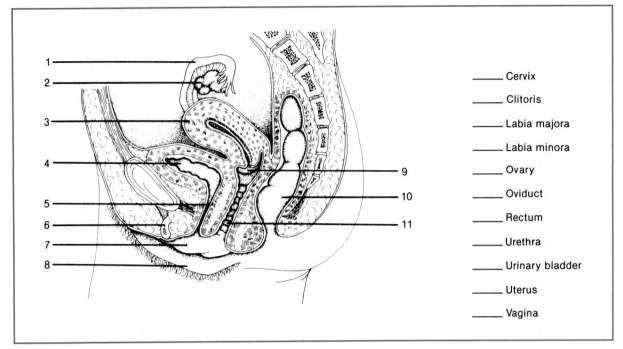

Figure 21.2. Female reproductive system

Labels (right side):
_____ Cervix
_____ Clitoris
_____ Labia majora
_____ Labia minora
_____ Ovary
_____ Oviduct
_____ Rectum
_____ Urethra
_____ Urinary bladder
_____ Uterus
_____ Vagina

cated laterally to the uterus, where they are supported by ligaments. One egg is released from alternate ovaries about every 28 days. The egg is picked up by the expanded end of the **oviduct** and carried toward the uterus by beating cilia of the ciliated epithelium lining the oviduct.

Ovaries secrete two female hormones. **Estrogen** is responsible for the development of the sex organs, the female sex drive, the secondary sex characteristics, and the buildup of the uterine lining. **Progesterone** prepares the uterine lining for the implantation of an early embryo. In turn, ovarian function is controlled by hormones released from the hypophysis (pituitary gland). Consult your text for a discussion of the ovarian and uterine cycles.

Materials

Per student
Colored pencils

Per lab
Anatomical charts of male and female reproductive systems
Models of male and female reproductive systems

Assignment 1

1. Label Figures 21.1 and 21.2. Color-code the organs of the reproductive systems.
2. Locate the organs of the male and female reproductive systems on the models and charts provided.
3. ***Complete item 1 on Laboratory Report 21 that begins on page 379.***

GAMETOGENESIS

The formation of gametes is called gametogenesis. It includes meiotic cell division, which reduces the chromosome number of gametes to one half that of somatic cells. For example, the diploid (2n) chromosome number in humans is 46, and the haploid (n) gametes contain only 23 chromosomes. If you need to review meiosis, see Exercise 8.

The patterns of gametogenesis described here and illustrated in Figures 21.3 and 21.4 are typical of mammals, although slight variations may occur among individual species.

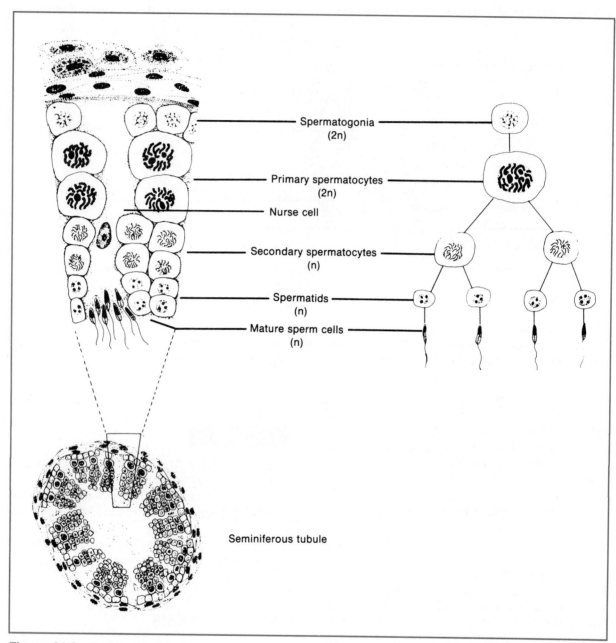

Figure 21.3. Spermatogenesis

Spermatogenesis

Sperm formation occurs in the seminiferous tubules of the testes. **Spermatogonia** (2n) are the outermost cells of the tubule. They divide mitotically to form a **primary spermatocyte** and a replacement spermatogonium. The primary spermatocyte divides meiotically to yield two secondary **spermatocytes** (n) after meiotic division I and four **spermatids** (n) after meiotic division II. The spermatids attach to "nurse cells" and mature into spermatozoa. Maturation includes the loss of most of the cytoplasm and the formation of a flagellum from a centriole. The sperm head consists mostly of the cell nucleus. Sperm are carried along the seminiferous

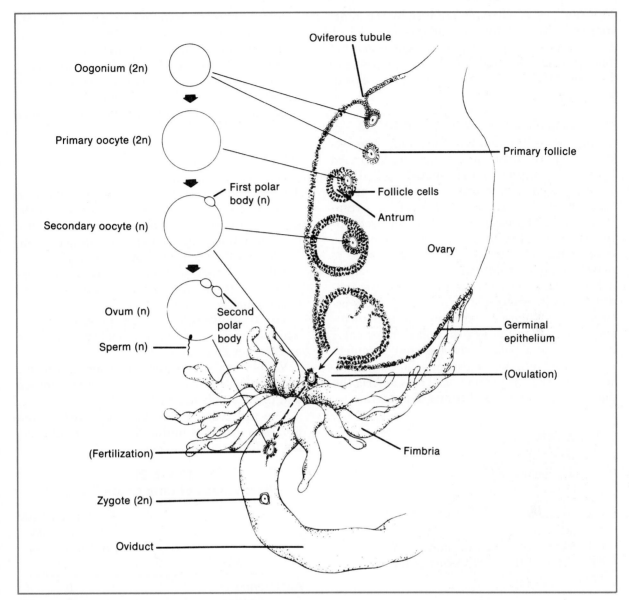

Figure 21.4. Oogenesis and fertilization

tubules and reach the epididymis in about 10 days.

Oogenesis

Prior to the birth of a female child, some **oogonia** of the germinal epithelium surrounding each ovary enlarge, become surrounded by follicular cells, and move into the ovary. These oogonia (2n) become the **primary oocytes** (2n) that enter prophase of meiosis I before oogenesis is arrested. The primary oocytes remain inactive at this stage until puberty.

At puberty, the **follicle stimulating hormone (FSH)** and the **luteinizing hormone (LH)** secreted by the pituitary gland stimulate primary oocytes and follicle cells to further growth. Each month one follicle develops more rapidly than the others to become a mature or **Graafian follicle** filled with fluid containing a large amount of **estrogen** secreted by the follicular cells. Meiotic division I proceeds to produce (1) a **secondary oocyte** (n) that receives most of the cytoplasm and (2) a much smaller **first polar body** that remains attached to the secondary oocyte.

Ovulation occurs when the mature follicle ruptures and ejects the follicular fluid and secondary oocyte through the ovary wall. The secondary oocyte enters the oviduct and is carried toward the uterus by the ciliated epithelium. Note that while it is common to speak of the "egg" as being released by the ovary in ovulation, a secondary oocyte is actually released. After ovulation, the empty follicle becomes the **corpus luteum** that produces progesterone to maintain the uterine lining.

No further division occurs unless the secondary oocyte is penetrated by a sperm. If this occurs, the secondary oocyte completes meiotic division II to form the **egg** (n) and another polar body. The first polar body also may complete meiosis II to form an additional polar body. Subsequently, egg and sperm nuclei fuse to form the diploid zygote. The polar bodies disintegrate.

Materials

Per student
Compound microscope

Per lab
Prepared slides of:
  cat testis, sectioned
  cat ovary, sectioned with mature follicle
  cat ovary, sectioned with corpus luteum
  human sperm

Assignment 2

1. Examine a prepared slide of cat testis. Locate the seminiferous tubules, which produce spermatozoa, and the interstitial cells, which produce testosterone, the male hormone. Compare your observations with Figure 21.3, and locate the cells involved in spermatogenesis.
2. Examine the prepared slide of human sperm. Note their small size. About 350 million sperm are released in an ejaculation.
3. Examine a prepared slide of cat ovary. Compare your slide with Figure 21.4. Locate the germinal epithelium, a primary follicle with a primary oocyte, and a Graafian (mature) follicle containing a secondary oocyte.
4. Examine a prepared slide of cat ovary showing a corpus luteum, which secretes progesterone to maintain the uterine lining.
5. ***Complete item 2 on the laboratory report.***

BIRTH CONTROL

The control of fertility is one of the major concerns of modern society, not only because of the desire to prevent unwanted pregnancies but also because of the great need to curb a human population growth rate that is out of control.

Table 21.1 indicates the effectiveness of common birth control methods. Descriptions of some of these methods follow.

Tubal ligation is a surgical procedure in which a small section of each oviduct is removed and the cut ends are tied.

Vasectomy is a surgical procedure in which a small section of each vas deferens is removed and the cut ends are tied.

The pill consists of synthetic estrogens and progesterones that inhibit development of ovarian follicles and ovulation.

TABLE 21.1
Effectiveness of Birth Control Methods

Method	Pregnancies per 100 Sexually Active Women per Year
Abstinence	0
Tubal ligation	0
Vasectomy	0.4
Pill	2
IUD only	5
Condom (high quality)	10
Diaphragm plus spermicide	13
Sponge plus spermicide	17
Rhythm method	24
Spermicide only	25
Withdrawal	26
Condom (poor quality)	30
Douche	60
No birth control method	90

Source: Cecie Starr, *Biology, Concepts and Applications* (San Francisco, Wadsworth), 1991; p. 484.

An **intrauterine device (IUD)** is a metal or plastic device placed in the uterus by a physician, and it remains there for long periods of time. An IUD prevents implantation of an embryo. Studies have shown that the use of an IUD increases the probability of pelvic inflammatory disease and the probability of female sterility.

A **diaphragm** is a dome-shaped device inserted into the vagina and placed over the cervix prior to sexual intercourse. A **cervical cap** is similar to a diaphragm. Both are used with spermicides.

A **sponge** is a spongelike device that is inserted into the vagina and placed against the cervix prior to sexual intercourse. The sponge contains a spermicide.

A **condom** is a thin, tight-fitting sheath of latex or lamb intestine that is worn over the penis during sexual intercourse. It is more effective when used with a spermicide. Only latex condoms provide some protection against sexually transmitted diseases.

Spermicides are chemicals that are lethal to sperm. They are marketed as foams, jellies, creams, and suppositories.

The **rhythm method** involves abstention from sexual intercourse during a woman's fertile period that extends from a few days before to a few days after ovulation. It requires a woman to take her temperature each morning before arising to detect the 0.5–1.0 °F rise in body temperature that occurs just prior to ovulation.

Withdrawal is the removal of the penis from the vagina just prior to ejaculation. Effectiveness is limited because some sperm are often emitted from the penis prior to ejaculation.

A **douche** is the rinsing out of the vagina after sexual intercourse. It is not very effective since sperm can enter the uterus within 1.5 min after being deposited in the vagina.

Materials

Per lab
Demonstration table of birth control devices and spermicides

Assignment 3

1. Examine the various birth control methods set up as a demonstration. Read the directions for use that accompany each device or spermicide.
2. *Complete item 3 on the laboratory report.*

22

FERTILIZATION AND DEVELOPMENT

OBJECTIVES

After completion of the laboratory session, you should be able to:
1. Describe activation and cleavage and identify activated eggs and cleavage stages when observed with the microscope.
2. Describe and identify a blastula and a gastrula.
3. Describe the formation of the germ layers and identify the germ layers in a late gastrula.
4. Indicate the adult tissues and organs formed from the germ layers.
5. Identify the extra-embryonic membranes and state their functions.
6. Correlate size and visible features with the age of human fetuses.
7. Define all terms in bold print.

The union of gametes and early embryological development are difficult to study in many animals, especially humans. Echinoderms are good subjects for such a study, however, because the gametes are easy to procure and minimal care is needed for the adults and embryos. In addition, embryological development in echinoderms is similar to that in chordates, including humans.

The penetration of an egg by a sperm is called **activation.** The subsequent fusion of egg and sperm nuclei is **fertilization,** and this process forms the diploid **zygote.** A series of mitotic divisions called **cleavage** begins and produces progressively smaller cells. As cleavage progresses, a solid ball of cells, the **morula,** is formed, and continued cleavage produces the **blastula,** a hollow ball of cells. This concludes the cleavage process. The blastula is not much larger than the zygote.

Mitotic divisions continue and transform the blastula into a **gastrula** by a process called gastrulation. This stage of early development is completed by the formation of the three **embryonic tissues** or **germ layers.**

In this exercise, you will study activation and early development in the sea urchin. Follow the directions carefully.

PROCUREMENT OF GAMETES

Your instructor will obtain gametes for the entire class. Several sea urchins may be needed to find a male and female since sex cannot be easily determined by external examination.

Materials

Per lab
Beakers, 50 ml and 100 ml
Dropping bottles
Finger bowls
Hypodermic needles, 22 gauge
Hypodermic syringes, 5 ml
Medicine droppers
Syracuse dishes
Potassium chloride solution, 0.5 M
Seawater at 20 °C
Sea urchins

Procedure

1. Inject 1 ml of the 0.5 M potassium chloride solution into each of three or four sea urchins. Insert the hypodermic needle through the membranous region around the mouth as shown in Figure 22.1.
2. Place the urchins on paper towels with the oral (mouth) side down. Watch for the release of the gamete secretions from the aboral surface. The sperm secretion is white, and the egg secretion is pale buff in color.
3. As soon as a female starts shedding (releasing eggs), place her on a beaker full of cold seawater, aboral side down so that the aboral surface is in the water. Release of all the eggs will take several minutes. See Figure 22.2.
4. After the eggs have been released, discard the female sea urchin. Swirl the eggs and water to wash the eggs and allow them to settle to the bottom of the beaker. Pour off the water and add fresh seawater. Repeat this washing procedure twice. It will facilitate the activation process.
5. After the final washing, swirl the water to disperse the eggs. Then pour about 25 ml of seawater and eggs into each of five or six finger bowls and keep them at 20 °C until used. The eggs will remain viable for 2–3 days at 20 °C when the bowls are stacked to reduce evaporation.
6. Allow several minutes for the sperm secretion to accumulate on the aboral surface of a male. Then remove the secretion with a medicine dropper and place it in a Syracuse dish. See Figure 22.3. Undiluted sperm secretion in a covered dish will be viable for 2–3 days at 20 °C. Prepare a sperm solution, just before use, by placing 2 drops of sperm secretion

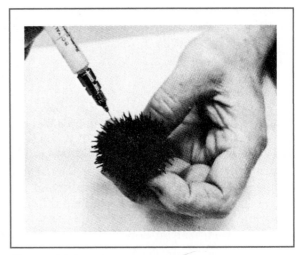

Figure 22.1. Injection of sea urchin

Figure 22.2. Shedding female on beaker of seawater

Figure 22.3. Removing sperm with a dropper

in 25 ml of seawater and dispense in a dropping bottle.

7. Keep gametes at 20 °C until used.

ACTIVATION

Chemicals called **gamones** are released by both the sperm and eggs of sea urchins. They serve to attract sperm to the eggs and also enable the attachment of a sperm to the egg membrane. Similar substances are released by human gametes. The first sperm to attach near the **animal pole** of the egg causes activation. The egg extends a **fertilization cone** (really an activation cone) through the egg membranes to engulf the sperm head and draw it into the egg. The sperm tail remains outside the egg. A rapid release of substances from cytoplasmic vesicles into the space between the **inner** and **outer egg membranes** immediately follows and results in the rapid inflow of fluid into this space. The accumulation of fluid pushes the outer membrane farther outward and prevents penetration of another sperm. The outer membrane is now called the **fertilization membrane** (really an activation membrane). Study Figure 22.4. Subsequently, the egg and sperm nuclei fuse to form the diploid nucleus of the zygote in the process of **fertilization.**

EARLY EMBRYOLOGY

After fertilization, the **zygote** begins a series of mitotic divisions that produce successively smaller cells. These divisions are called **cleavage,** and the first division occurs about 40–60 min after activation. Succeeding divisions occur at approximately 30-min intervals. The cells formed by the cleavage divisions are called **blastomeres.** See Figure 22.5.

The first cleavage division passes through the **animal** and **vegetal poles** of the zygote to yield cells of equal size. The second division also passes through both poles to form 4 cells of equal size.

The third division is perpendicular to the polar axis and forms 8 cells. The cells of the **animal hemisphere** are slightly smaller than those of the **vegetal hemisphere** due to the greater concentration of **yolk** in the vegetal hemisphere. The fourth division forms 16 cells; however, it forms 4 large cells and 4 tiny cells in the vegetal hemisphere. The tiny cells are called **micromeres.**

The **blastula** is formed about 9 hr after activation and is no longer enveloped by the fertilization membrane. The inner cavity, the **blastocoele,** is filled with fluid. Cilia develop on the outer surfaces of the cells, and their beating provides rotational motion of the blastula.

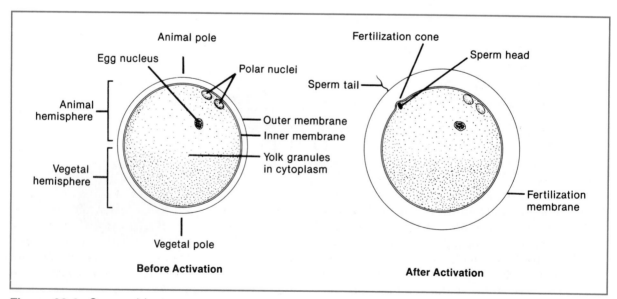

Figure 22.4. Sea urchin egg

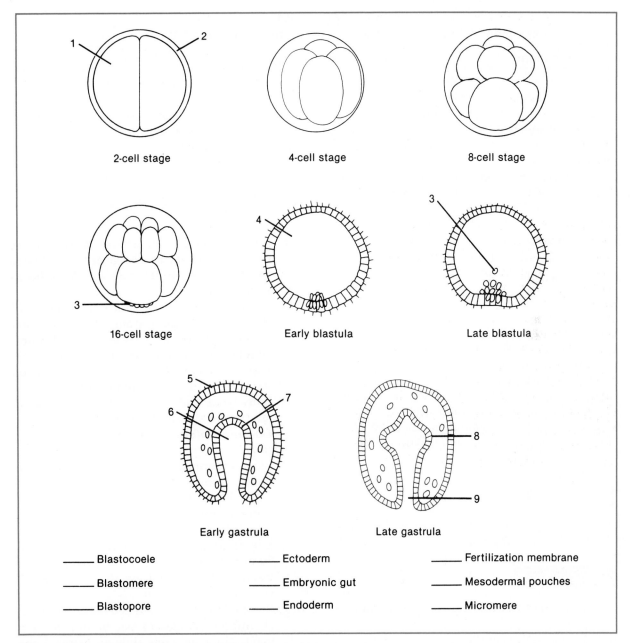

Figure 22.5. Early development in the sea urchin

Continued division of the cells results in the inward growth (invagination) of cells at the vegetal pole led by the micromeres. The **early gastrula** consists of two cell layers. The inner cell layer, the **endoderm,** forms the **embryonic gut** (archenteron), which opens to the exterior via the **blastopore.** The outer cell layer is the **ectoderm.** In the **late gastrula,** pouches bud off the endoderm to form the **mesoderm.** All three germ layers (embryonic tissues) are now present, and all later-appearing adult tissues and organs are derived from them. See Table 22.1. In both echinoderms and chordates, the blastopore becomes the anus, and a mouth forms later from another opening at the other end of the embryonic gut.

TABLE 22.1
Examples of Tissues and Organs Formed from the Germ Layers in Humans

Endoderm	Mesoderm	Ectoderm
Linings of the	Skeleton	Epidermis, including hair and nails
Urinary bladder	Muscles	Inner ear
Digestive tract	Kidneys	Lens, retina, and cornea of the eye
Respiratory tract	Gonads	Brain, spinal cord, nerves, and adrenal
Liver	Blood, heart, and blood ves-	medulla
Pancreas	sels	
Thyroid, parathyroids,	Reproductive organs	
and thymus	Dermis of the skin	

Materials

Per student
Colored pencils
Depression slides and cover glasses

Per lab
Glass marking pens
Medicine droppers
Toothpicks, flat
Ward's culture gum
Developing embryos at 3, 6, 12, 24, 48, and 96 hr
Unfertilized eggs in seawater at 20 °C
Sperm solution in dropping bottles at 20 °C
Prepared slides of sea urchin blastula, gastrula, and larval stages

Assignment 1

Complete item 1 on Laboratory Report 22 that begins on page 381.

Assignment 2

1. **Complete item 2a on the laboratory report.**
2. Place one drop of the egg and seawater mixture (8–12 eggs) in a depression slide and observe without a cover glass. Use the 10× objective. Compare the eggs with Figure 22.4. **Draw two or three eggs in the space for item 2b on the laboratory report.**
3. While the slide is on the microscope stage, add one drop of the sperm mixture at the edge of the depression, record the time, and quickly observe with the 10× objective. Note how the motile sperm cluster around the eggs. Why? Observe the rapid formation of the fertilization membrane. When most of the eggs have been activated, record the time. **Draw two or three activated eggs in the space for item 2b on the laboratory report.**
4. Use a toothpick to place a small amount of Ward's culture gum around the depression on the slide and add a cover glass. This prevents evaporation of water but allows passage of O_2 and CO_2. Write your initials on the slide with a glass marking pen.
5. Place your slide, egg mixture, and sperm mixture in the refrigerator at 20 °C.
6. **Complete item 2c on the laboratory report.**
7. Label Figure 22.5. Color the cells as follows to distinguish the embryonic tissue:
       ectoderm—blue
       mesoderm—red
       endoderm—yellow
8. Examine your slide of activated eggs at 15–20-min intervals, and try to observe the division of the zygote. Keep the slide at 20 ° C when not observing it.
9. Prepare and observe, in age sequence, slides of different stages of sea urchin development. Make only one slide at a time. Label it by age and with your initials. Keep it at 20 °C when not observing it. Examine these slides at 15–20-min intervals to observe a cell division. Compare your observations with Figure 22.5.
10. Compare the living blastula, gastrula, and larval stages with the prepared slides.
11. When finished, clean the slides, cover glasses, microscope stage, and objectives to remove all traces of seawater.
12. **Complete item 2 on the laboratory report.**

CHORDATE DEVELOPMENT

Examining the early development in a simple chordate, amphioxus, will help in understanding early development in humans. All chordates, including humans, exhibit a similar developmental pattern, and it is also similar to development in the sea urchin. See Figure 22.6.

Gastrulation forms ectodermal and endodermal layers as in the sea urchin, but mesoderm formation is a bit different. In amphioxus, two lateral mesodermal pouches bud off of the endoderm as in the sea urchin, but the dorsal portion of the endoderm between the pouches also becomes mesoderm that subsequently forms the **notochord.** A notochord is a flexible rod that provides support for the body, and it is an evolutionary forerunner of a vertebral column.

The formation of the notochord induces the overlying ectoderm to develop into a **neural tube.** In humans, as in other vertebrates, the neural tube subsequently develops into the brain and spinal cord. The lateral mesodermal pouches form the typical mesodermal structures as noted in Table 22.1.

Amniote Egg

Embryos of all chordates develop in an aqueous (water) environment. Among vertebrates, female fish and amphibians lay their eggs in water and males simultaneously deposit sperm

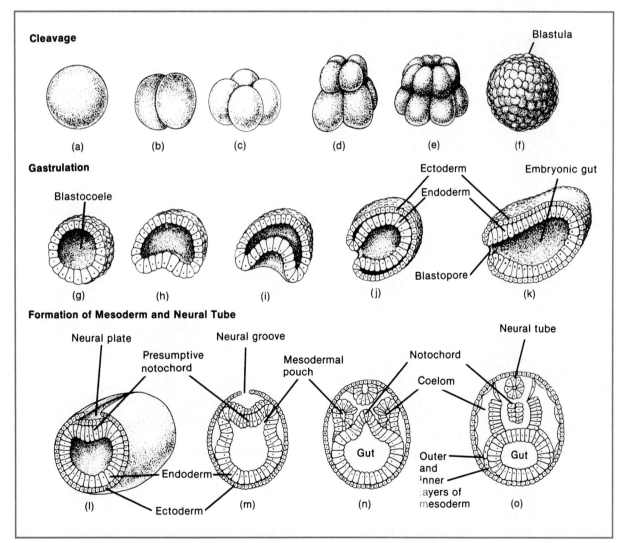

Figure 22.6. Early development in amphioxus

over them. Such organisms utilize external fertilization and external development of the young. It is a giant evolutionary step from the external fertilization and external development in fish and amphibians to internal fertilization and internal development found in humans and most mammals. This transition occurred in several stages and began long before humans appeared on the scene.

Reptiles were the first truly terrestrial vertebrates because they solved the problem of reproduction without returning to water. Their successful adaptations include internal fertilization and the **amniote egg.** The reptilian egg has a leathery shell that prevents excessive water loss and allows an exchange of oxygen and carbon dioxide between the developing embryo and the atmosphere. An adequate supply of stored nutrients (yolk and albumin) enables the development of the embryo to hatching. In addition, special protective membranes enclose the embryo.

The features of the amniote egg are shown in Figure 22.7. Note the reptilian **extra-embryonic membranes.** The **amnion** surrounds the embryo and contains the **amniotic fluid** that provides the embryo with its own "private pond" in which to develop. Instead of returning to water for embryonic development, reptiles

"brought the water to the embryo." The **yolk sac** envelops the yolk and absorbs nutrients for the embryo. The **allantois** is an embryonic urinary bladder, but it also spreads out against the outer membranes and serves as a gas-exchange organ. The **chorion** is the outermost membrane and encompasses all of the others.

Figure 22.7b shows how the extra-embryonic membranes are utilized in humans and most mammals. As in reptiles, the amnion envelops the embryo providing a "private pond" of amniotic fluid for the embryo, and the chorion surrounds all other membranes. The allantois carries embryonic blood vessels to the chorion. It forms the major part of the **umbilical cord** and, along with the chorion, forms the **placenta** that attaches the embryo to the uterine lining of the mother. There is no yolk, so the yolk sac plays a minor role as part of the umbilical cord. The amnion forms the outer covering of the umbilical cord.

The umbilical cord carries blood of the embryo to and from the placenta. Maternal and embryo bloods are separated by thin membranes in the placenta where an exchange of materials occurs:

$$\text{Maternal} \underset{\text{wastes and } CO_2}{\overset{\text{nutrients and } O_2}{\rightleftarrows}} \text{Embryo}$$
$$\text{blood} \qquad\qquad\qquad \text{blood}$$

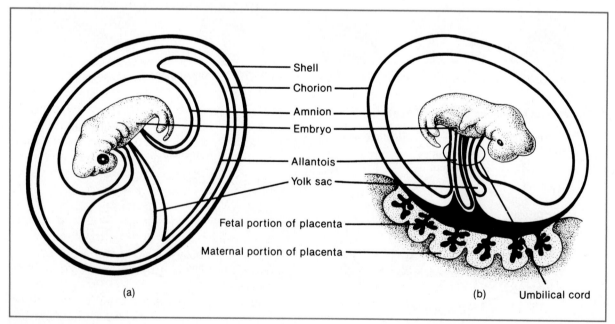

(a)

(b)   Umbilical cord

Shell — Chorion — Amnion — Embryo — Allantois — Yolk sac — Fetal portion of placenta — Maternal portion of placenta

Figure 22.7. Extra-embryonic membranes of the amniote egg. (a) Reptiles and birds. (b) Humans and most mammals.

Thus, mammals, including humans, have capitalized on the extra-embryonic membranes "invented" by reptiles.

Materials

Per student
Colored pencils

Per lab
Prepared slides of:
amphioxus, cleavage
amphioxus, gastrula
amphioxus, x.s., neurula stage

Assignment 3

1. Study Figure 22.6. Note the similarity of development in amphioxus with that of the sea urchin up to the gastrula. Note that (a) the mesoderm is formed from pouches that bud off the endoderm, (b) the mesoderm destined to be the notochord is located between the mesodermal pouches, and (c) the neural plate is formed by the dorsal part of the ectoderm and folds up to form the neural tube.
2. Color the embryonic tissues in Figure 22.6k, n, and o:
   ectoderm—blue
   neural tube—green
   mesoderm—red
   endoderm—yellow
3. Examine a prepared slide of amphioxus development and locate stages like those in Figure 22.6.
4. ***Complete item 3 on the laboratory report.***

HUMAN DEVELOPMENT

A human embryo (blastocyst) is implanted in the uterus about 5 days after fertilization, and germ layers and extra-embryonic membranes are evident by the 14th day. See Figure 22.8. Chorionic villi (projections) attach the embryo firmly to the uterine lining, and some of them will become part of the embryonic portion of the placenta. The first 8 weeks of development is known as the **embryonic period,** and the remainder of pregnancy is known as the **fetal period.**

At 8 weeks, the **fetus** is recognizably human with all organ systems in rudimentary form. It is about 3 cm (1¼″) in length and 1 g in weight. Some recognizable characteristics include: (1) The limbs are recognizable as arms and legs, and the fingers and toes are formed. (2) Bone formation begins, and internal organs continue to form. (3) The head is nearly as large as the body. All major brain regions are present, and the eyes are far apart with the eyelids fused. (4) The cardiovascular system is functional.

Subsequent development results in a baby being born about 280 days after fertilization (conception). When born prematurely, a fetus has about a 15% chance of survival at 24 weeks but nearly a 100% chance at 30 weeks.

Materials

Per student
Colored pencils

Per lab
Demonstration of pregnant cat or pig uterus
Model of a pregnant human female torso
Models of human developmental stages
Preserved human fetuses of various ages

Assignment 4

1. Study Figure 22.7 until you understand the location of the extra-embryonic membranes.
2. Study Figure 22.8a, noting the formation of the extra-embryonic membranes and germ layers. Color-code the germ layers: ectoderm—green; mesoderm—red; and endoderm—yellow.
3. Examine Figure 22.8b, noting the components of the umbilical cord, the placenta, and the amniotic fluid enveloping the fetus.
4. ***Complete items 4a and 4b on the laboratory report.***
5. Examine the pregnant human torso model and compare it with Figure 22.8c. Note how the amnion and chorion are pressed together, forming the amniochorion in late stages of

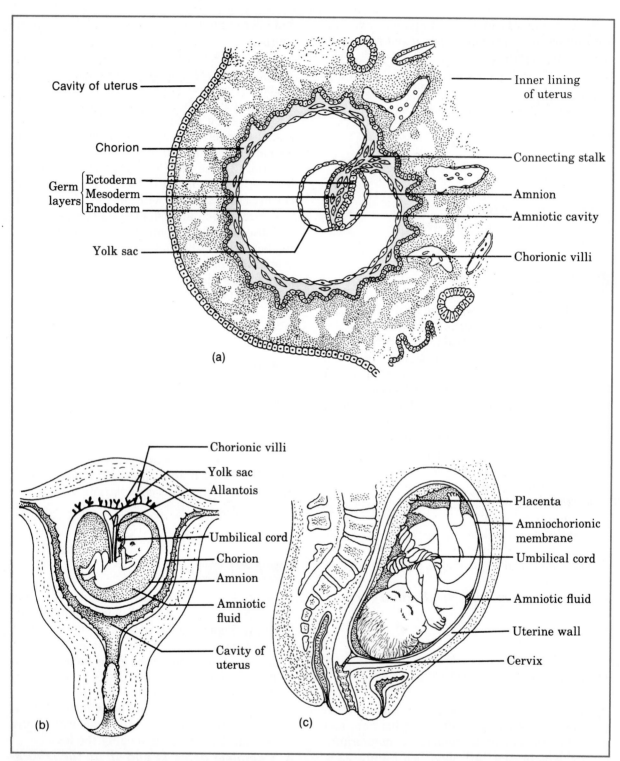

Figure 22.8. Selected stages in human development. (a) A human embryo, about 14 days old, implanted in the uterus. Note the germ layers and extra-embryonic membranes. (b) Uterus with a fetus about 10 weeks old. Note the extra-embryonic membranes and the umbilical cord. (c) A full-term fetus with head pressed against the cervix. Note the placenta, umbilical cord, and the fetal position.

pregnancy. Does the fetus fill all available space?

6. Examine the pregnant cat or pig uterus, noting the amniochorion, placenta, and umbilical cord.

7. Examine the series of models showing human development. Note the progression of development with the age of the fetus.

8. Examine the series of preserved human fetuses, noting the degree of development and the age of each fetus.

9. ***Complete the laboratory report.***

PART IV

ORGANISMIC DIVERSITY

23

MONERANS, PROTISTS, AND FUNGI

OBJECTIVES

After completion of the laboratory session, you should be able to:
1. Describe the distinguishing characteristics of monerans, protists, and fungi.
2. Identify representatives of these groups.
3. Describe the reproductive patterns of selected representatives.
4. Define all terms in bold print.

The next few exercises are included to acquaint you with the variety of organisms in the biotic world. The emphasis is on the distinguishing characteristics and life cycles of the major groups.

Taxonomy is the science of classifying organisms. Several **taxonomic categories** are used as shown in Table 23.1. The kingdom is the category containing the largest number of species (kinds of organisms), and the species category contains the fewest—only one. Organisms are classified according to their degree of similarity, which indicates the degree of evolutionary relationship. Therefore, a kingdom contains related phyla, a phylum contains related classes, and so forth.

The **phylum** category has been traditionally used by zoologists in classifying animals, while botanists have traditionally used the term **division** for this same category in classifying plants and plantlike organisms. These terms have persisted in the classification of organisms even though zoology and botany have now been unified in biology.

The scientific name of an organism is composed of both the genus and species names. For example, *Homo sapiens* is the scientific name for humans. Note that the genus name is capitalized while the species name is not. Also, the scientific name is always in italics.

Biologists now use a five-kingdom system of classification rather than the traditional "animals" and "plants." The five kingdoms are **Monera, Protista, Fungi, Plantae,** and **Animalia.** The major organismic groups composing each kingdom are shown in Figure 23.1.

KINGDOM MONERA

Monerans are the simplest organisms. They are either **unicellular,** composed of a single cell, or **colonial,** composed of a group of independently functioning cells joined together. All cells are **prokaryotic** because they lack membrane-bound organelles. A rigid, nonliving cell wall is

TABLE 23.1
Classification of the Human Species,
Homo sapiens

Taxonomic Category	Classification of Humans
Kingdom	Animalia
Phylum	Chordata
Class	Mammalia
Order	Primates
Family	Hominidae
Genus	*Homo*
Species	*sapiens*

formed external to the cell membrane. Review the structure of prokaryotic cells in Exercise 3 to refresh your understanding.

All monerans are haploid (n), and they reproduce by **binary fission,** a simple form of cell division. Review the process of binary fission in Exercise 8, if necessary. Binary fission can produce enormous numbers of bacteria or cyanobacteria in just a few hours.

Bacteria (Division Schizophyta)

Bacteria are the smallest organisms, with cells only 1–10 μm in length. Their cells exhibit

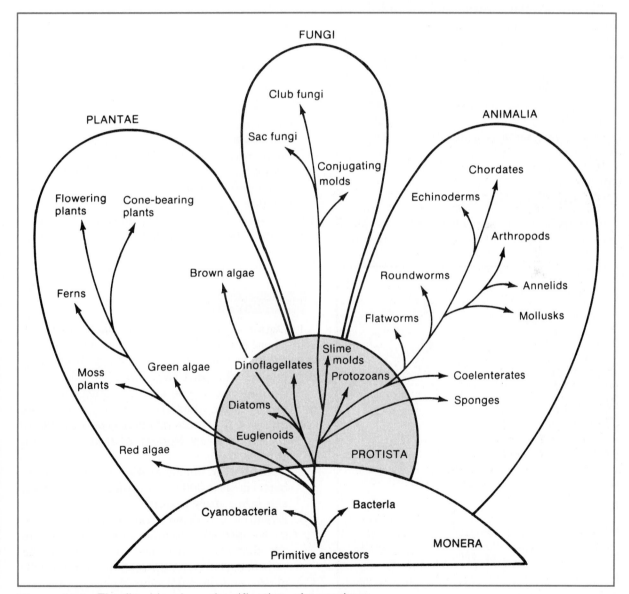

Figure 23.1. The five-kingdom classification of organisms

three distinctive shapes: rodlike (bacillus), spherical (coccus), or spiral (spirillum). Cells of some species possess flagella that enable them to swim about. Bacteria are identified by their shape and physiological characteristics. See Figure 23.2 and Plates 1.1–1.3.

Bacteria occur in all places on earth where life exists. Most bacteria are **heterotrophs** (feed on others) and must obtain nutrients from outside sources. Most heterotrophs are **saprotrophs,** meaning that they obtain nutrients by digesting nonliving organic matter. A few bacteria are disease-causing **parasites,** organisms that obtain nutrients from other living organisms.

Bacteria exhibit a sexual process called **conjugation.** During conjugation, two cells become connected by a cytoplasmic bridge, and a replicate strand of DNA passes from one cell to the other. The DNA combines with or replaces the DNA of the recipient cell, modifying its genetic composition. After separation, the cells reproduce by binary fission as before. Conjugation increases the genetic variability in a bacterial population.

Blue-Green Bacteria (Division Cyanobacteria)

All cyanobacteria are photosynthetic **autotrophs** (self-feeders), meaning that they pro-

duce their own organic nutrients by photosynthesis. They lack chloroplasts, but chlorophyll a and other pigments are concentrated in the **chromoplasm,** the outer region of a cyanobacterial cell. Each cell usually is enveloped in a gelatinous coat to prevent excess water from entering or leaving the cells. Most species are blue-green, but many other colors exist. Cyanobacteria occur in both freshwater and marine habitats. Some species are especially tolerant to high temperatures, and these forms are responsible for the colors often seen in hot springs. See Plate 1.4.

Materials

Per student
Compound microscope

Per lab
Cultures of cyanobacteria
   Gloeocapsa
   Oscillatoria
Demonstration microscope setups:
   motile bacteria (*Pseudomonas*), 1,000×
   Gloeocapsa, 1,000×
   Oscillatoria, 1,000×
Microscope slides and cover glasses
Prepared slides of bacterial types

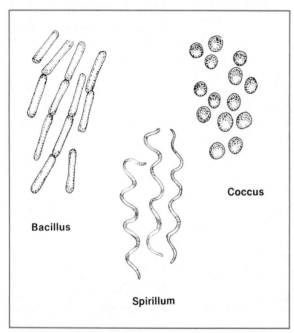

Figure 23.2. Bacterial cell types

1. Obtain a prepared slide of bacterial types. There are three stained smears of bacteria on the slide, and each consists of a different morphological type. Use reduced illumination to examine each type at 400×. Bacterial cells are smaller than you probably imagine. ***Draw a few cells of each type in item 1a on Laboratory Report 23 that begins on page 385.***
2. Examine the demonstration setup of motile bacteria at 1,000×. ***Complete item 1b on the laboratory report.***
3. Examine the demonstration setups of live *Gloeocapsa* and *Oscillatoria* at 1,000×. Note the shape of each colony, the shape of the cells in each colony, the color of their chromoplasm, and the gelatinous sheath, if present.
4. ***Complete item 1 on the laboratory report.***

KINGDOM PROTISTA

Protists are a heterogeneous group of unicellular or colonial organisms that exhibit animal-like, plantlike, or funguslike characteristics. The major groups probably are not closely related but are products of evolutionary lines that diverged millions of years ago. Protists and all higher organisms are composed of **eukaryotic cells** that contain membrane-bound organelles.

Mitotic cell division is the most common form of reproduction among protists. However, sexual reproduction does occur. In haploid (n) protists, only the zygote is diploid, and it promptly divides by meiotic division, forming four new haploid individuals. In diploid (2n) protists, haploid gametes are formed by meiotic division, and the fusion of gametes forms a new diploid individual.

Protozoans: Animal-like Protists

These animal-like protists lack a cell wall and are usually motile. Most forms engulf food into vacuoles, where it is digested. Some absorb nutrients through the cell membrane, and a few are parasitic. Protozoans occur in most habitats where water is available. Water tends to diffuse into freshwater forms, and these protozoans possess **contractile vacuoles** that repeatedly collect and pump out the excess water to maintain their water balance. Three groups of protozoans are shown in Figure 23.3.

Flagellated Protozoans (Phylum Mastigophora)

These primitive protozoans possess one or more **flagella** that provide a means of movement. Food may be engulfed and digested in vacuoles, or nutrients may be absorbed through the cell membrane.

Some species have established symbiotic relationships with other organisms. *Trypanosoma brucei* is a parasitic form that causes African sleeping sickness. It lives in the blood and nervous system of its vertebrate host and is transmitted by the bite of tsetse flies. *Trichonympha collaris* is a mutualistic symbiont that lives in the gut of termites. It digests the cellulose (wood) to produce simple carbohydrates that can be digested or utilized by the termite. See Plates 2.1–2.3.

Amoeboid Protozoans (Phylum Sarcodina)

These protists move by means of **pseudopodia,** flowing extensions of the cell. Prey organisms are engulfed and digested in food vacuoles. Some forms secrete a shell for protection. Calcareous shells of foraminiferans and siliceous shells of radiolarians are abundant in ocean sediments. *Entamoeba histolytica* is a parasitic form causing amoebic dysentery in humans. See Plates 2.4, 2.5, 3.1, and 3.2.

Ciliated Protozoans (Phylum Ciliophora)

Ciliates are the most advanced and complex of the protozoans. They are characterized by the presence of a **macronucleus,** one or more **micronuclei,** and movement by means of numerous **cilia,** hairlike processes extending from the cell. A flexible outer covering, the **pellicle,** is located exterior to the cell membrane. *Paramecium* is a common example that also possesses **trichocysts,** tiny dartlike weapons of offense and defense located just under the cell surface. Food organisms are swept down the **oral groove** by cilia and into food vacuoles, where digestion occurs. See Figures 23.3 and 23.4 and Plates 3.3 and 3.4.

Sporozoans (Phylum Sporozoa)

All species lack motility and are internal parasites of animals. *Plasmodium vivax,* a pathogen causing malaria, is a typical example. It is transmitted by the bite of *Anopheles* mosquitoes and infests red blood cells and liver cells of human hosts. Malaria has probably caused the death of more humans than any other disease. See Plate 3.5.

Plantlike Protists

The plantlike protists possess chloroplasts containing chlorophyll, and most of them have a cell wall. You will study representatives of three major groups.

Euglenoids (Division Euglenophyta)

These unicellular protists possess both plantlike and animal-like characteristics. They have **chlorophylls a** and **b** in chloroplasts, a flagellum for movement, and an "eyespot" (stigma)

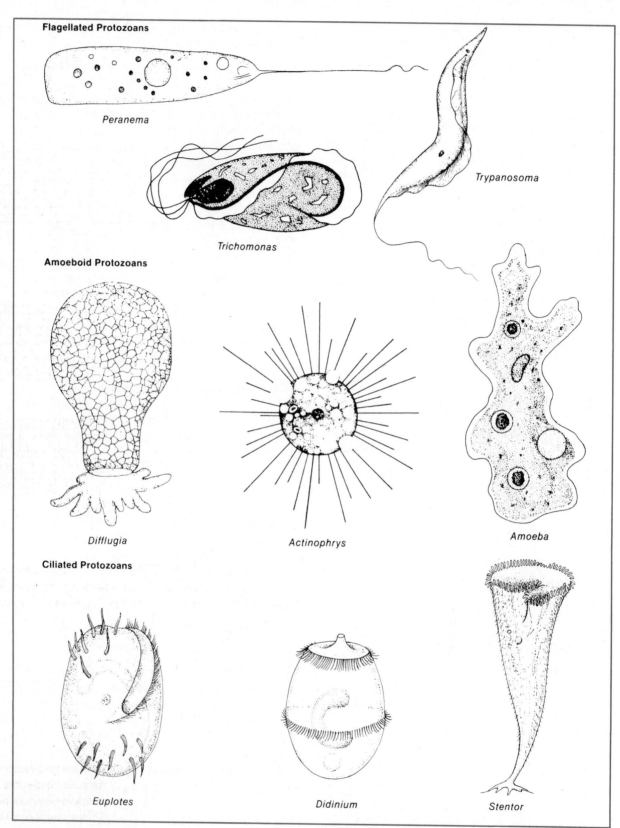

Figure 23.3. Examples of flagellated, ameoboid, and ciliated protozoans

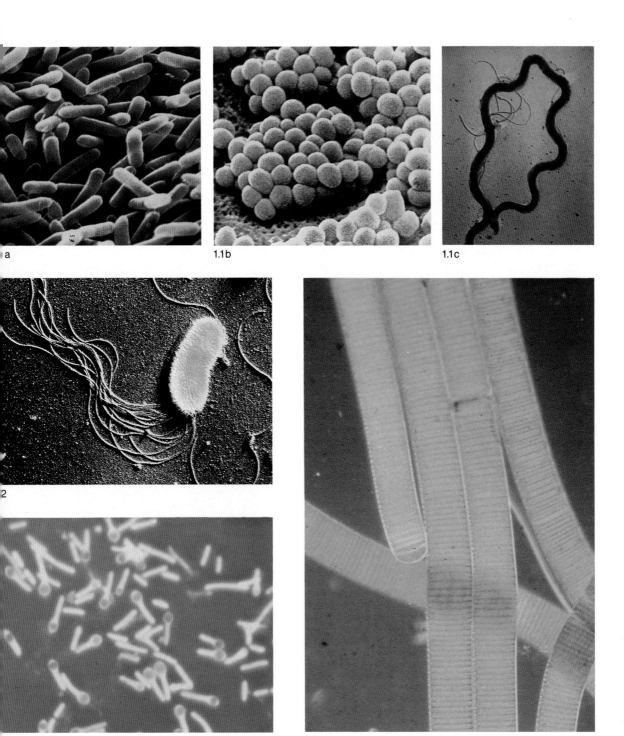

a   1.1b   1.1c

2

3   1.4

Plate 1.1   The three basic shapes of bacterial cells. (a) A false-color scanning electron micrograph (SEM) of the rod-shaped cells of *Pseudomonas aeruginosa*, a chlorine-resistant germ that sometimes causes skin infections among swimmers and users of hot tubs. 20,000X. (b) A false-color SEM of the spherical cells of *Staphylococcus aureus*, a pathogen causing boils and abscesses. 3,000X. (c) The spiral-shaped cell of *Treponema pallidum* which causes syphilis. 60,000X.

Plate 1.2   A false-color transmission electron micrograph (TEM) of *Pseudomonas fluorescens*, a motile soil bacterium with flagella. 10,000X.

Plate 1.3   *Clostridium tetani* causes tetanus (lockjaw). The enlarged end of some cells contains a spore that is resistant to unfavorable environmental conditions. 800X.

Plate 1.4   Filaments of *Oscillatoria*, a photosynthetic cyanobacterium, are composed of thin, circular cells joined together like stacks of coins. 100X.

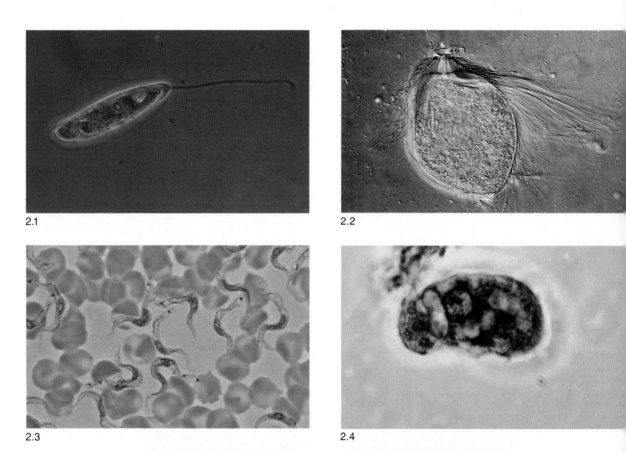

2.1

2.2

2.3

2.4

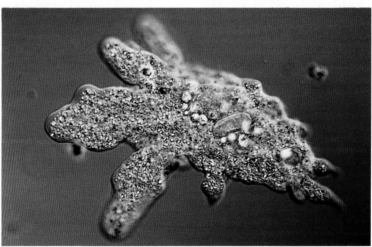

2.5

Plate 2.1 *Peranema trichophorum*, a freshwater flagellated protozoan, is propelled by a thick anterior flagellum held rigid and straight except for its tip. It engulfs prey through an opening near the base of the flagellum. A thin trailing flagellum is also present but is not visible here. 150X.

Plate 2.2 *Trichonympha* is a complex multiflagellated protozoan. It digests wood in the gut of termites. 135X.

Plate 2.3 *Trypanosoma brucei*, shown among human red blood cells, causes African sleeping sickness. It is transmitted by the bite of tsetse flies. 320X.

Plate 2.4 *Entamoeba histolytica* invades the lining of the intestine causing amoebic dysentery in humans. 600X.

Plate 2.5 *Amoeba proteus*, a common freshwater amoeboid protozoan. Food is digested within food vacuoles. Excess water is collected in contractile vacuoles and pumped out of the cell. 160X.

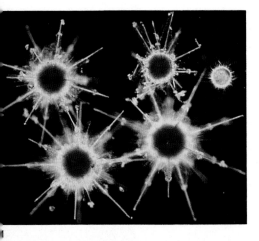

3.2

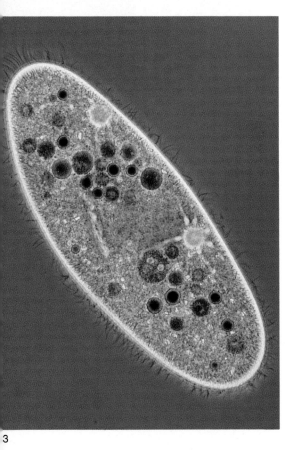

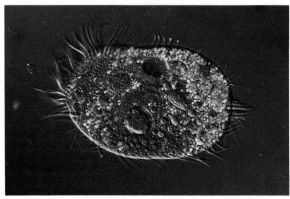

3.4

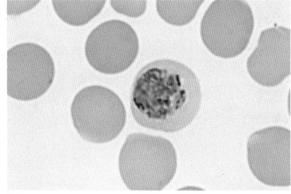

3

3.5

Plate 3.1 Living radiolarians. Radiolarians are marine amoeboid protozoans that secrete a siliceous shell. Slender pseudopodia extend through small openings in the shell.

Plate 3.2 A living foraminiferan. Foraminiferans are marine amoeboid protozoans that secrete a calcareous shell. Thin, ray-like pseudopodia extend through tiny openings in the shell.

Plate 3.3 *Paramecium* is a freshwater ciliate. Beating cilia function like little oars that move the cell through the water. Captured food organisms are digested in food vacuoles. Excess water is collected in contractile vacuoles and pumped out of the cell.

Plate 3.4 *Euplotes* is an advanced ciliate. Tufts of cilia unite to form cirri that function almost like legs as the ciliate "walks" over the bottom of a pond. Note the food vacuoles and contractile vacuole that appears like a cavity in this view.

Plate 3.5 *Plasmodium vivax*, a sporozoan, in a human red blood cell. It causes malaria and is transmitted by the bite of female anopheline mosquitoes.

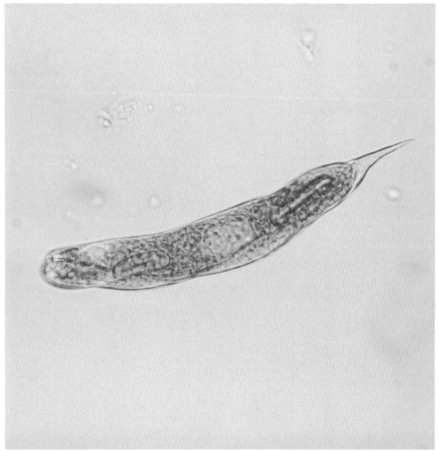

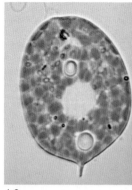

4.2

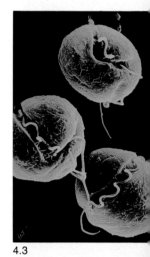

4.1

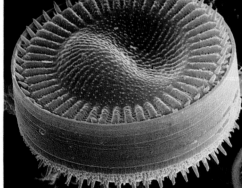

4.3

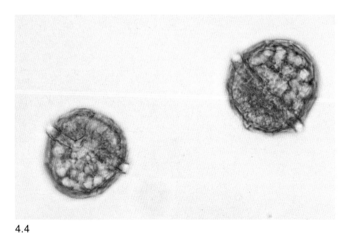

4.4

4.5

Plate 4.1 *Euglena* is a photosynthetic freshwater euglenoid protist. The eye spot detects light intensity. Food reserves are stored as paramylon, a complex carbohydrate. The flagellum is recurved along the underside of the cell in this photomicrograph.

Plate 4.2 *Phacus* is a freshwater euglenoid. Note the flagellum, eye spot, and chloroplasts.

Plate 4.3 Like all dinoflagellates, *Gymnodinium*, a marine form, possesses two flagella. One lies in an equatorial groove in the cell wall, and the other hangs from the end of the cell. (SEM 10,000X.)

Plate 4.4 *Peridinium* is a freshwater dinoflagellate common in acid-polluted lakes. Here, chloroplasts and equatorial groove are visible, but the flagella are not.

Plate 4.5 The siliceous walls of diatoms are formed of two parts that fit together like the top and bottom of a pillbox as in this SEM of *Cyclotella meneghinians*. 875X.

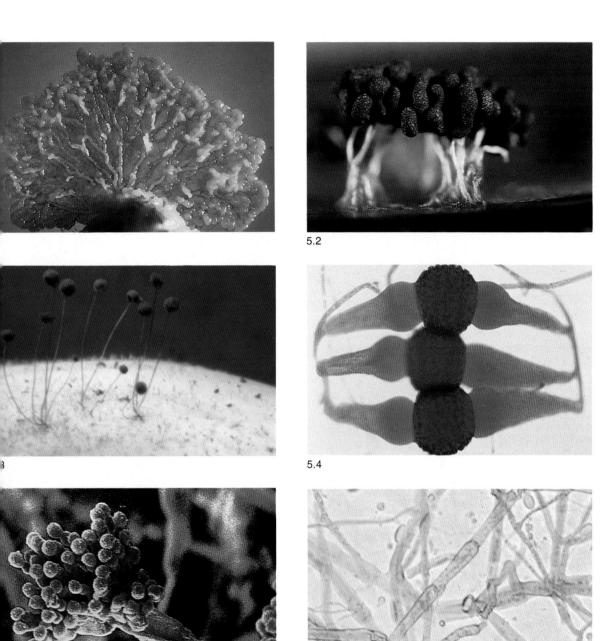

5.2

5.4

5.6

Plate 5.1 The plasmodium of the slime mold *Physarum* lives like a giant, multinucleate amoeba as it creeps along absorbing nutrients and engulfing food organisms.

Plate 5.2 When either food or moisture becomes inadequate, a *Physarum* plasmodium transforms into spore-forming sporangia supported by short stalks.

Plate 5.3 In *Rhizopus stolonifer*, black bread mold, sporangiophore hyphae support sporangia that form asexual spores. Each sporangiophore is attached to the substrate by rhizoid hyphae.

Plate 5.4 The fusion of (+) and (−) gametes in *Rhizopus stolonifer* results in the formation of a zygospore sandwiched between a pair of gametangia. Each zygospore later produces a spore-forming sporangium.

Plate 5.5 This false-color SEM of the asexual conidiospores of *Penicillium* shows their spherical shape and attachment to a sporangiophore. 5,750X.

Plate 5.6 The hyphae of *Nectria* are composed of separate cells joined end-to-end. Such hyphae are said to be septate because cross walls separate the cells.

6.1a

6.1b

6.2

6.3

6.4

Plate 6.1 (a) Ascocarps (fruiting bodies) of *Peziza aurantia*, a sac fungus. (b) Ascospores are formed in the sac-like asci after a complex sexual process.

Plate 6.2 Basidiocarps (fruiting bodies) of shelf fungi, a club fungus, contain spore-forming basidia. The basidiospores are released through tiny pores on the lower surface of the basidiocarps.

Plate 6.3 Each basidium of a club fungus produces four basidiospores as shown in this SEM. 1,100X.

Plate 6.4 A discharge of millions of basidiospores from a puffball, a club fungus.

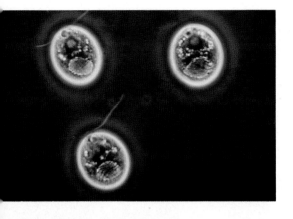

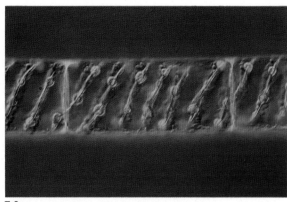

7.3

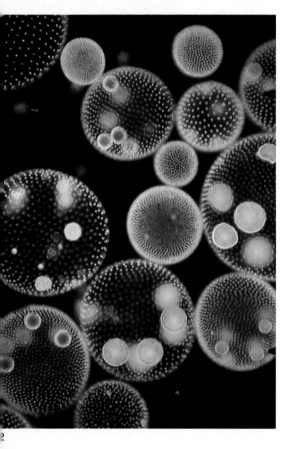

7.4

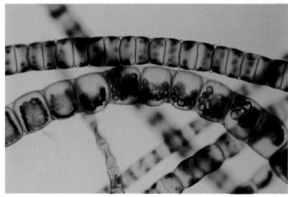

7.5

Plate 7.1 *Chlamydomonas* is a tiny unicellular green alga that moves by means of two flagella. Only the basal portion of the cup-shaped chloroplast is visible here. An eye spot, not visible here, enables detection of light intensity.

Plate 7.2 *Volvox* colonies may consist of up to 50,000 *Chlamydomonas*-like cells. *Volvox* may reproduce asexually by forming daughter colonies within the sphere. Some forms also have a few cells specialized for sexual reproduction.

Plate 7.3 The spiral chloroplast in a cell of *Spirogyra*, a filamentous green alga. The enlarged regions are pyrenoids, sites of starch storage.

Plate 7.4 Cellular specialization for sexual reproduction is evident in *Oedogonium*, a filamentous green alga. The enlarged, dark cells are oogonia which contain an oospore formed after union of sperm and egg cells.

Plate 7.5 Cellular specialization for reproduction also occurs in *Ulothrix*, another filamentous green alga. The cells in the lower filament are sporangia containing asexual zoospores.

8.1

8.2

8.3

8.4

8.5

Plate 8.1 *Ulva* or sea lettuce is a marine green alga that exhibits alternation of generations. Note the holdfast.

Plate 8.2 Brown marine algae are characterized by a robust body that can withstand strong wave action. Note the holdfast, stipe, and divided blade in *Durvillaea*, a brown alga from the Australian coast.

Plate 8.3 A forest of giant kelp, *Macrocystis*, along the California coast provides an important habitat for marine animals. The long stipes may be over 100 ft. in length.

Plate 8.4 The delicate, feathery body structure of red algae sharply contrasts with the robust brown algae. Red algae, like *Callithamnion*, live at ocean depths where wave action is minimal.

Plate 8.5 Gas-filled floats of *Macrocystis* lift the blades and stipe toward the water surface where more sunlight is available.

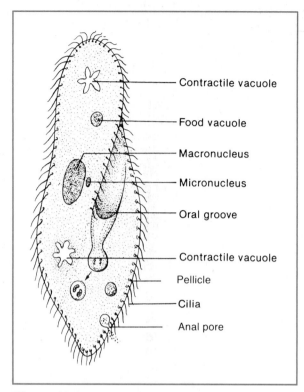

Figure 23.4. *Paramecium caudatum*

that detects light intensity. A cell wall is absent. See Figure 23.5 and Plates 4.1 and 4.2.

Dinoflagellates (Division Pyrrophyta)

Most of these unicellular forms are marine and photosynthetic with **chlorophylls a** and **c** in their chloroplasts. Most have a cell wall of cellulose, and all forms have two flagella. One lies in a groove around the equator of the cell, and the other hangs free from the end of the cell. When nutrients are abundant, certain marine species reproduce in such enormous numbers that the water turns a reddish color, a condition known as the red tide. Toxins produced by some species may cause massive fish kills and make shellfish unfit for human consumption. See Plates 4.3 and 4.4.

Diatoms (Division Chrysophyta)

These protists are unicellular and have a cell wall of silica, a natural glass. The cell wall consists of two halves that fit together like the top and bottom of a box. **Chlorophylls a** and **c** are found in their chloroplasts. When diatoms die,

the siliceous walls settle to the bottom. In some areas, they have formed massive deposits of diatomaceous earth that are mined and used commercially in a variety of products such as fine abrasive cleaners, toothpaste, filters, and insulation. Much of the atmospheric oxygen has been produced by photosynthesis in marine diatoms. See Plate 4.5.

Slime Molds: Funguslike Protists

The feeding stages of slime molds resemble amoeboid protozoans, but they reproduce by forming sporangia that produce spores.

Cellular Slime Molds (Division Acrasiomycota)

Members of this group live as single-celled amoeboid organisms engulfing bacteria in leaf litter and soil. Poor environmental conditions cause the cells to congregate in what is known as the "swarming stage" that results in the formation of a sporangium containing spores. *Dictyostelium* is an example of this group.

Plasmodial Slime Molds (Division Myxomycota)

Feeding stages of plasmodial slime molds consist of strands of protoplasm streaming along in amoeboid fashion and engulfing bacteria on leaf litter, rotting wood, and the like. The large plasmodium is multinucleate. Unfavorable conditions stimulate the plasmodium to migrate and ultimately to form a sporangium with spores. *Physarum* is an example of a plasmodial slime mold. See Plates 5.1 and 5.2.

Materials

Per student
Compound microscope

Per lab
Medicine droppers
Microscope slides and cover glasses
Protoslo
Toothpicks
Cultures of:
  Euglena
  Paramecium, containing yeast cells stained with congo red

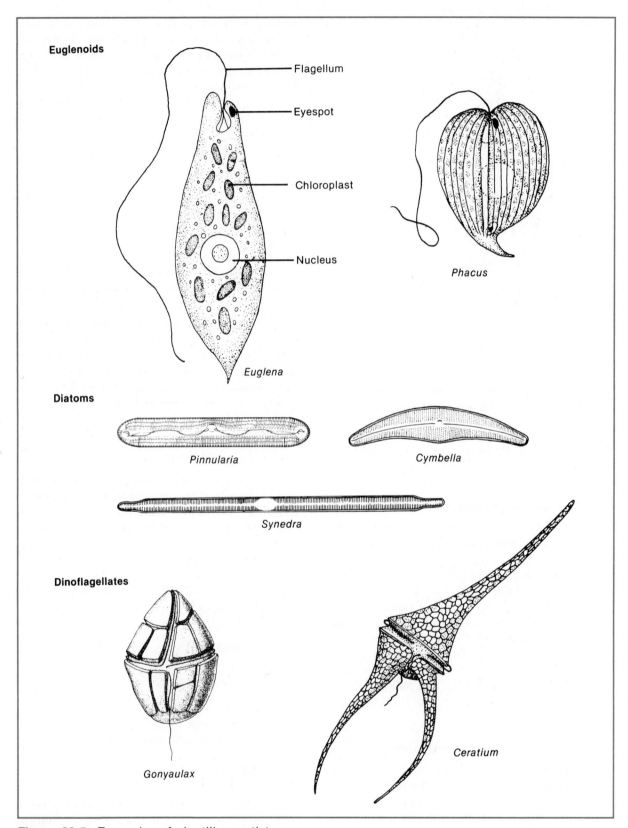

Figure 23.5. Examples of plantlike protists

Pelomyxa
Physarum
pond water
Diatomaceous earth
Demonstration setup showing effect of light on the distribution of *Euglena* in a partially shaded Petri dish
Demonstration microscope setups of:
   diatomaceous earth
   dinoflagellates
   foraminifera
   radiolaria
Termites, living
Prepared slides of:
   diatoms
   dinoflagellates
   foraminifera
   radiolaria

Assignment 2

1. Obtain a live termite and separate the abdomen from the thorax. Squeeze the abdominal contents into a drop of water on a microscope slide. Add a cover glass and examine at 100×. Locate a multiflagellated *Trichonympha* and examine it at 400×. **Complete items 2a–2c on the laboratory report.**
2. Examine the *Pelomyxa* culture set up under a dissecting microscope. Locate a specimen on the bottom of the jar, remove it with a dropper, and place it, along with a drop of water, on a microscope slide. Examine this large, multinucleate amoeboid protozoan at 40× and 100×. Locate food and contractile vacuoles and observe the flowing motion of cytoplasm as pseudopodia form during **amoeboid movement. Complete item 2d on the laboratory report.**
3. Examine the slides of foraminifera and radiolaria "shells" set up under demonstration microscopes. How do they differ?
4. Place a drop of the *Paramecium* culture containing stained yeast cells on a microscope slide and examine it at 40× or 100× without a cover glass. Observe how *Paramecium* moves and what happens when it bumps into an object.
5. Add a drop of Protoslo to the drop of *Paramecium* culture on your slide and mix it with a toothpick. Add a cover glass and at 40×

locate a *Paramecium* stuck in the Protoslo. Examine it at 100× and 400×. Locate the structures shown in Figure 23.4. Since *Paramecium* has been feeding on stained yeast cells, the red-stained yeast cells will be visible in food vacuoles. As digestion proceeds, the yeast cells will change from bright red to purple to blue.
6. **Complete item 2 on the laboratory report.**

Assignment 3

1. Make a slide of a drop of the *Euglena* culture and examine it at 100× and 400×. Note the color, manner of movement, and lack of a cell wall. **Complete item 3a on the laboratory report.**
2. Observe the distribution of *Euglena* in the partially shaded Petri dish set up as a demonstration. *Do not move the dish.* Lift the black-paper light shield to make your observations, and return it to its prior position. **Complete item 3b on the laboratory report.**
3. Examine the demonstration microscope setup of dinoflagellates. **Complete item 3c and draw 1–2 cells in item 3d on the laboratory report.**
4. Examine a prepared slide of diatoms. Note the symmetry of their cell walls. Examine the sample of diatomaceous earth and the demonstration microscope setup of diatomaceous earth. **Complete item 3 on the laboratory report.**

Assignment 4

Examine the demonstration setups of slime molds. Note the organization of the plasmodium stage and the sporangia of the reproductive stage. Compare them with Plates 5.1 and 5.2. **Complete item 4 on the laboratory report.**

Assignment 5

Make and examine several slides of pond water and observe the monerans and protists present. **Complete item 5 on the laboratory report.**

FUNGI

The **kingdom fungi** contains a large and diverse group of heterotrophic organisms that occur in freshwater, marine, and terrestrial habitats. Terrestrial fungi reproduce by **spores,** dormant reproductive cells that are dispersed by wind and that germinate to form a new fungus when conditions are favorable. Most fungi are multicellular; only a few are unicellular.

Fungi are either **saprotrophs** or **parasites.** Most species are saprotrophs and play a beneficial role in decomposing nonliving organic matter, but some saprotrophs cause serious damage to stored food products. Rusts and mildews are important plant parasites. Ringworm and athlete's foot are common human ailments caused by fungi.

The vegetative (nonreproductive) body of a multicellular fungus is called a **mycelium,** and it is composed of threadlike filaments, the **hyphae.** Hyphae are formed of cells joined end to end. The cells of hyphae may be separated by cell walls (septate hyphae), or the cell walls may be incomplete or lacking (nonseptate hyphae). The cell walls are formed of chitin. See Plate 5.6.

Nutrients are obtained by hyphae secreting digestive enzymes into the surrounding substrate, which is digested extracellularly. The resulting nutrients are then absorbed into the hyphae.

Spores of fungi are formed either from terminal cells of reproductive hyphae or in **sporangia,** enlarged structures at the ends of specialized hyphae. Some fungi form **fruiting bodies** that contain the spore-forming hyphae.

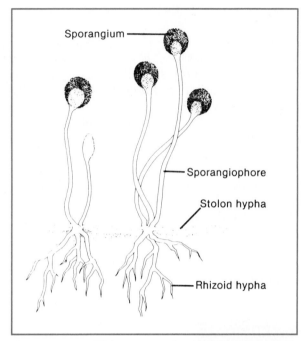

Figure 23.6. *Rhizopus stolonifer,* black bread mold

mycelium. See Figure 23.6 and Plates 5.3 and 5.4.

Asexual reproduction by mitospores occurs continuously. Sexual reproduction occurs only when opposite mating types (designated + and −) come in contact. Then, as shown in Figure 23.7, special cells become **gametes** (n) that fuse forming a **zygote** (2n). The zygote develops a resistant cell wall forming a **zygospore,** the characteristic giving the name to this group of fungi. Subsequently, the zygospore germinates and forms a sporangiophore whose sporangium produces both + and − haploid spores.

Conjugating Molds (Division Zygomycota)

The common black bread mold, *Rhizopus stolonifer,* is an example of this group. It produces three types of hyphae. **Stolon hyphae** spread over the surface of bread as the mycelium grows. **Rhizoid hyphae** penetrate the bread to digest it and to anchor the mycelium. **Sporangiophores** are upright hyphae that form a **sporangium** at their tips. Asexual **mitospores** (spores formed by mitosis) develop within the sporangia and are released when mature. Germination of these spores forms the haploid hyphae of a new

Sac Fungi (Division Ascomycota)

Yeasts, mildews, most molds, and cup fungi belong to this group. Many members of this group are important parasites of plants and animals, while others provide benefits to humans. Yeasts are used in brewing and wine-making and to release bubbles of CO_2 to make dough rise in the baking industry. *Penicillium* species produce antibiotics.

Asexual reproduction occurs by **budding** in yeasts and by **conidiospores** in mildews and molds. Conidiospores are formed by mitotic divi-

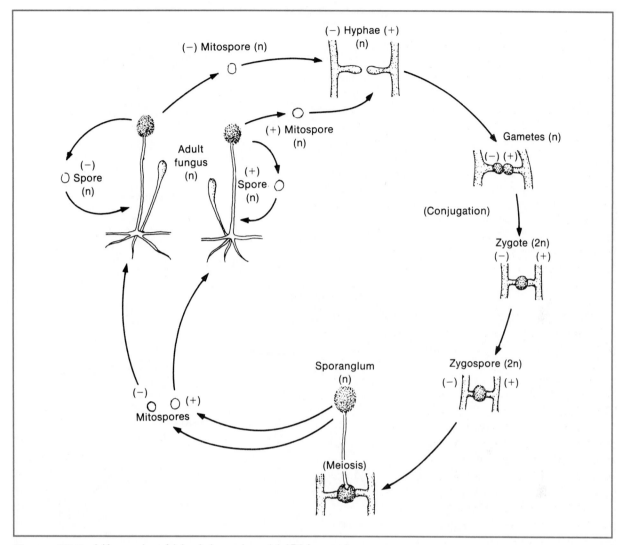

Figure 23.7. Life cycle of black bread mold (*Rhizopus*)

sion at the tips of reproductive hyphae. See Plate 5.5. Sexual reproduction occurs as well. Some forms, like cup fungi, undergo a complex sexual process that culminates in the production a **fruiting body** that contains the spore-forming hyphae. The tips of these hyphae enlarge to form tiny sacs, **asci,** within which **ascospores** are formed by meiotic division. See Plates 6.1a and 6.1b.

Club Fungi (Division Basidiomycota)

Mushrooms, puffballs, and shelf fungi belong to this group. See Plates 6.2–6.4. Many forms are beneficial saprotrophs, but some are serious plant parasites, such as rusts and smuts, that cause enormous losses in wheat, corn, and other cereal crops.

The common mushroom, *Coprinus,* is a suitable example of this group. The mushroom mycelium derives nutrients from nonliving organic substances in the soil. It is haploid and occurs in two mating types, + and −. Whenever opposite mating types come in contact with each other, cells (analogous to gametes) from each mating type of hyphae grow toward each other and fuse, forming a single cell containing separate nuclei of each mating type. The nuclei do not fuse, so mitotic division of the cell forms **dikaryotic hyphae** whose cells contain two nuclei (n + n), one of each mating type. See Figure 23.8.

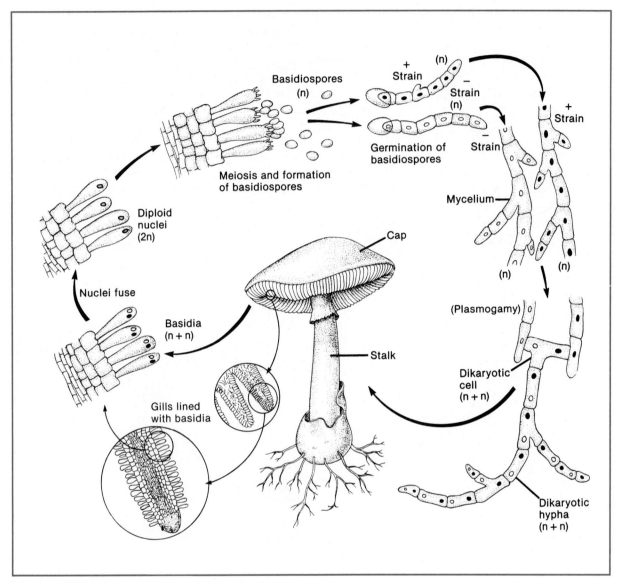

Figure 23.8. Life cycle of a mushroom (*Coprinus*)

Growth of many dikaryotic hyphae results in the formation of a **fruiting body,** the structure that you think of as a "mushroom." Spore-forming hyphae of the gills develop terminal enlarged cells called **basidia,** and then the two nuclei in each basidium fuse, forming a zygote nucleus (2n). By meiotic division, basidia form four **basidiospores,** two of each mating type, that are released and dispersed by air currents. Basidiospores germinate to form haploid hyphae by mitotic division.

Materials

Per lab
Demonstration cultures under dissecting microscopes:
  Penicillium on citrus fruit or agar
  Rhizopus on bread or agar
  Rhizopus mating types with zygospores on agar
Dropping bottles of methylene blue, 0.01%
Medicine droppers

Microscope slides and cover glasses
Mushrooms
Yeast culture in 5% glucose
Prepared slides of:
   Rhizopus zygospores
   Coprinus gills, x.s.
"Unknown" monerans, protists, and fungi

Assignment 6

1. ***Complete items 6a–6c on the laboratory report.***
2. Examine the *Rhizopus* colony set up under a dissecting microscope. Note its growth pattern and the location of mature and immature sporangia. Locate the three types of hyphae.
3. Examine (a) *Rhizopus* mating colonies and the gametes and zygospores that have formed and (b) a prepared slide of *Rhizopus* zygospores. ***Complete items 6d–6f on the laboratory report.***
4. Place a drop of the yeast culture on a microscope slide. Add a drop of methylene blue and a cover glass, and examine the stained cells at 100× and 400×. Note how buds form on the cells, and locate the nuclei. ***Draw a few budding yeast cells in item 6g on the laboratory report.***
5. Examine the *Penicillium* mold set up under a dissecting microscope. Note the growth pattern and the color of mature and immature conidiospores. ***Complete items 6h–6j on the laboratory report.***
6. Examine a mushroom. Locate the stalk, cap, and gills.
7. Examine a prepared slide of *Coprinus* gill, x.s. Locate the hyphae forming the gill, the basidia, and basidiospores. ***Complete item 6 on the laboratory report.***

Assignment 7

Examine the "unknown" specimens set up by your instructor. ***Complete item 7 on the laboratory report.***

24

PLANTS

The kingdom **Plantae** includes the algae as well as the more familiar and more complex terrestrial plants. Plants are photosynthetic autotrophs. Their cells have photosynthetic pigments contained in plastids, cell walls of cellulose, and a large central vacuole in mature cells.

ALGAE

Algae occur primarily in freshwater and marine habitats, but some forms occur in moist terrestrial areas. They lack vascular tissue and true leaves, stems, and roots, although complex forms may have structures that resemble them. There are three groups of algae based on their coloration due to pigments in their chloroplasts: green, brown, and red algae.

Green Algae (Division Chlorophyta)

Green algae may be unicellular, colonial, or multicellular. Most species occur in freshwater or marine habitats, but a few occur in moist areas on land. The presence of (1) **chlorophylls a** and **b** in chloroplasts, (2) cellulose cell walls, (3) starch as the nutrient storage form, and (4) whiplash flagella on motile cells suggests that green algae are ancestral to higher plants. See Figure 24.1 and Plates 7.1–7.5 and 8.1.

Brown Algae (Division Phaeophyta)

These multicellular algae are almost exclusively marine and are often called seaweeds because of their abundance along rocky seacoasts. Their brownish color results from the presence of the pigment **fucoxanthin** in addition to **chlorophylls a** and **c** in the chloroplasts. The typical structure of a brown alga includes a **holdfast,** which anchors the alga to a rock, a **stipe** (stalk), and **blades,** which are leaflike structures. See Figure 24.1. All parts of the plant carry on photosynthesis, but the blades are the most important photosynthetic organs. Blades of large brown algae often have air-filled bladders associated with them to hold them near the surface

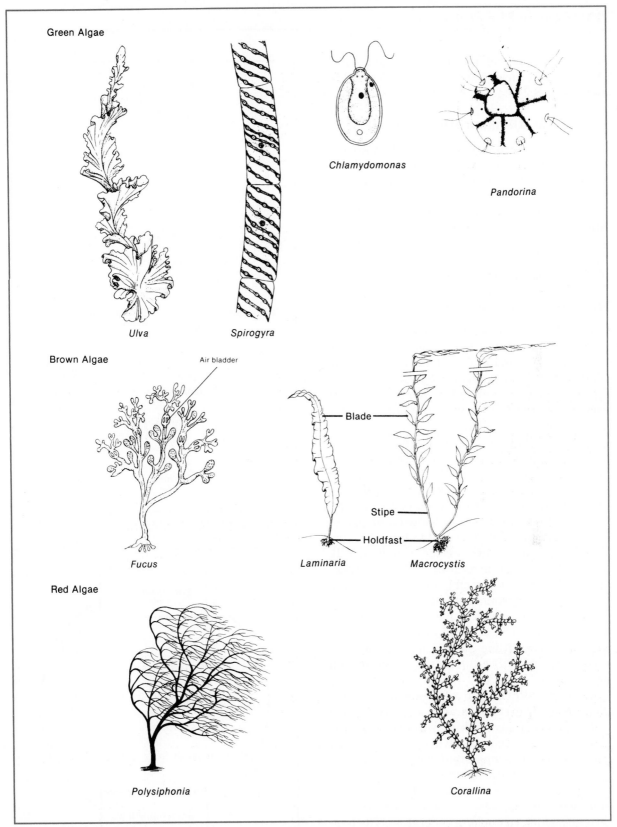

Figure 24.1. Examples of green, brown, and red algae. The drawings are not to scale. *Chlamy-domonas* and *Pandorina* are microscopic in size, while *Macrocystis* is a giant kelp that may be over 100 ft in length.

of the water. Flagellated cells have one whip-lash and one tinsel flagellum. See Figure 24.1 and Plates 8.2–8.4.

Red Algae (Division Rhodophyta)

Most red algae are marine and occur at greater depths than brown algae. They have a delicate body structure. Like brown algae, they are attached to rocks by holdfasts. See Figure 24.1 and Plate 8.5. The presence of red pigments, **phycobilins,** in addition to **chlorophylls a** and **d,** allows them to absorb the deeper-penetrating wavelengths of light for photosynthesis. Flagellated cells are absent.

Reproduction in Algae

Green algae reproduce vegetatively by mitotic division in unicellular forms and by fragmentation in colonial and multicellular forms. Sexual reproduction ranges from conjugation in colonial green algae to the more complex alternation of generations in multicellular forms of green, brown, and red algae.

Consider **conjugation** in *Spirogyra,* a filamentous, green alga, shown in Figure 24.2. When opposite mating types are in contact with each other, a tube forms between cells of the two filaments. The contents of each cell condense, forming a **gamete.** One gamete migrates through the tube from one cell to the other and fuses with the gamete in the receiving cell, forming a diploid (2n) **zygote.** It forms a **zygospore** that can withstand unfavorable conditions and that germinates under favorable conditions. Germination is by meiosis, forming four haploid nuclei. Three nuclei disintegrate, leaving the cell with one haploid nucleus. Mitotic division of this cell results in the formation of a new haploid filament.

Although **alternation of generations** in a plant life cycle first appeared in advanced multicellular algae, it is most highly developed in terrestrial plants, where you will study it in some detail. Here, it is important for you to understand its basic characteristics.

Examine Figure 24.3. Note that there are two adult generations: a gametophyte generation alternates with a sporophyte generation. The haploid (n) **gametophyte** produces gametes (eggs and sperm) by mitosis. Union of sperm and egg forms a diploid (2n) **zygote** that grows to become

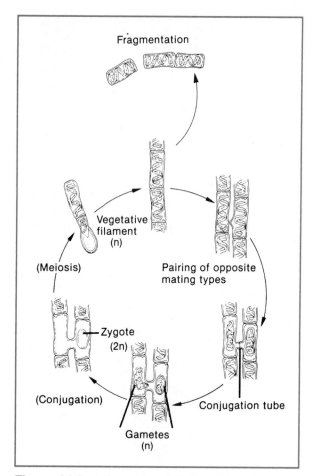

Figure 24.2. Life cycle of *Spirogyra*

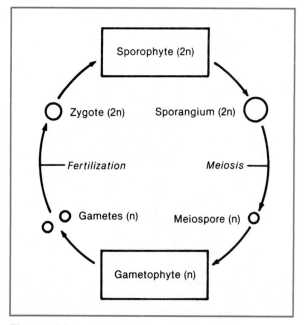

Figure 24.3. Alternation of generations

a diploid (2n) **sporophyte.** The sporophyte produces haploid (n) **spores** by meiotic division that germinate and grow into gametophytes of the next generation.

Materials

Per student
Colored pencils

Per lab
Medicine droppers
Microscope slides and cover glasses
Cultures of:
   Chlamydomonas
   Spirogyra
   Volvox
Demonstration microscope setups of:
   Chlamydomonas, 1,000×
   Volvox, 40×
Demonstration of the effect of light on *Chlamydomonas* distribution in a partially shaded Petri dish
Prepared slides of *Spirogyra* conjugation showing gametes and zygospores
Representative brown, green, and red algae

Assignment 1

1. **Complete item 1a on Laboratory Report 24 that begins on page 389.**
2. Examine *Chlamydomonas* set up under a demonstration microscope. Compare your observations with Figure 24.1 and Plate 7.1.
3. Examine the distribution of *Chlamydomonas* in the partially shaded Petri dish. *Do not move the dish.* Lift the black-paper light shield to make your observations, then replace it in its prior position.
4. Examine *Volvox* set up under a demonstration microscope and note its manner of movement. *Volvox* is a colony formed of many *Chlamydomonas*-like cells. Compare your observations with Plate 7.2. **Complete items 1b–1e on the laboratory report.**
5. Make a water-mount slide of a few strands of *Spirogyra,* a nonmotile, filamentous alga, and examine them at 100× and 400×. Locate the cellular parts shown in Figure 24.1. Compare your specimen with Plate 7.3. **Complete items 1f and 1g on the laboratory report.**

6. Examine the demonstration specimens of brown, green, and red algae. Compare the specimens with Figure 24.1 and Plates 8.1–8.5. **Complete items 1h and 1i on the laboratory report.**
7. Study the life cycle of *Spirogyra* in Figure 24.2. Color the diploid stages of the cycle green. Examine a prepared slide of *Spirogyra* conjugation. Note the gametes and zygospores. **Complete item 1 on the laboratory report.**

MOSS PLANTS (Division Bryophyta)

Mosses, liverworts, and hornworts compose the bryophytes. They occur in habitats that are moist for at least part of the year because surface water is required for sperm to swim to the eggs. Most are not well adapted for terrestrial life because they lack vascular tissues (xylem and phloem) that conduct water, minerals, and nutrients. They also lack true leaves, stems, and roots.

Study the moss life cycle in Figure 24.4. Locate the gametophytes and the sporophyte. In this moss, there are separate male and female gametophytes, but some moss species have a single gametophyte that contains both male and female reproductive organs.

The dominant generation is the **gametophyte,** meaning that the gametophyte is larger and lives longer than the sporophyte. A gametophyte has leaflike photosynthetic organs that are attached to a stemlike stalk from which **rhizoids** extend, anchoring it to the soil. Rhizoids do not absorb water and minerals from the soil like roots.

A **sporophyte** consists of a stalk (seta) attached to a gametophyte and a terminal **sporangium.** The sporophyte is partially parasitic on the gametophyte since it has limited photosynthetic capabilities.

Materials

Per student
Colored pencils
Compound microscope

Per lab
Moss gametophytes and sporophytes, living
Representative bryophytes, living or plastomounts

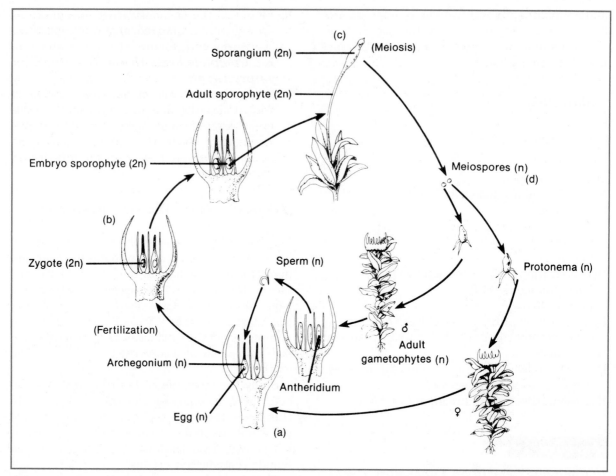

Figure 24.4. Life cycle of a moss. (a) A male gametophyte (n) forms sperm-forming antheridia at its tip, and a female gametophyte (n) forms archegonia, each forming one egg. When released, a flagellated sperm swims into an archegonium and fertilizes the egg cell. (b) Union of sperm (n) and egg (n) forms a zygote (2n) that develops into an embryo sporophyte (2n). (c) The embryo sporophyte grows to become an adult sporophyte that remains attached to the base of the archegonium. (d) Meiospores (n) are formed by meiosis within the sporangium and released. Upon germination, they grow into protonemas (n) that subsequently develop into adult male and female gametophytes (n).

Prepared slides of *Mnium:*
    antheridia with sperm, l.s.
    archegonia with eggs, l.s.
    protonema
    sporangium (capsule) with spores, l.s.

Assignment 2

1. Study Figure 24.4 and color all diploid phases of the life cycle green.

2. Examine a living gametophyte with sporophyte attached. Locate the rhizoids, stalk, and leaflike organs of the gametophyte, and the stalk and sporangium (capsule) of the sporophyte. **Complete item 2a on the laboratory report.**

3. Examine prepared microscope slides of antheridia with sperm, archegonia with eggs, sporangium with spores, and protonema at 40× and 100×. **Complete item 2 on the laboratory report.**

Ferns (Division Pterophyta)

Ferns are the most common and best known of the nonseed vascular plants. Vascular plants possess two vascular tissues: xylem and phloem. **Xylem** conducts water and dissolved minerals upward from roots to stem and leaves. **Phloem** conducts organic nutrients either upward or downward within the plant. Plants that possess vascular tissue have true **roots, stems,** and **leaves.** Without vascular tissue, plants would not be much bigger than mosses and could not live in drier habitats.

Examine the fern life cycle in Figure 24.5. A fern sporophyte is the dominant generation, and it is what you recognize as a fern plant. A sporophyte has an underground stem, a **rhizome,** that is anchored by numerous **roots. Leaves** arising from the stem are usually large and subdivided into many leaflets. Leaflets of

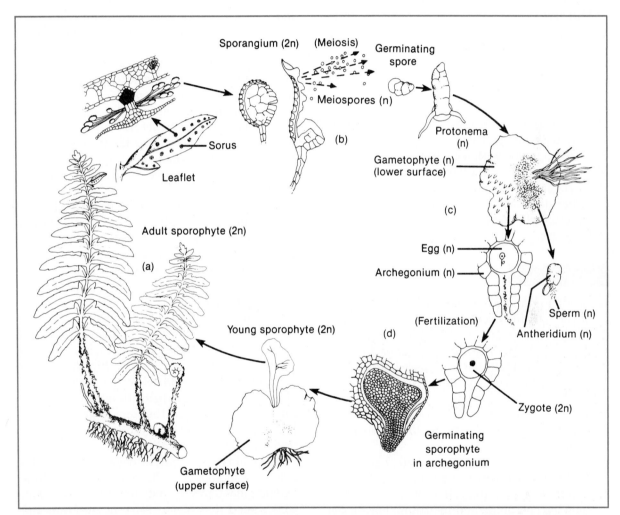

Figure 24.5. Life cycle of a fern. (a) A mature fern sporophyte (2n) forms sporangia on the undersurface of the leaflets of some leaves. Sporangia are clustered in groups called sori. (b) Meiospores (n), formed by meiosis, are released from sporangia. The spores germinate and develop into protonemas (n) that grow to become gametophytes (n). (c) Sperm-forming antheridia and egg-containing archegonia are located on the lower surface of a gametophyte. When released, a flagellated sperm swims into an archegonium to fertilize the egg. (d) Union of sperm (n) and egg (n) forms a zygote (2n) that develops into an embryo sporophyte within the archegonium. Growth of the embryo sporophyte ultimately produces an independent adult sporophyte (2n).

some leaves have small brown spots, **sori,** on their undersurface. A sorus consists of many spore-forming sporangia that are usually protected by a thin cover, the indusium.

The separate gametophyte is small, flat, and roughly heart-shaped. It is anchored to the soil by rhizoids. Archegonia are clustered together on its undersurface near the notch, while antheridia are more widely scattered closer to the bases of the rhizoids.

Materials

Per student
Colored pencils
Compound microscope
Dissecting microscope

Per lab
Demonstration microscope setup of archegonium with an egg cell
Fern sporophytes with sori, living
Microscope slides and cover glasses
Prepared slides of fern:
   antheridium with sperm
   gametophyte, w.m.
   sorus, x.s.

Assignment 3

1. Study Figure 24.5. Color the diploid portions of the fern life cycle green.
2. Examine a prepared slide of a fern gametophyte. Locate the rhizoids, antheridia, and archegonia. Examine a prepared slide of an antheridium and the demonstration microscope setup of an archegonium. How many eggs are located within its swollen base? **Complete items 3a–3c on the laboratory report.**
3. Examine a living fern sporophyte. Locate the rhizome, roots, leaves, and sori.
4. Examine a prepared slide of a sorus, x.s., showing sporangia and indusium. Locate the spores within the sporangia.
5. Remove a leaflet with sori and examine a sorus with a dissecting microscope. If an indusium is present, remove it with a dissecting needle to expose the sporangia. Place the leaflet on a microscope slide without water or cover glass, and examine the exposed sporangia at 40× with your compound microscope.

Watch what happens as the sporangia dry out. **Complete item 3 on the laboratory report.**

SEED PLANTS

Seed plants are better adapted to terrestrial life than ferns but maintain a life cycle involving alternation of generations. The dominant sporophyte—the generation that you recognize as a plant—consists of a root system and a shoot system. The **root system** usually lies below ground. It provides anchorage, and it absorbs water and minerals. The **shoot system** consists of stems, leaves, and reproductive organs. It produces organic nutrients via photosynthesis.

A seed plant contains well-developed vascular tissue that provides structural support for roots, stems, and leaves as well as the transport of materials. Leaves, the main photosynthetic organs, are supported by **veins** composed of vascular tissue.

Sporophytes of seed plants form two different types of meiospores. **Microspores** are formed in **microsporangia** and mature to form **pollen grains** that, in turn, develop into microscopic **male gametophytes. Megaspores** are formed by **megasporangia** and develop into **female gametophytes.**

Seed plants are not dependent on water for sperm transport. Instead, pollen grains are transferred to the female gametophyte by wind or insects, a process called **pollination.** Then the pollen grain develops into a male gametophyte with a **pollen tube** that carries **sperm nuclei** to the **egg** in the female gametophyte. Fertilization results in a zygote, the first cell of a new sporophyte generation.

The embryo sporophyte develops within the female gametophyte. The embryo sporophyte, female gametophyte, and stored nutrients compose a **seed** that has a resistant seed coat enabling dormancy during unfavorable conditions. When conditions are favorable, a seed germinates, producing a new sporophyte generation.

Seed plants are divided into two large groups. **Gymnosperms** have seeds borne exposed on the surface of modified leaves. **Angiosperms** produce flowers and have their seeds enclosed within a **fruit.**

Cone-Bearing Plants
(Division Coniferophyta)

Conifers are the best known and largest group of gymnosperms. The cones contain the reproductive organs, and two types of cones are formed. **Male** (pollen) **cones** are small with paper-thin scales. **Female** (seed) **cones** are large with woody scales. Pollen is transferred from male cones to female cones by wind. Seeds are borne exposed on the upper surface of the scales of mature female cones. The leaves are either needlelike or scalelike. Conifers may attain considerable size and may live in rather dry habitats because their vascular tissue is well developed, their leaves restrict water loss, and water is not required for sperm transport.

Study the life cycle of the pine illustrated in Figure 24.6.

Materials

Per student
Colored pencils
Compound microscope
Dissecting instruments

Per lab
Cones and leaves of representative conifers
Demonstration microscope setups:
    pine female gametophyte with archegonium and egg
    sectioned pine seed showing embryo sporophyte
Female pine cones with seeds
Male pine cones with pollen
Pine seeds
Prepared slides of pine microsporangium with pollen

Assignment 4

1. Study the pine life cycle in Figure 24.6 and color the diploid stages green. *Complete item 4a on the laboratory report.*
2. Examine the male and female pine cones. Compare their size and weight. Locate the microsporangia of a male cone and the seeds on the upper surfaces of the scales of a female cone.
3. Examine the cones and scalelike and needlelike leaves of representative conifers.
4. Examine a prepared slide of pine microsporangia with pollen. Note the structure of pollen. *Draw a few pollen grains in item 4b on the laboratory report.*
5. Examine a pine female gametophyte (megasporangium) with archegonium set up under a demonstration microscope. Locate the egg in an archegonium. *Draw a female gametophyte with archegonium and egg in item 4b on the laboratory report.*
6. Examine a sectioned pine seed set up under a demonstration dissecting microscope. Locate the embryo sporophyte embedded in tissue of the female gametophyte.
7. *Complete item 4 on the laboratory report.*

Flowering Plants
(Division Anthophyta)

Flowering plants are the most advanced plants. Their success in colonizing the land is due to well-developed vascular tissues and **flowers** (reproductive organs) that greatly enhance reproductive success. Most flowers attract insects that bring about pollination, a process leading to the fertilization of egg cells by sperm nuclei. Seeds are enclosed within **fruits** that facilitate dispersal. The reproductive patterns of flowering plants show some major adaptations over gymnosperms.

1. Reproductive structures are grouped in **flowers** that usually contain both microsporangia and megasporangia.
2. Pollination is usually by wind in grasses, but it is by insects in most flowering plants. Insects are attracted to flowers by color and nectar.
3. Portions of the flower develop to form a **fruit** that encloses the seeds and enhances seed dispersal by wind in some plants but by animals in most.

Study the life cycle of a flowering plant illustrated in Figure 24.7.

Flower Structure

The basic structure of a flower is shown in Figure 24.8, but this fundamental organization has many variations.

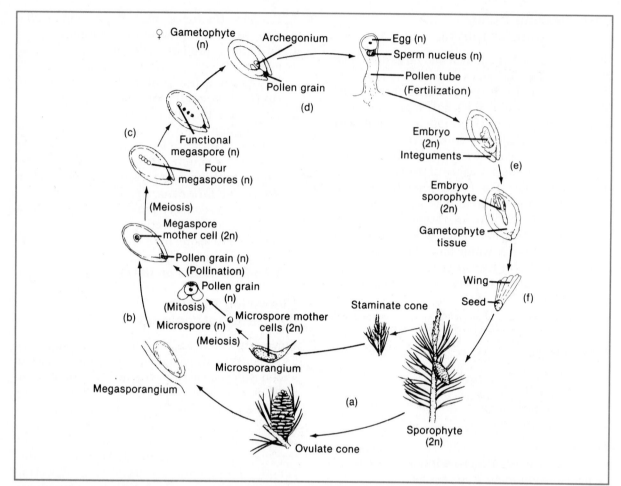

Figure 24.6. Life cycle of a pine. (a) The sporophyte (2n) produces staminate (male) and ovulate (female) cones. Each "leaf" of a staminate cone contains two microsporangia, and each "leaf" of an ovulate cone contains two megasporangia. (b) Microsporangia form microspores (n) by meiosis, and microspores develop into pollen grains. Pollen grains are released and carried by wind to the megasporangia, where they begin to develop into male gametophytes. (c) Megasporangia form four megaspores (n) by meiosis. Three disintegrate, and one megaspore grows to form a female gametophyte. (d) The pollen tube of a male gametophyte grows into an archegonium and discharges two sperm nuclei (n), one of which fertilizes the egg (n). (e) The resultant zygote (2n) grows to become an embryo sporophyte embedded within the female gametophyte. The wall of the megasporangium forms the seed coat of the new seed. (f) The seed is released and dispersed by wind. It germinates to grow into a new sporophyte (2n) generation.

The **receptacle** supports the flower on the stem, and the reproductive structures are enclosed within two whorls of modified leaves. The inner whorl consists of **petals** that are usually colored to attract pollinating insects. The outer whorl consists of **sepals** that are typically smaller than the petals and are usually green in color.

The **stamens** are the male portions of the flower. Each stamen consists of an **anther** supported by a **filament.** Anthers contain the microsporangia that produce pollen. The **pistil** is

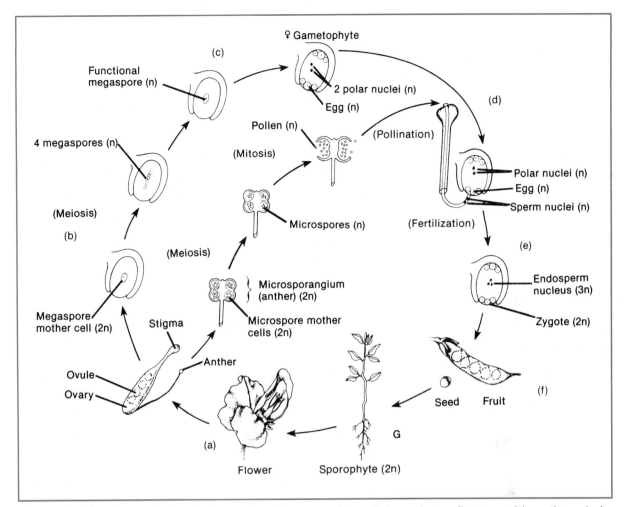

Figure 24.7. Life cycle of a pea plant. (a) The sporophyte (2n) produces flowers with anthers (microsporangia) and ovules (megasporangia). (b) Anthers form microspores (n) by meiosis. Each microspore develops into a pollen grain. Ovules form four megaspores (n), but only one remains functional. (c) The functional megaspore grows to form a female gametophyte (n) containing an egg and two polar nuclei within the ovule. (d) Pollen grains are transferred to the stigma of a flower and develop to become male gametophytes (n) that grow pollen tubes down the style. A pollen tube carries two sperm nuclei (n) to a female gametophyte. (e) One sperm nucleus fertilizes the egg, forming a zygote (2n), and the other fuses with the two polar nuclei, forming the endosperm nucleus (3n). The zygote grows to become an embryo sporophyte, and the endosperm develops to provide stored nutrients for the embryo. (f) The ovule, endosperm, and embryo sporophyte form the seed that is contained within a fruit, an enlarged, ripened ovary. Fruit and/or seeds are disseminated by wind or animals. Upon germination, seeds grow to form a new sporophyte (2n) generation.

the female portion, and it consists of three parts. The basal portion is the **ovary,** which contains **ovules** (megasporangia). The tip of the pistil is the **stigma,** which receives pollen and secretes enzymes promoting pollen germination. The **style** is a slender stalk that joins stigma and ovary. Nectar is secreted near the base of the ovary.

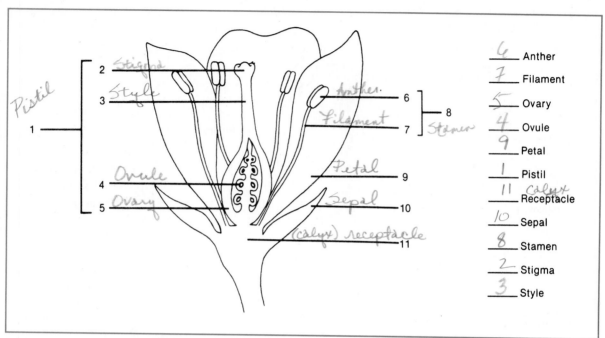

Figure 24.8. Flower structure

Labels on diagram (left side):
- Pistil
- 2 — Stigma
- 3 — Style
- 1
- 4 — Ovule
- 5 — Ovary
- 6 — Anther
- 8 — Stamen
- 7 — Filament
- 9 — Petal
- 10 — Sepal
- 11 — (calyx) receptacle

Answer list (right side):
- 6 Anther
- 7 Filament
- 5 Ovary
- 4 Ovule
- 9 Petal
- 1 Pistil
- 11 calyx / Receptacle
- 10 Sepal
- 8 Stamen
- 2 Stigma
- 3 Style

Fruits and Seeds

As the seeds develop in the ovary, the ovary grows and ripens to form a fruit that provides protection for the seeds and facilitates seed dispersal. The three basic types of fruits are shown in Figure 24.9.

1. **Dry dehiscent fruits** split open when sufficiently dry to cast out the seeds, sometimes with considerable force. Pea and bean pods are examples.
2. **Dry indehiscent fruits** do not open, and the ovary wall tightly envelops the seed. Acorns and fruits of corn and other cereals are examples.
3. **Fleshy fruits** remain moist for a considerable period of time and are usually edible and colored. Animals scatter the seeds by feeding on the fruits.

A seed (Figure 24.10) consists of a protective **seed coat** that is derived from the wall of the ovule, stored nutrients, and a dormant **embryo sporophyte.** In monocots and some dicots, the stored nutrients compose the endosperm. In most dicots, the embryonic leaves, **cotyledons,** contain many of the stored nutrients, and the endosperm is reduced. Seeds are able to with-

stand unfavorable conditions and tend to germinate only when favorable conditions exist.

Materials

Per student
Colored pencils
Compound microscope
Dissecting instruments
Dissecting microscope

Per lab
Dropping bottles of iodine solution
Microscope slides and cover glasses
Model of a flower
Bean seeds, soaked
Corn fruits, soaked
Lily or *Gladiolus* flowers
Pea pods, fresh
Representative flowers, fruits, and seeds
Prepared slides of lily anthers, x.s.

Assignment 5

1. Study the life cycle of a flowering plant in Figure 24.7. Color the diploid stages green.

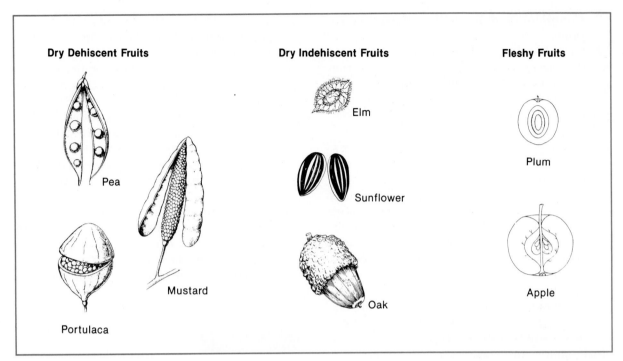

Figure 24.9. Types of fruits

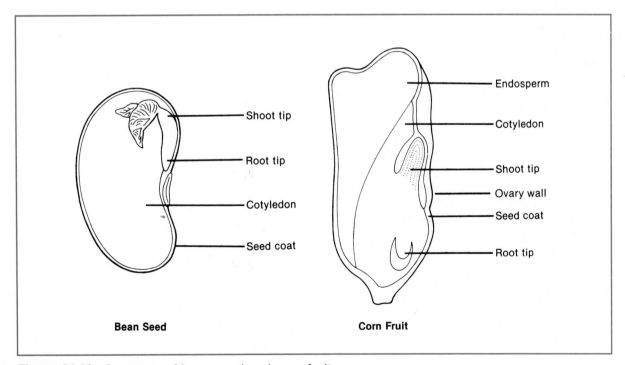

Figure 24.10. Structure of bean seed and corn fruit

Complete item 5a on the laboratory report.

2. Label Figure 24.8. Color the male portion of the flower yellow and the female portion green. *Complete items 5b and 5c.*

3. Obtain a lily flower and compare it with Figure 24.8 and the flower model to identify its parts.

4. Examine a prepared slide of an anther, x.s. Note the microsporangia and enclosed pollen. Remove an anther from your flower and observe it with a dissecting microscope. Note the pollen. Place a few pollen grains on a microscope slide without water and observe them at 40× and 100× with a compound microscope. *Complete item 5d on the laboratory report.*

5. Use a scalpel to make a cross section of the ovary and examine it with a dissecting microscope. Locate the ovules. *Complete item 5e on the laboratory report.*

6. Examine the representative flowers and note the variations in their structure. Locate the male and female portions of each.

7. Examine the representative seeds and fruits. Use Figure 24.9 to classify the fruits according to type. Obtain and examine a pea pod. *Complete items 5f and 5g on the laboratory report.*

8. In Figure 24.10, color the embryos green and the cotyledons blue. Obtain a soaked bean seed and corn kernel. Open the "halves" of the bean seed as in Figure 24.10 to expose the embryo. Use a scalpel to section the corn kernel and examine its cut surface. Examine both with a dissecting microscope to locate their components shown in Figure 24.10.

9. Add a drop of iodine solution to the bean cotyledons and embryo and to the cut surface of the corn kernel. After 3 min, rinse, blot dry, and examine.

10. *Complete item 5 on the laboratory report.*

Assignment 6

Examine the numbered "unknown" plants and identify the group to which they belong. *Record your responses in item 6 on the laboratory report.*

25

STRUCTURE OF FLOWERING PLANTS

OBJECTIVES

After completion of the laboratory session, you should be able to:
1. Identify the external structure of a flowering plant.
2. Identify monocots and dicots.
3. Identify types of root systems and leaves.
4. Identify tissues composing roots, stems, and leaves.
5. Define all terms in bold print.

Flowering plants are the most advanced vascular plants. They have well-developed vascular tissues providing support as well as material transport, and, as you learned in Exercise 24, their life cycle involves flowers, pollination, and seeds enclosed in fruits. In this exercise, you will study the external and internal structure of roots, stems, and leaves.

Flowering plants are subdivided into two major classes: **monocotyledonous** (monocots) **plants** and **dicotyledonous** (dicots) **plants.** Their distinguishing characteristics are shown in Figure 25.1.

Monocots include grasses, palms, and lilies. Most flowering plants are dicots, and they fall into one of two major categories: herbaceous or woody. **Herbaceous** (not woody) **dicots** are usually annuals or biennials. Annuals complete their life cycle in a single growing season like most wildflowers, beans, and tomatoes. Biennials require two growing seasons; flowers are produced only in the second season. **Woody dicots** are usually perennials that live several years and produce flowers each year, such as oaks and roses.

GENERAL EXTERNAL STRUCTURE

Figure 25.2 shows the basic parts of a flowering plant. The shoot system consists of a **stem** that supports the **leaves, flowers,** and **fruits.** Leaves branch from the stem at sites called **nodes.** A section of stem between nodes is an **internode.** Leaves are the primary photosynthetic organs of the plant, and they exhibit two types of venation: net venation and parallel venation. Leaves with **net venation** have a central vascular bundle (vein) called a midrib from which smaller lateral veins branch. Such leaves consist of a thin, expanded portion called a **blade** and a leaf stalk called a **petiole.** Leaves with **parallel venation** lack a midrib and have veins running parallel to each other for the length of the blade. Such leaves usually lack a petiole.

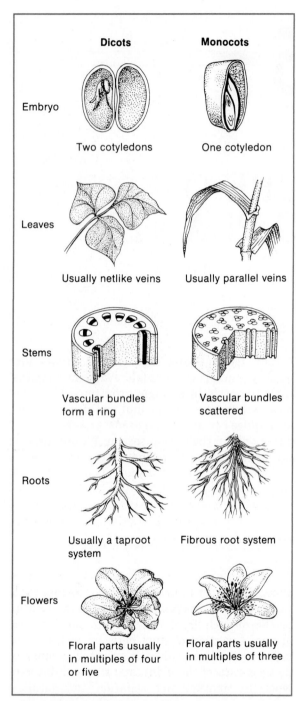

Figure 25.1. Distinguishing characteristics of monocots and dicots

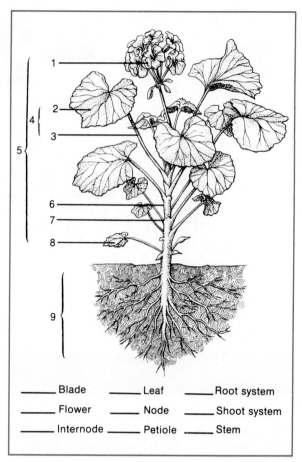

Figure 25.2. External structure of a flowering plant

_____ Blade   _____ Leaf   _____ Root system

_____ Flower   _____ Node   _____ Shoot system

_____ Internode   _____ Petiole   _____ Stem

Materials

Per student
Dissecting microscope

Per lab
Coleus seedlings
Coleus and corn stems in solution of red food coloring
Corn seedlings
Representative dicots and monocots
Razor blades, single-edged

Assignment 1

1. Label Figure 25.2. ***Complete items 1a and 1b on Laboratory Report 25 that begins on page 393.***
2. Obtain corn and *Coleus* seedlings. Gently wash the soil from their roots and lay them

The root system is located in the ground, and it may be even more highly branched than the shoot system. Roots not only anchor the plant but they absorb water and nutrients as well.

on a paper towel for observation. Compare the roots, stems, and leaves with Figures 25.1 and 25.2.

3. Stems of several *Coleus* and corn seedlings have been placed in a water-soluble dye that stains vascular tissue as it is carried up the stem. Use a razor blade to cut thin cross sections from the *Coleus* and corn stems and examine them with a dissecting microscope. Observe the arrangement of the vascular bundles and compare them with Figure 25.1.

4. Identify the "unknown" plants as dicots or monocots. ***Complete item 1 on the laboratory report.***

ROOTS

Roots perform three important functions: (1) anchorage and support, (2) absorption and transport of water and minerals, and (3) storage and transport of organic nutrients.

The Root Tip

The basic structure of a root tip is shown in Figure 25.3. Cells formed in the **region of cell division** become either part of the root proper or form the root cap. The **root cap** is composed of rather large cells that protect the region of cell division. It also provides a sort of lubrication as its cells are eroded by the root growing through the soil. Cells of the root are enlarged in the **region of elongation,** and this accounts for the greatest increase in the linear growth of the root. As the cells become older, they develop their specialized characteristics in the **region of differentiation** where the **primary root tissues** are formed. Note that the cells of the root are arranged in columns. The column in which a cell is located determines the type of cell it will become. For example, cells in the outermost columns become epidermal cells, and those in the innermost columns become xylem cells. **Root hairs** are extensions of epidermal cells in the region of differentiation. They greatly increase the surface area of the root tip and are the primary sites of water and mineral absorption.

Growth in length of both roots and stems results from the formation of new cells in the region of cell division and their subsequent enlargement in the region of elongation. Thus,

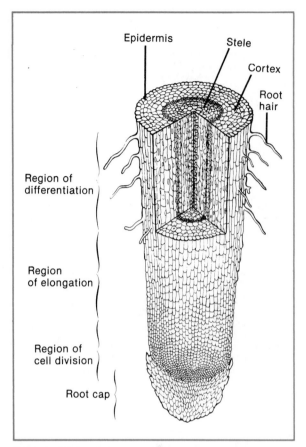

Figure 25.3. A dicot root tip

growth of both roots and stems occurs at their tips, and this growth is continuous throughout the life of the plant.

Root Tissues

Figure 25.4 shows the tissues found in a root of a herbaceous dicot (*Ranunculus*) as viewed in cross section. Note the three major divisions: epidermis, cortex, and stele.

The **epidermis** is the outermost layer of cells that provides protection for the underlying tissues and reduces water loss.

The **cortex** composes the bulk of the root and consists mostly of large, thin-walled cells used for food storage. The **endodermis,** the innermost layer of the cortex, is composed of thick-walled, water-impermeable cells and a few water-permeable cells with thinner walls. This ring of cells controls the movement of water and minerals into and out of the xylem.

The **stele** is that portion of the root within the

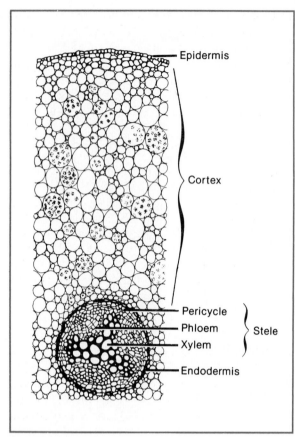

Figure 25.4. Herbaceous dicot (*Ranunculus*) root, x.s.

endodermis. It is sometimes called the central cylinder. The outer layer of the stele is composed of thin-walled cells, the **pericycle,** from which branch roots originate. The thick-walled cells in the center of the stele are the **vessel cells** of the **xylem.** Between the rays of the xylem are the **sieve tubes** and **companion cells** of the **phloem.**

Root Types

There are three types of roots. **Tap root systems** have a single dominant root from which branch roots arise. **Fibrous root systems** consist of a number of similar-sized roots that branch repeatedly. Fibrous roots are characteristic of monocots. Both types of root systems may be shallow or deep, depending on the species of plant, but tap roots are capable of the deepest penetration. **Adventitious roots** are unique in that, unlike other roots, they do not grow from

the primary root of the embryo. They originate from stems or leaves and are typically fibrous in nature.

Materials

Per student
Colored pencils
Compound microscope
Dissecting microscope
Dissecting instruments

Per lab
Dropping bottles of methylene blue, 0.01%
Microscope slides and cover glasses
Examples of adventitious, fibrous, and tap roots
Germinated grass (or radish) seeds
Prepared slides of:
  Allium (onion) root tip, l.s.
  Ranunculus (buttercup) root, x.s.

Assignment 2

1. Color-code the root cap and regions of cell division, cell elongation, and cell differentiation in Figure 25.3 and the xylem and phloem in Figure 25.4.
2. Obtain a germinated grass seed. Place it on a microscope slide in a drop of water. Examine the young root with a dissecting microscope and at 40× with a compound microscope. Note the root hairs. Locate the oldest (longest) and youngest root hairs. Use a scalpel to cut the seed from the root. Add a drop of methylene blue and a cover glass. Examine the root at 100×. Note the attachment of a root hair to an epidermal cell. Try to locate a cell nucleus in a root hair. ***Complete items 2a–2c on the laboratory report.***
3. Examine a prepared slide of *Allium* root tip, l.s., with a compound microscope at 40×. Locate the root cap and the regions shown in Figure 25.3., except for the region of differentiation, which is not shown on your slide. Compare the size of the cells in each region. ***Complete item 2d on the laboratory report.***
4. Examine a prepared slide of *Ranunculus* root, x.s., with a compound microscope at 40×. Locate the tissues shown in Figure 25.4. Note the starch granules in the cells of the

cortex and the arrangement of xylem and phloem.

5. Examine the examples of root types. Note their characteristics. ***Complete item 2 on the laboratory report.***

STEMS

The stem serves as a connecting link between the roots and the leaves and reproductive organs. It also may serve as a site for food storage. The arrangement of vascular tissue in stems varies among the subgroups of vascular plants.

Monocot Stems

Figure 25.5. illustrates the cross-sectional structure of part of a corn stem. Note the *scattered* **vascular bundles** surrounded by large thin-walled cells, a characteristic of monocots. Each vascular bundle has thick-walled, fibrous cells around the edges surrounding the large xylem vessels and the smaller sieve tubes and companion cells of phloem. Most of the support for the stem is provided by the xylem and fibrous cells.

Herbaceous Dicot Stems

Figure 25.6 shows the structure of a portion of an alfalfa stem in cross section. Note that the vascular bundles are arranged in a *broken ring* just interior to the **cortex.** This pattern is characteristic of herbaceous dicots. Each vascular bundle is composed of three tissues (from outside): **phloem, vascular cambium,** and **xylem;** the central portion of the stem, the **pith,** is composed of large, thin-walled cells.

Woody Dicot Stems

The structure of a young woody stem in cross section is shown, in part, in Figure 25.7. Note the characteristic *continuous ring* of primary vascular tissue and the arrangement of the tissues within the stem. The **wood** of woody plants is actually xylem tissue.

In each growing season, the vascular cambium forms new (secondary) xylem and phloem. Each growing season is identifiable by an **annual ring** of xylem, which is composed of large-celled **spring wood** (lighter color) and small-

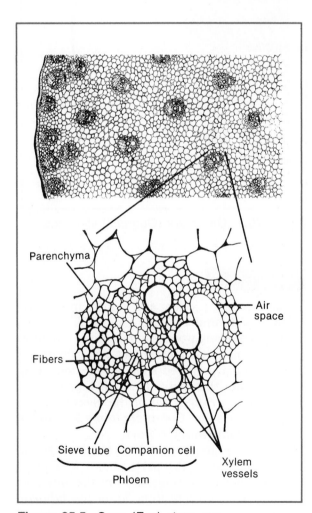

Figure 25.5. Corn (*Zea*) stem, x.s.

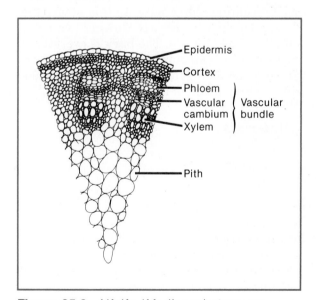

Figure 25.6. Alfalfa (*Medicago*) stem, x.s.

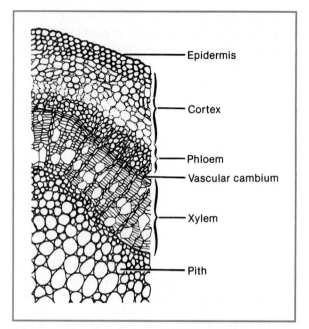

Figure 25.7. Young oak (*Quercus*) stem, x.s.

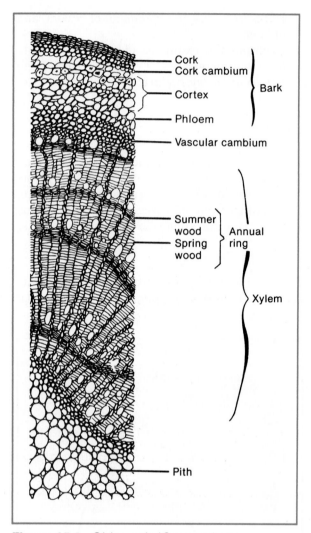

Figure 25.8. Older oak (*Quercus*) stem, x.s.

celled **summer wood** (darker color). No cells are formed in fall or winter in temperate climates. The stem grows in diameter by the formation and enlargement of new xylem and phloem. Since the growth is actually from the inside of the stem and since the epidermis and cortex cannot grow, they tend to fracture and slough off. The **cork cambium** forms in the cortex, and it produces cork cells that assume the function of protecting the underlying tissues and preventing water loss. At this stage, the stem is subdivided into (1) bark, (2) vascular cambium, and (3) wood. The **bark** is everything exterior to the vascular cambium: phloem, cortex, cork cambium, and cork. Figure 25.8 illustrates the structure of an older woody stem.

Materials

Per student
Colored pencils
Compound microscope

Per lab
Tree stem, x.s.
Prepared slides of:
   Quercus (oak) stem, 1 yr, x.s.
   Quercus stem, 4 yr, x.s.
   Medicago (alfalfa) stem, x.s.
   Zea (corn) stem, x.s.

Assignment 3

1. Color-code the xylem and phloem in Figures 25.5, 25.6, and 25.7. In Figure 25.8, color-code the spring wood, summer wood, vascular cambium, phloem, cork, and cork cambium.
2. Examine a prepared slide of *Zea* stem, x.s., and locate the parts shown in Figure 25.5. Note the arrangement of the vascular bundles in the stem. *Complete items 3a and 3b on the laboratory report.*
3. Examine a prepared slide of *Medicago* stem, x.s., and locate the parts shown in Figure 25.6. Note the arrangement of the vascular bundles. *Complete item 3c on the laboratory report.*

4. Examine prepared slides of *Quercus* stems, 1 yr and 4 yr, x.s. Locate the parts shown in Figures 25.7 and 25.8.
5. Determine the age of the demonstration section of tree stem. ***Complete item 3 on the laboratory report.***

LEAVES

Leaves are the primary organs of photosynthesis, and they are sites of gas exchange, including water loss by evaporation. The major part of a leaf is the broad, thin **blade** that is attached to a stem by a leaf stalk, the **petiole,** in most dicots. Some leaves, especially in monocots, lack a petiole, so the blade is attached directly to a stem.

Venation

The blade is supported by many **veins** that are composed of vascular tissue plus a fibrous vascular bundle sheath. Leaves of monocots have **parallel venation** in which veins are arranged in parallel rows running the length of the leaf. See Figure 25.9e. Leaves of dicots have **net venation** in which a large central vein, the **midrib,** extends the length of the leaf. Veins branching from the midrib branch repeatedly into smaller and smaller veins that form a network throughout the leaf. See Figure 25.9a.

The pattern of veins in net-veined leaves follows one of two major types. In **pinnate** leaves, there is a single midrib with several lateral branches forming a pattern resembling the structure of a feather. In **palmate** leaves, there are several major veins branching from the petiole resembling the fingers extending from the hand. If a leaf is divided into a number of leaflets (parts), it is said to be **compound.** If not, it is said to be **simple.** The combination of these patterns determines the type of venation, and it is used in classifying plants. See Figure 25.9.

Internal Structure

The internal structure of a lilac (*Syringa*) leaf is shown in Figure 25.10. An upper and lower **epidermis** form the surfaces of the leaf, and each is coated with a waxy **cuticle** that retards evaporative water loss through the epidermis. Epidermal cells do not contain chlorophyll, except for the scattered **guard cells** that surround tiny openings called **stomata** (stoma is singular). Gas exchange between leaf tissues and the atmosphere occurs through the stomata.

Stomata are often found only on the lower epidermis. This location prevents direct exposure to sunlight and helps to reduce evaporative water loss. When guard cells are photosynthesizing, the accumulation of sugars causes them to swell with water that enters by osmosis, and the swelling bends the guard cells, opening the sto-

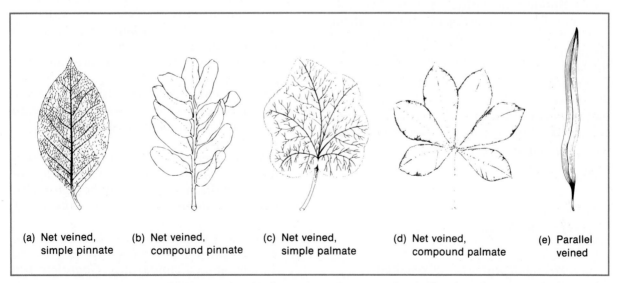

(a) Net veined, simple pinnate   (b) Net veined, compound pinnate   (c) Net veined, simple palmate   (d) Net veined, compound palmate   (e) Parallel veined

Figure 25.9. Examples of leaf venation in flowering plants. a–d are dicots; e is a monocot.

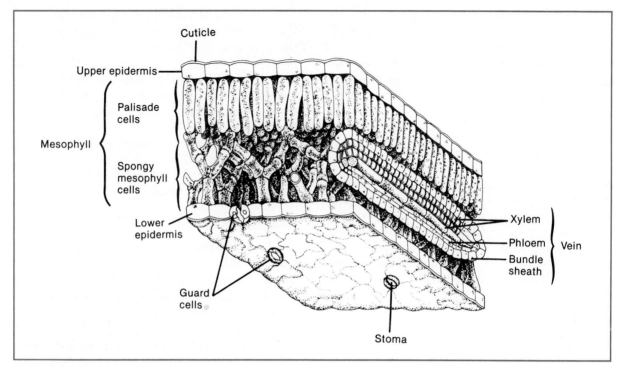

Figure 25.10. Lilac (*Syringa*) leaf, x.s.

mata, which increases gas exchange. At night, water leaves the guard cells, so they straighten out and close the stomata.

The **mesophyll** is the main photosynthesizing tissue in a leaf. It consists of two types. **Palisade mesophyll** consists of closely packed, upright cells located just under the upper epidermis, and it carries out most of the photosynthesis. **Spongy meosphyll** consists of a meshwork of cells with many spaces between them, and it occurs in the lower half of a leaf. This arrangement aids the movement of gases within the leaf.

The leaf veins not only provide support but also movement of materials. Xylem brings water and minerals from roots to the leaf cells, and phloem carries away sugar to stem and root cells for use or storage.

Materials

Per student
Colored pencils
Compound microscope

Per lab
Leaves with different types of venation
Leaves with "unknown" venation

Prepared slides of:
Syringa (lilac) leaf, x.s.
Syringa leaf epidermis with stomata
Zea (corn) leaf, x.s.

Assignment 4

1. Examine the leaves with different types of venation. Identify the types of venation for the "unknowns." ***Complete items 4a and 4b on the laboratory report.***
2. Study Figure 25.10. Color the mesophyll and guard cells green.
3. Examine a prepared slide of *Syringa* leaf, x.s. Locate the cells and tissues shown in Figure 25.10. ***Complete items 4c–4e on the laboratory report.***
4. Examine a prepared slide of *Syringa* epidermis with stomata. Note the density of the stomata and the arrangement of the guard cells. ***Complete items 4f–4i on the laboratory report.***
5. Examine a prepared slide of *Zea* (corn) leaf, x.s. Compare it to the *Syringa* leaf, noting differences and similarities. ***Complete the laboratory report.***

SIMPLE ANIMALS

OBJECTIVES

After completion of the laboratory session, you should be able to:

1. Compare the animal groups studied as to their distinguishing characteristics.
2. Classify to phylum members of the groups studied.
3. Identify the major structural components of organisms studied and describe their functions.
4. Define all terms in bold print.

Members of the kingdom **Animalia** are characterized by (1) a **multicellular body** formed of different types of **eukaryotic cells** that lack a cell wall and plastids, (2) **heterotrophic nutrition,** and (3) movement by means of **contractile fibers.** Animals are diploid (2n), and they reproduce sexually by haploid (n) **gametes,** large nonmotile eggs and small motile sperm, that are formed by meiotic division. A few simple animals also reproduce asexually.

Figure 26.1 shows the presumed evolutionary relationships of the major phyla of animals. These relationships are based on fundamental similarities and differences that will become clear as you work through the next three exercises. You will study the major phyla from simplest to most complex. Your primary goal is to learn the recognition characteristics of each group and to understand the major evolutionary advances exhibited by each group.

In this exercise, you will study the simpler phyla of animals. Table 26.1 indicates a few basic characteristics that distinguish sponges, coelenterates, flatworms, and roundworms. Refer to it as you watch for the gradual increase in complexity among these groups.

SPONGES (Phylum Porifera)

Sponges are the simplest animals. They probably evolved from colonial flagellated protozoans. Most sponges are marine; only a few live in fresh water. The nonmotile adults live attached to rocks or other objects, but the larvae are ciliated and motile. The cells of an adult are so loosely organized that the level of organization is intermediate between cellular and tissue levels, a **cellular-tissue level.** Sponges are categorized according to which one of three types of **spicules,** or skeletal elements, they possess: chalk (calcium carbonate), glass (silica), or fibrous (protein). Note their basic characteristics in Table 26.1.

Figure 26.2 illustrates the basic structure of a simple sponge. Note the numerous pores in the body wall. The beating flagella of **collar cells** create a steady current of water flowing through the **incurrent pores** into **incurrent canals,** through **pore cells** into **radial canals,** on into the **spongocoel** (central cavity), and out the **osculum.** Sponges are filter feeders. Tiny food par-

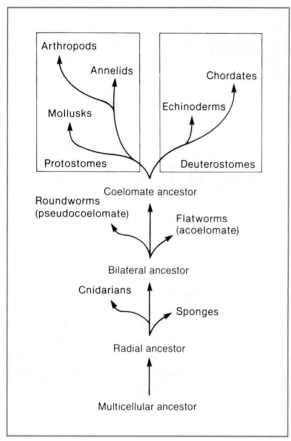

Figure 26.1. Presumed evolutionary relationships of the major animal phyla

ticles in the water are engulfed by collar cells and digested intracellularly within food vacuoles. Nutrients are then distributed to other cells by diffusion that is aided by wandering amoebocytes. Gas exchange and removal of metabolic wastes occur by diffusion.

Sponges release their gametes into the water, where fertilization occurs. The zygote develops into a motile, ciliated larva that ultimately settles to the bottom, attaches, and grows into an adult sponge.

Materials

Per student
Colored pencils
Dissecting microscope

Per lab
Syracuse dishes
Demonstration microscope setups of:
    Grantia, l.s., at 40×
    Grantia, l.s., showing collar cells at 400×
Preserved *Grantia*

Assignment 1

1. Color-code the collar cells, epidermis, spicules, and amoebocytes in the enlargement of the body wall in Figure 26.2.
2. Examine the demonstration skeletons of representative sponges, noting the type of each one.
3. Place a preserved *Grantia* in a Syracuse dish and add sufficient water to cover it. Examine it with a dissecting microscope and locate the osculum, spicules around the osculum, and incurrent pores.
4. Examine a slide of sectioned *Grantia* set up under demonstration microscopes. At 40×,

TABLE 26.1
Distinctive Characteristics of Sponges, Coelenterates, Flatworms, and Roundworms

	Sponges	*Coelenterates*	*Flatworms*	*Roundworms*
Symmetry	Asymmetrical or radial symmetry	Radial symmetry	Bilateral symmetry	Bilateral symmetry
Level of organization	Cellular-tissue level	Tissue level	Organ level	Organ system level
Germ layers	None	Ectoderm and endoderm	Ectoderm, mesoderm, and endoderm	Ectoderm, mesoderm, and endoderm
Coelom	N/A	N/A	Acoelomate	Pseudocoelomate
Body plan	N/A	Saclike	Saclike	Tube within a tube

N/A = Not applicable.

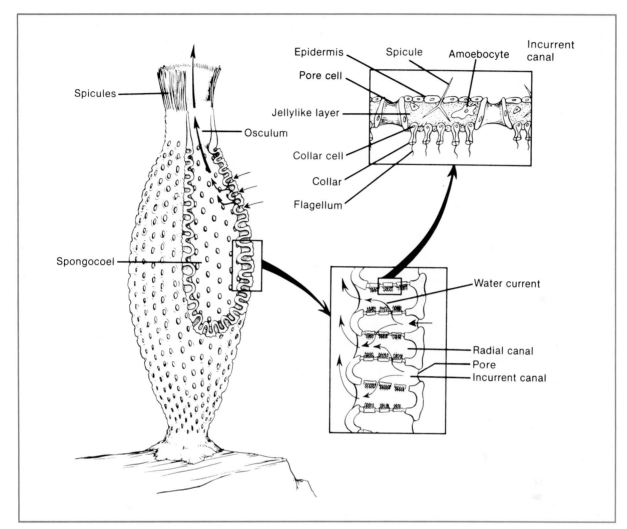

Figure 26.2. *Granitia,* a simple sponge

locate the incurrent pores, radial canals, spongocoel, and osculum. At 400×, observe the collar cells. ***Complete item 1 on Laboratory Report 26 that begins on page 397.***

COELENTERATES
(Phylum Coelenterata)

Coelenterates (cnidarians) include jellyfish, sea anemones, corals, and hydroids. Most forms are marine, but a few forms occur in fresh water. Note their basic characteristics in Table 26.1.

Coelenterates exhibit two distinct body forms, as shown in Figure 26.3. The tubular body of a **polyp** is attached to the substrate at one end and has a mouth surrounded by tentacles at the other. A **medusa** is a free-swimming form. Medusae have bell-shaped bodies with tentacles hanging from the edge of the bell and a mouth located at its center. Pulsating contractions of the bell produce a swimming motion. Most species occur as either polyps or medusae, but some species exhibit both forms. In such species, a sexually reproducing medusa alternates with an asexually reproducing polyp, as shown in Figure 26.4.

The structure of *Hydra,* a freshwater form, illustrates the basic coelenterate structure. See Figure 26.5. The body wall consists of two cell layers, **epidermis** (ectoderm) and **gastrodermis** (endoderm), separated by a noncellular **mesoglea**. Contraction of **myofibrils** in cells of each layer enables movement. A **nerve net,**

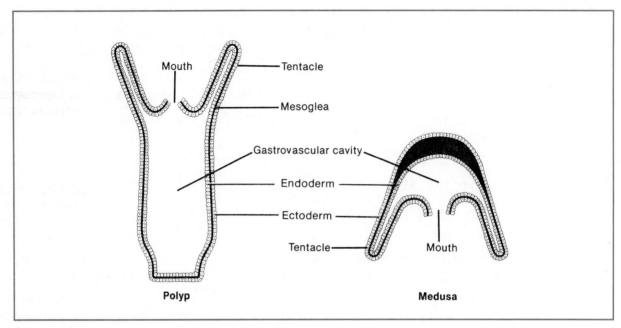

Figure 26.3. Comparison of polyp and medusa

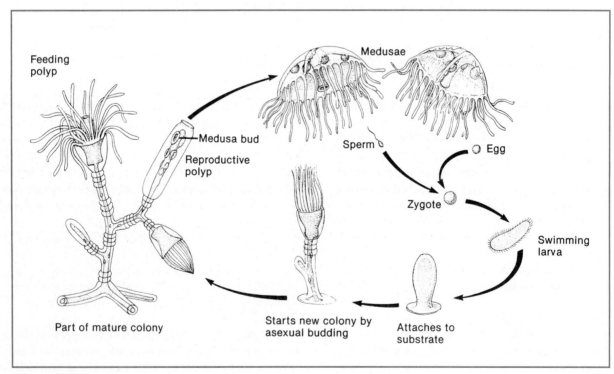

Figure 26.4. Life cycle of *Obelia*

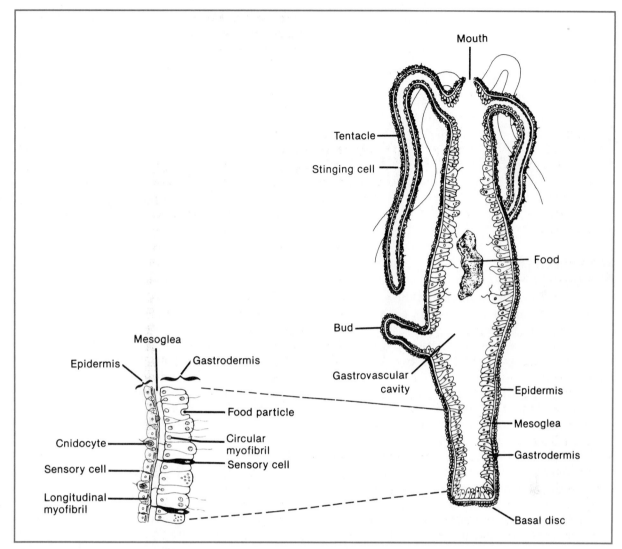

Figure 26.5. Longitudinal section of *Hydra* with detail of body wall

formed of interconnected neurons, lies within the mesoglea and coordinates body functions.

Tentacles of all coelenterates contain **stinging cells** (cnidocytes) that eject dartlike projectiles (nematocysts) of offense and defense. Prey organisms are paralyzed by the stinging darts and engulfed through the mouth into the **gastrovascular cavity,** where partial digestion occurs. Then the partially digested particles are engulfed by gastrodermal cells, and digestion is completed within food vacuoles. The distribution of nutrients, gas exchange, and removal of metabolic wastes occur by diffusion.

Hydra reproduces asexually by budding as well as sexually by gametic reproduction. A *Hydra* develops an egg within an ovary and sperm within a testis. Sperm use their single flagellum to swim to the egg through the water. After fertilization, the zygote develops into a new individual.

Materials

Per student
Colored pencils
Compound microscope

Per lab
Cultures of:
  Daphnia, small
  Hydra
Demonstration setups of:
  Hydra in small aquarium feeding on small
    Daphnia
  representative coelenterates
Demonstration microscope setup of a tentacle of
  a living *Hydra,* showing stinging cells
Prepared slides of:
  Hydra, l.s.
  Obelia medusae, w.m.
  Obelia polyps, w.m.

Assignment 2

1. Color the epidermis green and the gastrodermis yellow in the enlargement of the *Hydra* body wall in Figure 26.5.
2. Examine a prepared slide of *Hydra,* l.s., and locate as many of the structures labeled in Figure 26.5 as you can.
3. Examine prepared slides of *Obelia* polyp and medusa. Locate the tentacles and mouth in each body form. Note the buds forming on the asexually reproducing polyps.
4. Observe the living *Hydra* in the small aquarium (or beaker) as they feed on *Daphnia.* Note their feeding strategy and behavior. Do some *Hydra* have buds?
5. Examine the representative coelenterates. Note which are polyps and which are medusae.
6. Examine a tentacle of a living *Hydra* set up under a demonstration microscope. Locate a stinging cell and its hairlike trigger. ***Complete item 2 on the laboratory report.***

FLATWORMS
(Phylum Platyhelminthes)

Flatworms have a saclike body plan like coelenterates, but their dorsoventrally flattened body is **bilaterally symmetrical** with anterior and posterior ends. All three embryonic tissues are present, and they develop into an adult with an **organ level of development.** See Table 26.1.

All flatworms are **hermaphroditic.** Sperm is exchanged during copulation, but self-fertiliza-tion may occur, especially in parasitic species. Internal fertilization produces fertile eggs laid by the female. Many flatworms are free-living, and they feed on nonliving organic material. Some are important parasites of humans and other vertebrates.

Free-Living Flatworms

A common freshwater planarian, *Dugesia,* is typical of nonparasitic forms. See Figures 26.6 and 26.7. When feeding, the **pharynx** is extended through a midventral opening, the **mouth,** to engulf food into a branched **gastrovascular cavity.** Digestion begins within the gastrovascular cavity and is completed within gastrodermal cells, as in coelenterates. Nutrients are distributed by diffusion.

Unlike coelenterates, neurons in flatworms are more specialized and are organized into specific structures that are concentrated at the anterior end. **Eyespots** are photoreceptors, and chemical and tactile receptors are located on the auricles. Body movements are coordinated by the anterior **cerebral ganglia** ("brain") and two **ventral nerve cords** that extend posteriorly. Movement in *Dugesia* results from the beating cilia of epidermal cells on the ventral body surface.

Specialized excretory organs first appear in flatworms. Metabolic wastes are collected via diffusion into **excretory canals.** The beating of long cilia of numerous flame cells creates a current propelling the liquid wastes out of the body through **excretory pores** scattered over the body surface.

Parasitic Flatworms

Tapeworms and **flukes** are flatworms that parasitize humans and other vertebrates. Their life cycles include one or two intermediate hosts and a final host in which the adults live.

Tapeworm adults live in the small intestine of their hosts, where they attach to the intestinal wall by hooks and/or suckers of a **scolex.** Nutrients are absorbed through the body wall; a gastrovascular cavity is absent.

A tapeworm body consists of a series of repeating units, called **proglottids,** that are continuously produced by the scolex. When mature, proglottids are self-fertilizing reproductive factories. When proglottids become filled with fer-

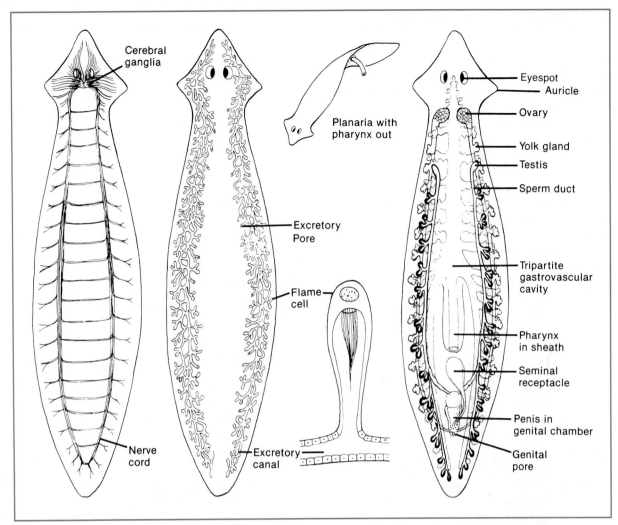

Figure 26.6. Neural, excretory, and reproductive structures in *Dugesia*

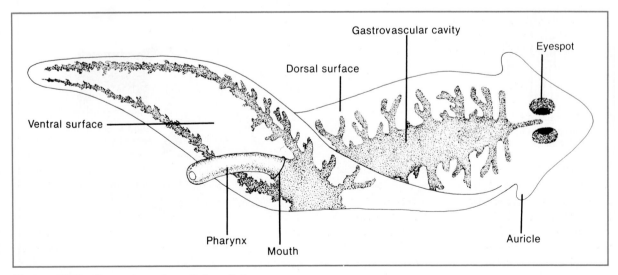

Figure 26.7. *Dugesia,* a free-living flatworm

tile eggs, they separate from the posterior end of the tapeworm and pass from the host in feces.

Intermediate hosts (herbivores) are infected by accidentally eating the eggs. Larvae hatch from the eggs in the intermediate host's intestine, burrow into blood vessels, and are carried by blood to skeletal muscles, where they encyst. When a carnivore or omnivore eats muscles of the intermediate host infested with encysted larvae (bladder worms), the larvae are activated and attach to the new final host's intestine. See Figure 26.8.

Adult flukes infest the lungs, liver, and blood of vertebrates, including humans. Intermediate forms in the life cycle often include snails and fish. For example, the human liver fluke is prevalent in portions of Asia where sanitation is poor and the eating of raw or undercooked freshwater fish is practiced.

Materials

Per student
Compound microscope
Dissecting microscope

Per lab
Camel's hair brushes
Watch glasses
Culture of *Dugesia*
Representative flatworms
Demonstration setup of planaria's response to light

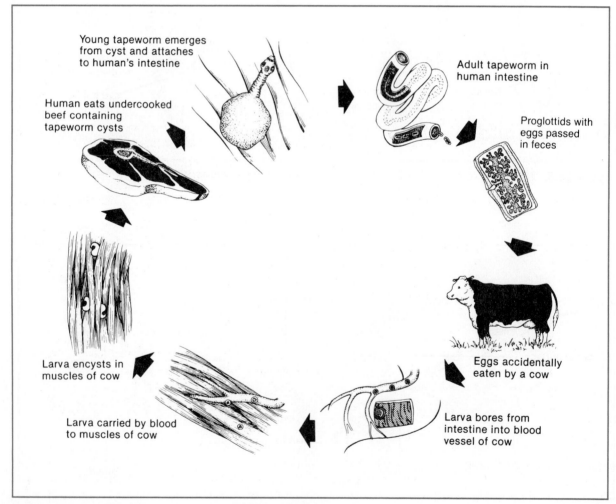

Young tapeworm emerges from cyst and attaches to human's intestine

Human eats undercooked beef containing tapeworm cysts

Adult tapeworm in human intestine

Proglottids with eggs passed in feces

Eggs accidentally eaten by a cow

Larva encysts in muscles of cow

Larva carried by blood to muscles of cow

Larva bores from intestine into blood vessel of cow

Figure 26.8. Life cycle of the beef tapeworm (*Taenia*)

Preserved specimens:
  Ascaris
  Tapeworms
Prepared slides of:
  Dugesia, w.m., injected gastrovascular cavity
  Taenia sp. scolex, w.m.
  Taenia sp. larvae (cysticerci) encysted in muscle
  Taenia sp. gravid proglottid

Assignment 3

1. Use a camel's hair brush to place a living planarian in a watch glass with a small amount of water. Examine it with a dissecting microscope. Note the eyespots, auricles, and manner of movement. Touch the posterior end gently with a pencil and note the response. Touch an auricle with a pencil and note the response.
2. Examine a prepared slide of *Dugesia* and note the highly branched gastrovascular cavity.
3. Your instructor has placed a few *Dugesia* in water in a partially shaded Petri dish. Lift the light shield to observe their distribution. Do they prefer shade or light? ***Complete items 3a–3d on the laboratory report.***
4. Examine prepared slides of *Taenia* encysted larvae, scolex, and gravid proglottids. Note the hooks and suckers on the scolex and the dark eggs in the expanded uterus of the gravid proglottid. Study the life cycle of the beef tapeworm in Figure 26.8. ***Complete items 3e and 3f on the laboratory report.***
5. Observe the demonstration flatworms. Estimate the lengths of the tapeworms. Note the infestation sites of the adult flukes. ***Complete item 3 on the laboratory report.***

ROUNDWORMS (Phylum Nematoda)

Roundworms are advanced over flatworms by having an **organ system level** of development, a **tube-within-a-tube body plan,** and a **false coelom.** Animals with a tube-within-a-tube body plan have a complete digestive tract that has two openings: a **mouth** at one end and an **anus** at the other. The space between the digestive tract and the body wall is the **coelom** or body cavity. It is a **false coelom** (pseudocoelom)

in roundworms, because it is not completely lined with mesodermal tissue. See Figure 26.9. Sexes are separate. Most roundworms are free-living, but some are important parasites of plants and animals.

Ascaris is a large roundworm parasite of swine and humans. Adults live in the small intestine, where they feed on partially digested food. An external cuticle protects *Ascaris* from digestive juices of the host. Dorsal and ventral nerve cords coordinate body movements. Metabolic wastes are collected in laterally located excretory canals and excreted through an excretory pore near the anterior end of the body. See the internal structure of a female in Figure 26.10. Note that the reproductive system is the dominant feature, a common trait of endopara-

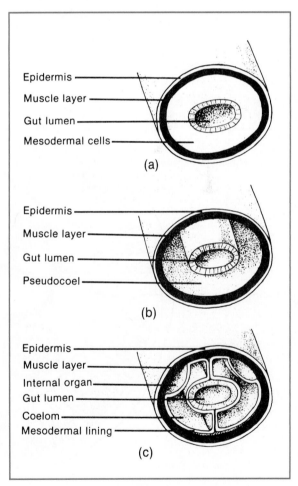

Figure 26.9. Acoelomate (a), pseudocoelomate (b), and eucoelomate (c), conditions in bilaterally symmetrical animals

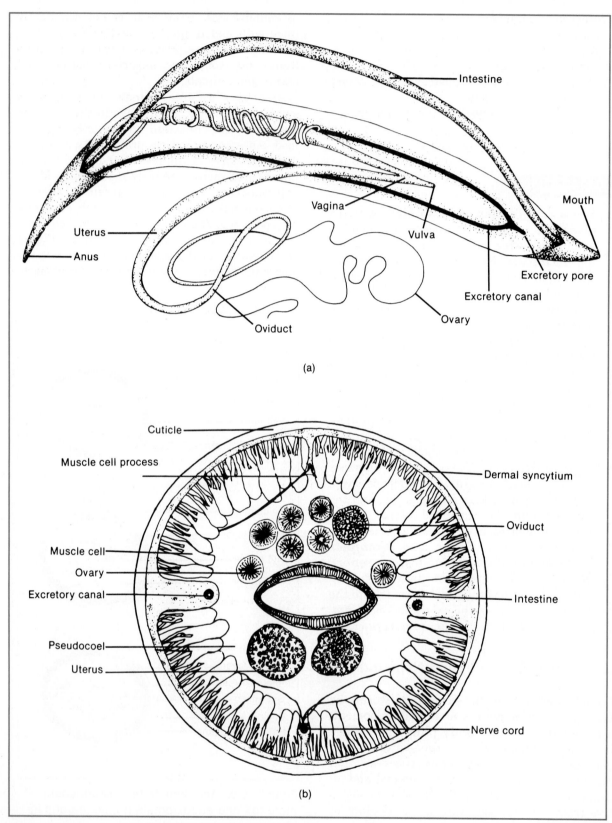

Figure 26.10. Female *Ascaris*. (a) Dissection. (b) Cross section.

sites. Each tubular half of the bipartite uterus is continuous with an oviduct and ovary.

A female *Ascaris* is about 30 cm in length. A male is smaller and has a curved posterior end. After mating in the small intestine, the female releases eggs that are passed in the host's feces. New hosts are infected by accidentally eating the eggs. Note the life cycle in Figure 26.11.

There are a number of important roundworm parasites infesting humans. Infestation by one of these, *Trichinella spiralis,* occurs by eating undercooked pork or wild game in which microscopic encysted larvae are present. The larvae emerge in the intestines and attach to the intes-

tinal wall and grow into adults that feed by sucking blood from the intestinal wall. After copulation, females release larvae that burrow into blood vessels of the intestine. The larvae are carried by the blood to skeletal muscles, where they burrow through the vessel walls and encyst within muscle tissue. Note that each host is infested with both adult and larval forms.

Materials

Per student
Compound microscope
Dissecting instruments

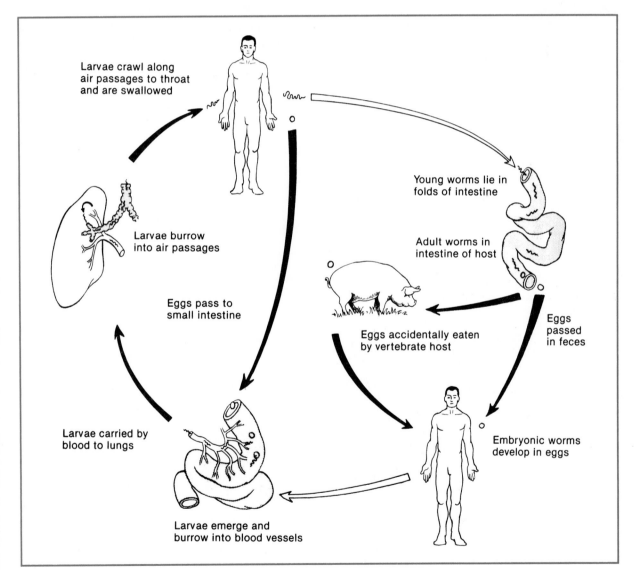

Larvae crawl along air passages to throat and are swallowed

Larvae burrow into air passages

Eggs pass to small intestine

Larvae carried by blood to lungs

Larvae emerge and burrow into blood vessels

Young worms lie in folds of intestine

Adult worms in intestine of host

Eggs accidentally eaten by vertebrate host

Eggs passed in feces

Embryonic worms develop in eggs

Figure 26.11. *Ascaris* life cycle

Per lab
Dissecting pans, wax bottomed
Dissecting pins
Medicine droppers
Microscope slides and cover glasses
Culture of *Turbatrix*
Demonstration setups of:
  Ascaris, dissected female
  representative roundworms
Preserved *Ascaris*
Prepared slides of:
  Ascaris, x.s.
  Trichinella, encysted larvae
Representative roundworms

Assignment 4

1. Make a slide of a drop of *Turbatrix* culture and observe these free-living roundworms that can live in vinegar. Note their swimming motion. ***Complete item 4a on the laboratory report.***

2. Examine the preserved male and female *Ascaris*. Note their sizes and distinctive shapes.
3. Examine the demonstration dissected female *Ascaris*. Locate the parts shown in Figure 26.10a.
4. Examine a prepared slide of *Ascaris* female, x.s., and locate the structures labeled in Figure 26.10b. ***Complete items 4b–4d on the laboratory report.***
5. Examine a prepared slide of *Trichinella* larvae encysted in muscle tissue. Can you see the larvae with the naked eye? ***Complete item 4 on the laboratory report.***

Assignment 5

Examine the numbered "unknown" specimens of these animal groups and identify the phylum to which each belongs. ***Complete item 5 on the laboratory report.***

Materials

Per student
Colored pencils

Per lab
Demonstration dissections of a freshwater clam
Representative mollusks that are numbered but
not labeled
Squid, fresh or frozen

Assignment 1

1. Color-code the digestive tract, heart and
blood vessels, and kidney of the clam in Figure 27.1.
2. Examine the demonstration dissected clam
and locate as many of the structures labeled
in Figure 27.1 as possible.
3. Examine a squid and note how it is adapted
for a predaceous mode of life. The foot is modified into arms and tentacles with suction
cups. The body is streamlined, and swimming
is powered by water-jet propulsion as water is
ejected forcefully from the siphon. The shell
is reduced to a thin supporting membrane,
and the image-forming eyes resemble vertebrate eyes but are formed differently.
4. Examine the representative mollusks and
use Table 27.1 to determine the class of each.
***Complete item 1 on Laboratory Report 27
that begins on page 399.***

ANNELIDS (Phylum Annelida)

The major characteristics of this phylum is a
segmented body formed of repeating units
called **somites.** Most somites are basically similar, including the arrangement of many internal
structures. Most annelids are marine, but some
occur in freshwater and terrestrial habitats. Table 27.2 lists the characteristics of the major
classes of annelids.

The most familiar annelids are earthworms
(class Oligochaeta) that literally eat their way
through the soil, digesting whatever organic
material is ingested. Each somite has four pairs
of **setae** (bristles) that provide traction for a
crawling worm. The smooth band around the
body is the **clitellum.** It secretes a mucus band
around copulating worms and later secretes an
egg case for the fertile eggs.

Figures 27.2, 27.3, and 27.4 show the internal
structures of an earthworm. Note the specializations of the anterior portion of the digestive
tract. For example, the **pharynx** aids the ingestion of food, the crop temporarily holds food, and
the **gizzard** grinds food into tiny particles. Digestion is completed in the **intestine,** and nutrients are absorbed into the blood for transport.
Annelids have a **closed circulatory system,** so
blood is always contained within blood vessels.

Longitudinal and circular muscles occur in
the body wall and in the wall of the digestive
tract. Contractions are coordinated by the ner-

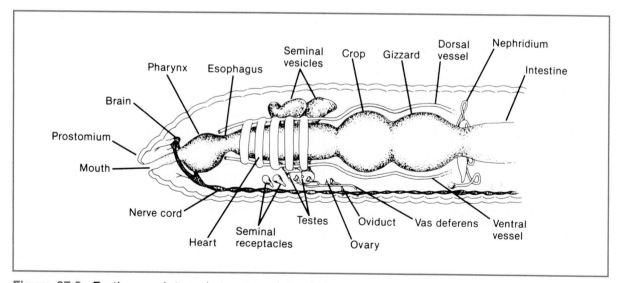

Figure 27.2. Earthworm internal structure, lateral view

TABLE 27.2
Distinguishing Characteristics of Annelid Classes

Class	Characteristics
Polychaeta	Head with simple eyes and tentacles; segments with lateral extensions (parapodia) and many bristles (setae); sexes usually separate; predaceous; marine
Oligochaeta	No head; segments without extensions and with few, small bristles; hermaphroditic; terrestrial or freshwater
Hirudinea (leeches)	No head; segments with superficial rings and without lateral extensions or bristles; anterior and posterior suckers; hermaphroditic; parasitic, feeding on blood; terrestrial or freshwater

vous system. Metabolic wastes are removed by a pair of **nephridia** in each somite. Gas exchange occurs by diffusion through the moist epidermis.

Earthworms are hermaphroditic, and they exchange sperm during copulation. **Seminal vesicles** hold sperm formed by testes until they are released during copulation. **Seminal receptacles** receive sperm during copulation and later release sperm to fertilize eggs as they are released from the oviducts.

The nervous system includes a "brain" (supraesophageal ganglia) and a double **ventral nerve cord** with **segmental ganglia** and lateral nerves.

Materials

Per student
Colored pencils
Dissecting instruments

Per lab
Dissecting pans, wax bottomed
Dissecting pins
Representative annelids that are numbered but not labeled
Preserved earthworms

Assignment 2

1. Study the internal structure of an earthworm in Figures 27.2–27.4. Color-code the digestive tract, blood vessels, and nephridia in Figures 27.2 and 27.4.
2. *Complete items 2a-2c on the laboratory report.*
3. Obtain an earthworm and place it in your dissecting pan. Locate the somites, mouth, anus, setae, and clitellum.
4. Dissect the earthworm as follows. Locate the structures shown in Figure 27.2.
   a. Pin the earthworm dorsal (convex) side up to the bottom of the dissecting pan by placing a pin through the prostomium, a small projection over the mouth, and another pin through a somite posterior to the clitellum.
   b. Use a sharp scalpel to cut through the body wall, *only,* along the dorsal midline from the prostomium to the clitellum. *Be careful not to damage underlying organs.*
   c. Use a dissecting needle to break the septa

Figure 27.3. Dorsal view of dissected earthworm

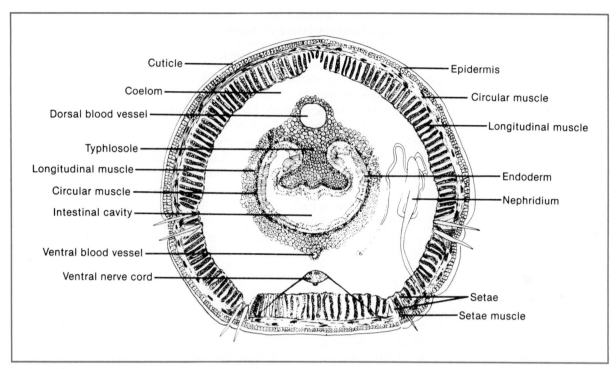

Figure 27.4. Earthworm (*Lumbricus*), x.s.

separating the somites as you spread out and pin down the body wall to expose the internal organs.

d. Add water to cover the worm to prevent it from drying out.

e. Locate the structures labeled in Figures 27.2–27.4. The ovaries and testes and their associated ducts are tiny and will be difficult to see.

f. Carefully use a dissecting needle and forceps to remove tissue to expose the brain.

g. Remove the crop, gizzard, and part of the intestine to expose the ventral nerve cord with its segmental ganglia.

h. Cut open the crop and gizzard and compare the thickness of their walls.

i. Dispose of the worm as directed by your instructor.

5. Examine the representative annelids and use Table 27.2 to determine the class of each. ***Complete item 2 on the laboratory report.***

ARTHROPODS (Phylum Arthropoda)

There are more species of arthropods than all other animals combined. The characteristics en-abling such success include a well-developed nervous system, complex sense organs, and a fusion of body segments to form specialized body regions. In addition, all arthropods exhibit these distinguishing characteristics: (1) **jointed appendages,** (2) an **exoskeleton** of chitin, (3) **compound** or **simple eyes,** and (4) an **open circulatory system.** Table 27.3 lists the characteristics of the major classes of arthropods.

Crayfish (Crustacea)

The freshwater crayfish exhibits the basic arthropod and crustacean characteristics. Note its external structure in Figure 27.5. The segmented body is divided into a **cephalothorax,** covered by the carapace, and an **abdomen.** Each body segment has a pair of jointed appendages, and most are modified for special functions. For example, **antennae** contain chemical and tactile sensory receptors, **compound eyes** are photoreceptors, **mouthparts** are used in feeding, **walking legs** enable movement and grasping, and **uropods** are used in swimming. The first pair of walking legs, the **chelipeds,** is modified for capturing food. **Swimmerets** on abdominal appendages are relatively unmodified,

TABLE 27.3
Distinguishing Characteristics of Arthropod Classes

Class	Characteristics
Arachnida (spiders, scorpions)	Cephalothorax and abdomen; four pairs of legs; simple eyes; no antennae; mostly terrestrial
Crustacea (crabs, shrimp)	Cephalothorax and abdomen; compound eyes; two pairs of antennae; five pairs of legs; mostly marine or freshwater
Chilopoda (centipedes)	Elongate, dorsoventrally flattened body; each body segment with a pair of legs; one pair of antennae; simple eyes; terrestrial
Diplopoda (millipedes)	Elongate, dorsally convex body; segments fused in pairs, giving *appearance* of two pairs of legs per segment; one pair of antennae; simple eyes; terrestrial
Insecta (insects)	Head, thorax, and abdomen; three pairs of legs on thorax; wings, if present, on thorax; simple and compound eyes; one pair of antennae; freshwater or terrestrial

but the first two pairs are modified for sperm transfer in males.

The exoskeleton is hardened for protection, but it remains flexible at joints, enabling movement. It is shed periodically to allow growth. The carapace covers the **gills** (gas exchange organs) that are located laterally attached to either the body wall or the bases of the walking legs.

Figure 27.6 shows the internal structure of a crayfish. Note the **heart** and **arteries** of the open circulatory system. Blood is emptied into

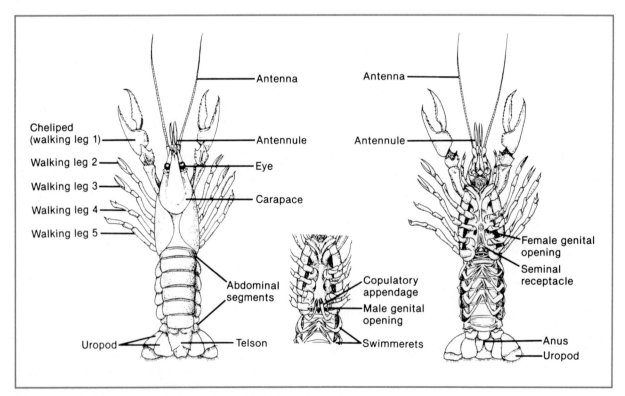

Figure 27.5. External structure of crayfish (*Cambarus*)

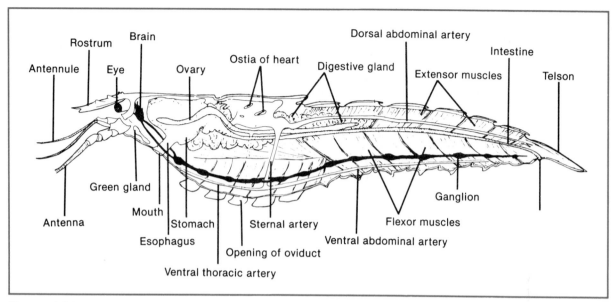

Figure 27.6. Internal structure of female crayfish

the coelom and slowly returns to reenter the heart through tiny valved openings, the **ostia.** A pair of elongated **gonads,** either ovaries or testes, lies just ventral to the heart. A large **digestive gland** surrounds the gonads and **stomach,** filling much of the cephalothorax. Metabolic wastes are removed by **green glands** located near the bases of the antennae. The highly developed nervous system includes a **"brain"** (supraesophageal ganglia) and a double ventral nerve cord with segmental ganglia and lateral nerves.

Grasshopper (Insecta)

A grasshopper illustrates typical insect characteristics. Note in Figure 27.7 that the body is divided into three distinct parts: **head, thorax,** and **abdomen.** The head contains both simple and compound eyes, only one pair of antennae, and the **mouthparts** that are modified appendages. The three pairs of legs and two pairs of wings are attached to the thorax. (All insects do not have wings.) Visible on the abdomen are the **tympanum,** a sound receptor comparable to your eardrum, and **spiracles** that are openings into the **tracheal system,** a complex of tubules that carry air directly to body tissues.

Internally, the digestive tract is divided into crop, stomach, and intestine. **Gastric caeca** are glands that aid digestion. Metabolic wastes are

removed from the hemolymph by **Malpighian tubules** into the intestine. Hemolymph is pumped by a series of "hearts" anteriorly through the "aorta" and emptied into the coelom. It then slowly returns through the coelom to the "hearts." Note the double **ventral nerve cord** with **segmental ganglia** and the **"brain"** (supraesophageal ganglia).

Most insects undergo **metamorphosis,** a hormonally controlled process that changes the body form at certain stages of the life cycle. In **complete metamorphosis,** as in butterflies, there are four life stages: egg; larva, a wormlike feeding stage; pupa, a nonfeeding stage; and adult. In **incomplete metamorphosis,** as in grasshoppers, the pupal stage is absent. The egg hatches into a nymph (in terrestrial forms) or naiad (in aquatic forms) that grows through several molts, with the adult emerging from the last molt.

Materials

Per student
Colored pencils

Per lab
Representative arthropods that are labeled
Representative arthropods that are numbered but not labeled
Preserved crayfish and grasshoppers

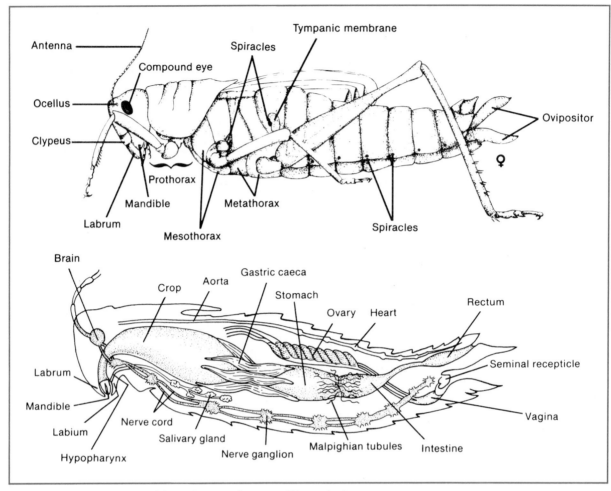

Figure 27.7. Anatomy of female grasshopper (*Romalea*)

Assignment 3

1. **Complete item 3a on the laboratory report.**
2. Examine a preserved crayfish and locate the external structures shown in Figure 27.5. Examine an eye with a dissecting microscope. Note the individual facets forming this compound eye. Note the differences of the walking legs. **Complete item 3b on the laboratory report.**
3. Color-code the digestive tract, heart and blood vessels, and green gland of the crayfish in Figure 27.6 and study its internal structure.
4. Dissect a crayfish as described below and locate the parts labeled in Figure 27.6.

a. Hold the crayfish dorsal side up with its head facing away from you. Use scissors to make incision 1 as shown in Figure 27.8. This removes the left side of the carapace, exposing the gills.
b. Note the gills. How are they arranged? To what are they attached? Note the thin, transparent body wall medial to the gills.
c. Make incision 2 by extending incision 1 from the cervical groove to the eye and downward to remove the left anterior lateral part of the carapace. This exposes the lateral wall of the stomach and a large mandibular muscle.
d. Make incision 3 as shown and *gently* lift off the median strip of carapace to expose

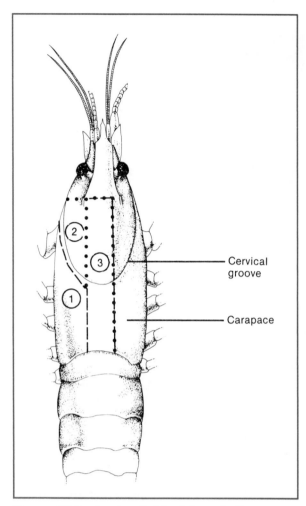

Figure 27.8. Sequential incisions in the crayfish dissection

the heart. Note the arteries extending from the heart and the ostia.

e. Carefully remove organs as you work your way from dorsal to ventral, identifying each organ encountered. When the nerve cord is exposed, locate the segmental ganglia.

f. Make a transverse cut through the abdomen and examine the cut surface. Locate the muscles, intestine, and nerve cord.

g. Dispose of your crayfish as directed by your instructor.

5. Obtain a preserved grasshopper. Locate the external structures shown in Figure 27.7. *Complete item 3c on the laboratory report.*

6. Examine the labeled representative arthropods. Compare them with Table 27.3.

7. Use Table 27.3 to determine the class of the "unknowns." *Complete item 3 on the laboratory report.*

ECHINODERMS AND CHORDATES

Echinoderms and chordates are called **deuterostomates** because the embryonic blastopore becomes the anus, and the mouth is formed from the second embryonic opening to appear. This embryonic characteristic suggests that echinoderms are more closely related to chordates than to protostomates.

ECHINODERMS
(Phylum Echinodermata)

These "spiny-skinned" animals are named after the spines that protrude from their calcareous **endoskeleton** that lies just under a thin epidermis. All echinoderms are marine. A unique characteristic is their **water vascular system.** Although larval stages are bilaterally symmetrical, adults are radially symmetrical. Adults have an **oral surface,** where the mouth is located, and an **aboral surface,** where the anus is located. There are no anterior or posterior ends nor dorsal or ventral surfaces. The characteristics of the classes of echinoderms are listed in Table 28.1.

Note the echinoderm characteristics in the dissected sea star in Figure 28.1. The typical pentamerous (five-part) structure is obvious by the five **arms** that radiate from a **central disc.** The water vascular system consists of interconnecting tubes filled with seawater. Seawater is filtered as it enters through the **madreporite. Radial canals** extend into each arm from the **ring canal,** and the **lateral canals** end in **tube feet** that have suction cups at their ends. Contraction and relaxation of an **ampulla** extends and retracts a tube foot. The oral surface of each arm has an **ambulacral groove** that is bordered by tube feet.

A very short digestive tract of mouth, stomach, intestine, and anus is housed in the central disc, but a pair of digestive glands fills most of each arm. A pair of gonads is also located in each arm. Sexes are separate. Nervous and circulatory systems are greatly reduced.

TABLE 28.1
Classes of Echinoderms

Class	Characteristics
Asteroidea (sea stars)	Star-shaped forms with broad-based arms; ambulacral grooves with tube feet that are used for locomotion
Ophiuroidea (brittle stars)	Star-shaped forms with narrow-based arms that are used for locomotion; ambulacral grooves covered with ossicles or absent
Echinoidea (sea urchins)	Globular forms without arms; movable spines and tube feet for locomotion; mouth with five teeth
Holothuroidea (sea cucumbers)	Cucumber-shaped forms without arms or spines; tentacles around mouth
Crinoidea (sea lilies)	Body attached by stalk from aboral side; five arms with ciliated ambulacral grooves and tentaclelike tube feet for food collection; spines absent

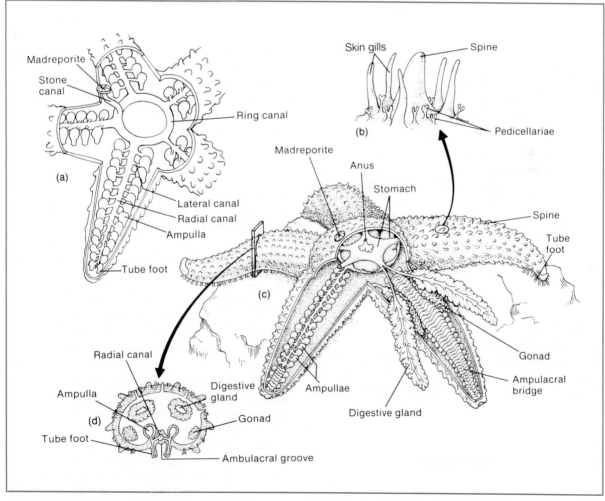

Figure 28.1. A sea star (*Asterias*). (a) Water vascular system. (b) Skin gills and pediceillariae. (c) Dissection. (d) Arm cross section.

Materials

Per student
Colored pencils
Dissecting instruments
Dissecting pans
Preserved sea star

Per lab
Representative echinoderms that are numbered
but not labeled

Assignment 1

1. **Complete items 1a and 1b on Laboratory
   Report 28 that begins on page 401.**
2. Study the structure of a sea star in Figure 28.1 and color-code the water vascular
   system.
3. Obtain a sea star and note its external structure. Perform the dissection as follows to locate the structures labeled in Figure 28.1.
   a. Use scissors to cut off one of the arms. Examine the cut end and locate the structures labeled in Figure 28.1d.
   b. Locate the anus and madreporite on the aboral surface of the central disc and the mouth on the oral surface.
   c. Use scissors to remove the aboral surface of the central disc. Locate the short intestine, stomach, and attachment of the digestive glands.
   d. Use scissors to remove the aboral surface of one of the arms and locate the digestive glands and gonads. Then locate the ambulacral bridge and the ampulla of the tube feet.
   e. Remove the stomach to locate the ring canal.
   f. Dispose of the sea star as indicated by your instructor.
4. Examine the representative echinoderms and use Table 28.1 to identify the classes to which they belong. **Complete item 1 on the laboratory report.**

CHORDATES (Phylum Chordata)

The four distinctive characteristics of chordates are present in the embryo but may not persist in the adult: (1) a **dorsal tubular nerve cord,** (2) **pharyngeal gill slits,** (3) a **notochord,** and (4) a **post-anal tail.** There are three major groups of chordates: tunicates, lancelets, and vertebrates.

Tunicates (Subphylum Urochordata)

All chordate characteristics are present in the motile larvae but absent in the nonmotile adults. See Figure 28.2. Both larval and adult tunicates (sea squirts) are filter feeders with pharyngeal gills providing both gas exchange and food collection. Collected food particles are passed by cilia into the stomach.

Lancelets (Subphylum Cephalochordata)

In contrast, all chordate characteristics are present in adult lancelets, such as amphioxus shown in Figure 28.3. The notochord, an evolutionary forerunner of a vertebral column, provides support for the body, and the segmentally arranged muscles enable a weak swimming motion. Note the simple dorsal and caudal fins.

Amphioxus lives buried in mud with its anterior end sticking out. Water is brought in through the buccal cavity, passes over the gills into the atrium, and flows out the atriopore. Like tunicates, lancelets are filter feeders. The pharyngeal gills serve as both gas exchange organs and food collectors. Food particles are moved by cilia into the intestine.

Vertebrates (Subphylum Vertebrata)

You are most familiar with this group of animals. Adult vertebrates may not exhibit all of the four chordate characteristics, although they are present in their embryos. A vertebrate body is usually divided into **head, trunk,** and **tail,** but a tail is not present in all adults. Most forms have **pectoral** and **pelvic appendages.** An **endoskeleton** of cartilage or bone provides support, including a **vertebral column** that usually completely replaces the notochord.

The **brain** is encased in a cranium, and the **nerve cord** is surrounded by the vertebral column. Metabolic wastes are removed by **kidneys,** and gas exchange occurs by either **pharyngeal gills** or **lungs.** The **closed circulatory system** includes a ventral heart of 2–4 cham-

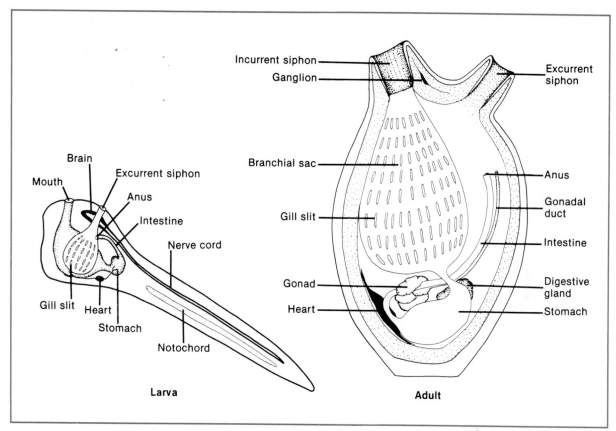

Figure 28.2. A tunicate (*Ciona*)

bers. The distinctive characteristics of vertebrate classes are listed in Table 28.2.

Jawless Fish (Class Agnatha). Present day agnaths (lampreys and hagfish) are degenerate forms living as scavengers and parasites on other fish, which probably explains the absence of scales and paired fins and the presence of a cartilaginous skeleton. External features of a lamprey are shown in Figure 28.4.

Cartilaginous Fish (Class Chondrichthyes). A side branch of jawed bony fish that appeared about 425 million years ago, sharks and rays have an endoskeleton of cartilage. Thick **pectoral** and **pelvic fins** provide stabilization when swimming. **Dermal scales,** a basic fish characteristic, are formed from the dermis of the skin. They are thick and heavy and provide protection without hindering body flexibility.

Bony Fish (Class Osteichthyes). Modern bony fish are better adapted for a swimming lifestyle. They have **lightweight dermal scales** that provide protection while reducing body weight and a **swim bladder** that provides controllable buoyancy. The **bony endoskeleton** is also usually lightweight. As in all fish, a **lateral line system** detects underwater vibrations.

Amphibians (Class Amphibia). Amphibians (frogs, toads, and salamanders) usually live in moist environments near ponds and streams. They must return to an aquatic habitat to reproduce since fertilization is external and embryonic and larval development requires free water. The larvae have gills as gas exchange organs and a fishlike tail for swimming. These features are absorbed and legs and lungs develop as amphibians mature into air-breathing adults. Adults lack scales, so excessive water loss through the skin is avoided by living in a moist habitat.

Reptiles (Class Reptilia). Reptiles were the first truly land animals because they do not have to return to water to reproduce. Fertilization is internal, and **leathery shelled eggs** (amniote eggs) contain nutrients and a "private

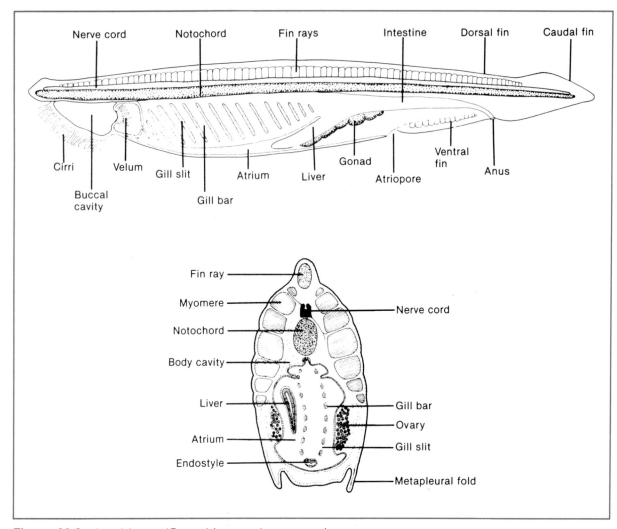

Figure 28.3. Amphioxus (*Branchiostoma*), w.m. and x.s.

pond" for the developing embryo. Further, they have **epidermal scales** that retard water loss through the skin, so they can live in dry habitats. **Claws** on their toes are modified epidermal scales. Ribs join to the sternum to form a **rib cage** that provides protection for the heart and lungs. Like fish, amphibians, and all lower animals, reptiles are **poikilothermic** (body temperature fluctuates with the ambient temperature) and **ectothermic** (heat for body temperature comes from external sources).

Birds (Class Aves). Birds possess many adaptations for flight, including (1) **feathers** (derived from epidermal scales) providing a lightweight airfoil, (2) hollow, lightweight bones, (3) air sacs associated with the lungs, enhancing buoyancy, (4) a fused skeleton except for neck, tail, and appendages, and (5) pectoral appendages modified into **wings.** Jaws are modified into a **horny beak** lacking teeth. Fertilization is internal, and hard-shelled eggs contain nutrients and an aqueous habitat for embryonic development. Parental care is greatly increased.

Unlike lower forms, birds are **homeothermic** (body temperature is relatively constant) and **endothermic** (heat from cellular respiration maintains body temperature).

Mammals (Class Mammalia). The key characteristics of mammals are **hair, mammary glands,** and a **diaphragm** that separates

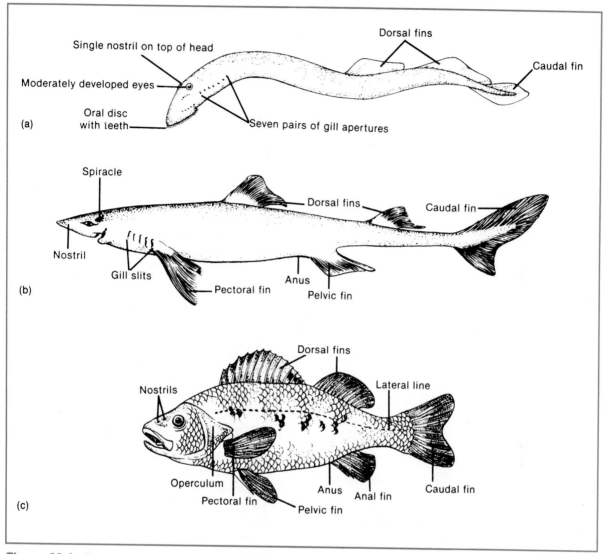

Figure 28.4. Representative fish. (a) Lamprey (*Petromyzon*). (b) Dogfish shark (*Squalus*). (c) Perch (*Perca*).

the abdominal and thoracic cavities. Like birds, mammals are homeothermic and endothermic.

Fertilization is internal and, except for primitive forms, offspring develop in the uterus of the female and receive nutrients from the mother via a placenta. Female mammals nurse their young for variable lengths of time after birth.

Materials

Per student
Colored pencils
Dissecting microscope

Per lab
Demonstration specimens of lamprey, shark, perch, and frog life cycle
Demonstration dissections
   lamprey, cross section through the tail
   shark, cross section through the tail
Demonstration microscope setups of:
   tunicate (ammocoete) larva
   amphioxus, w.m.
Preserved amphioxus and lamprey
Representative chordates that are numbered but not labeled

TABLE 28.2
Classes of Vertebrates

Class	Characteristics
Agnatha (jawless fish)	No jaws; scaleless skin; median fins only; cartilaginous endoskeleton; persistent notochord; two-chambered heart
Chondrichthyes (cartilaginous fish)	Jaws; subterminal mouth; placoid dermal scales; median and paired fins; cartilaginous endoskeleton; two-chambered heart
Osteichthyes (bony fish)	Jaws; terminal mouth; membranous median and paired fins; bony endo-skeleton; lightweight dermal scales; two-chambered heart; operculum covers gills; swim bladder
Amphibia (amphibians)	Aquatic larvae with gills; air-breathing adults with lungs; scaleless skin; paired limbs; bony endoskeleton; three-chambered heart
Reptilia (reptiles)	Bony endoskeleton with rib cage; well-developed lungs; epidermal scales; claws; paired limbs in most; leathery-shelled eggs; three-chambered heart
Aves (birds)	Wings; feathers, epidermal scales; claws; endoskeleton of lightweight bones; rib cage; horny beak without teeth; hard-shelled eggs; four-chambered heart
Mammalia (mammals)	Hair; reduced or absent epidermal scales; claws or nails; mammary glands; diaphragm; bony endoskeleton with rib cage; young develop in uterus and are nourished via placenta; four-chambered heart

Assignment 2

1. Color-code the nerve code green, the noto-chord blue, and pharyngeal slits red in the tunicates in Figure 28.2 and in the amphi-oxus in Figure 28.3.
2. Examine the slide of a tunicate larva set up under a demonstration microscope. Compare it with Figure 28.2 and locate the four chor-date characteristics.
3. Obtain a preserved amphioxus and examine it with a dissecting microscope. Locate the opening to the buccal cavity, the dorsal and caudal fins, and the segmentally arranged body muscles.
4. Examine a slide of amphioxus, w.m., set up under a demonstration microscope. Compare it with Figure 28.3 and locate the four chor-date characteristics. **Complete items 2a and 2b on the laboratory report.**

5. Compare a lamprey, shark, and perch as to: (a) type of fins, (b) number of gill slits, (c) type of dermal scales, and (d) location of jaws, if present. Locate the lateral line. **Complete items 2c–2f on the laboratory report.**
6. Examine cross sections through the tail of a lamprey and a shark. Locate the cartilagi-nous vertebrae, nerve cord, and notochord, if present. **Complete item 2g on the labora-tory report.**
7. Examine the demonstration of the frog life cycle. Why are amphibians considered to be poorly adapted to terrestrial life? **Complete items 2h–2m on the laboratory report.**
8. Examine the representative vertebrates and use Table 28.2 to identify the class to which each belongs. **Complete the laboratory re-port.**

PART V

EVOLUTION AND ECOLOGY

EVOLUTION

After completion of the laboratory session, you should be able to:
1. Identify each major organismic group and the geologic era and period in which it appeared.
2. Identify and compare the prehuman and human fossils believed to be in the evolutionary pathway leading to modern man.
3. Perform a comparative skull analysis of fossil and modern hominids.
4. Define all terms in bold print.

In 1858, Darwin presented his theory of evolution by natural selection, supported by a mass of evidence from various sources. Much of his evidence is still considered valid today. It was drawn from the fields of **paleontology, comparative morphology, comparative embryology,** and **biogeography.** Important new evidence has since been obtained from the fields of **biochemistry, physiology,** and **genetics.**

The fact that evolution has occurred since the first appearance of living organisms is well established, although biologists may quibble over the exact mechanisms producing it. Two current hypotheses are gradualism (microevolution) and punctuationalism (macroevolution). The **gradu-**alistic hypothesis proposes that new species result from an accumulation of continual, successive, small genetic changes. The **punctuational hypothesis** proposes that species are rather static except during relatively brief periods when substantive genetic changes occur to form new species and higher taxonomic categories. These two processes may both play a role in evolution.

THE FOSSIL RECORD

The fossil record shows changes that have occurred over millions of years and establishes the **phylogeny** (evolutionary relationships) in the descent from simpler to more complex organisms. Of course, the fossil record is incomplete. Considering that extremely few organisms become fossilized, that relatively few fossils survive normal geologic processes, and that the discovery of such fossils is limited, the fossil record is surprisingly good.

A fossil may be prints or casts left by organisms or an organism itself whose tissues have been replaced by minerals. The age of fossils is determined by establishing the age of the sedimentary rock in which they are found. Age-dating methods using the normal decay of radioactive elements have been especially useful.

A summary of the fossil record for a few major organismic groups is shown in Figure 29.1. The

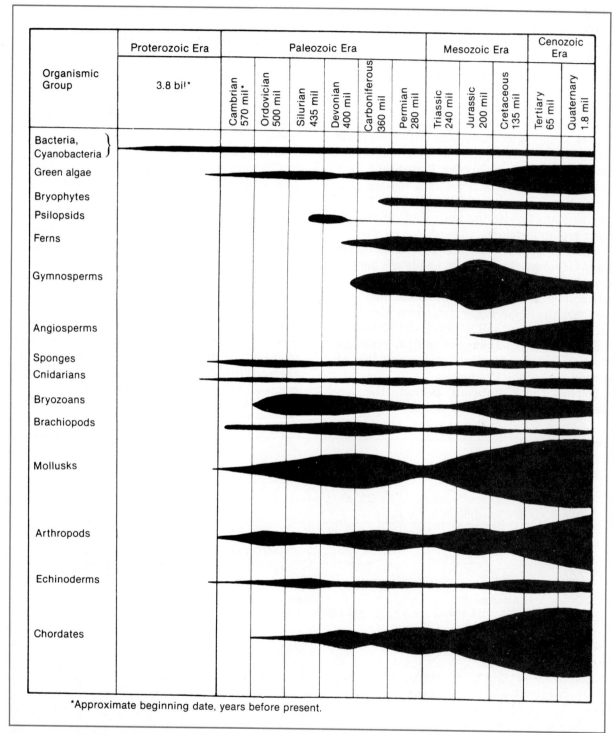

*Approximate beginning date, years before present.

Figure 29.1. Fossil record of major organismic groups

length of the bars indicates the appearance and persistence of the groups, and the width of the bars shows the number of species present.

The geographic distribution of fossil and modern organisms is largely explained by the movement of the continents. The earth's crust is divided into seven major plates and a number of smaller plates. The relative movement of these plates accounts for movement of the continents and variations in climate of a particular land mass during geologic history. Figure 29.2 shows the relative positions of the land masses at selected times in geologic history. Refer to this figure as you consider the key events of the geologic era and periods.

The **Proterozoic era** was dominated by prokaryotic forms. The earliest fossils are bacteria from 3.5 billion years ago. Unicellular green algae and fungi appeared about 900 million years ago. Multicellular algae date from about 650 million years ago, and multicellular soft-bodied animals appeared about 600 million years ago. Note the number of organismic groups that appeared during the Proterozoic.

During the **Cambrian period,** the land masses were located in tropical latitudes.

Gondwana was a very large land mass that wrapped around about half of the earth. **Laurentia** was the smaller land mass that later became part of North America. The warm, shallow seas flourished with multicellular algae and marine arthropods (crustaceans), especially trilobites. Other marine animals included sponges, coelenterates, brachiopods, mollusks, and echinoderms.

In the **Ordovician,** marine algae and invertebrates were dominant. Trilobites, primitive marine arthropods, were abundant, and the radiation of brachiopods and bryozoans was especially rapid. Cephalopods and jawless fish, the first vertebrates, appeared. Later in this period, Gondwana began moving toward the South Pole.

In the **Silurian,** Gondwana was concentrated near the South Pole, and later began its movement northward. The colonization of land by plants began. The psilopsids, which include club mosses and horsetails, were probably the first vascular plants. Jawed fish appeared, and arthropods invaded the land. Many marine crustaceans and cephalopods became extinct.

The **Devonian** was dominated by fish. The

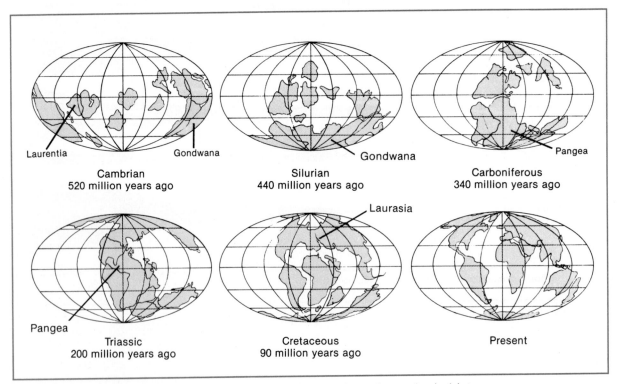

Figure 29.2. Positions of the land masses at selected times in geologic history

various fish groups appeared, including lungfish and lobe-finned fish. By the end of the period, the first amphibians had evolved from lobe-finned fish, and the vertebrate invasion of the land began. Among land plants, club mosses and horsetails were common, and ferns made their appearance, apparently evolving from psilopsids. Later in the period, seed ferns and gymnosperms appeared, apparently evolving from ferns.

The **Carboniferous** was the age of amphibians, when immense swamp forests of club mosses, horsetails, and ferns dominated the low-lying land masses. The covering of these coastal forests by sand and silt has formed the coal deposits of today. Insects radiated rapidly, and the reptiles arose from amphibians. The appearance of bryophytes in this period, long after the evolution of vascular plants, is an exception to the general pattern of evolutionary events.

About 300 million years ago, the land masses joined together to form a giant, unstable land mass called **Pangea** that remained intact until late in the Mesozoic era. At the beginning of the **Triassic,** a period of mountain building and reduction of the land-sea interface (due to the land masses joining together) produced cooler climates and resulted in the extinction of many species. Over 75% of the marine invertebrate species disappeared with the reduction of the shallow seas, and many land forms became extinct as well. Species of reptiles increased, but species of amphibians decreased. This large-scale extinction is known as the Permo-Triassic Crisis.

The **Mesozoic era** was the age of dinosaurs. They appeared in the **Triassic,** reached their peak in the early **Cretaceous,** and then rather quickly became extinct. The radiation of reptiles produced many diverse groups. Therapsids gave rise to the first mammals in the late Triassic. Birds evolved from thecodont reptiles in the **Jurassic.** Gymnosperms and ferns were the dominant land plants. Angiosperms appeared in the Triassic, possibly evolving from seed ferns, and became common in the Cretaceous as gymnosperms and ferns declined.

In the late Mesozoic, Pangea broke up, and the land masses began to disperse toward their present positions. The **Laramide Revolution,** a time of volcanic activity and mountain building, contributed to a cooler climate. The oceans re-ceded from the land masses. Fragmentation of the extensive tropical forests led to the emergence of grasslands in the mid-Cenozoic.

The **Cretaceous extinction** occurred at this time (65 million years ago). Over half of the marine species and many land forms, including dinosaurs, became extinct. Evidence suggests that an asteroid collided with the earth at about this time, but just how this contributed to the mass extinction is not clear.

The climatic change and the extinctions opened up new vistas for surviving forms. Angiosperms became the dominant land plants, and mammals and birds became the dominant vertebrates in the **Tertiary.**

Materials

Per lab
Representative fossils

Assignment 1

1. Study Figures 29.1 and 29.2. Note the initial appearance of each group and the variation in the number of fossil species. *Complete items 1a and 1b on Laboratory Report 29 that begins on page 403.*
2. Study Figure 29.3, and compare it with the preceding discussion.
3. Examine the fossils available. Mark the position of each fossil on Figure 29.1 and *complete item 1 on the laboratory report.*

HUMAN EVOLUTION

The pattern of human evolution is not clear, and the fossil record is sparse. Considerable disagreement exists among scientists regarding the position of some of the fossil forms, but they do agree on the following points:

1. Modern human's closest living relatives are the chimpanzee and gorilla.
2. The great apes and hominids (humans and humanlike forms) evolved from a common ancestor.
3. Hominid evolution is characterized by adaptive radiation and a branching evolutionary tree.

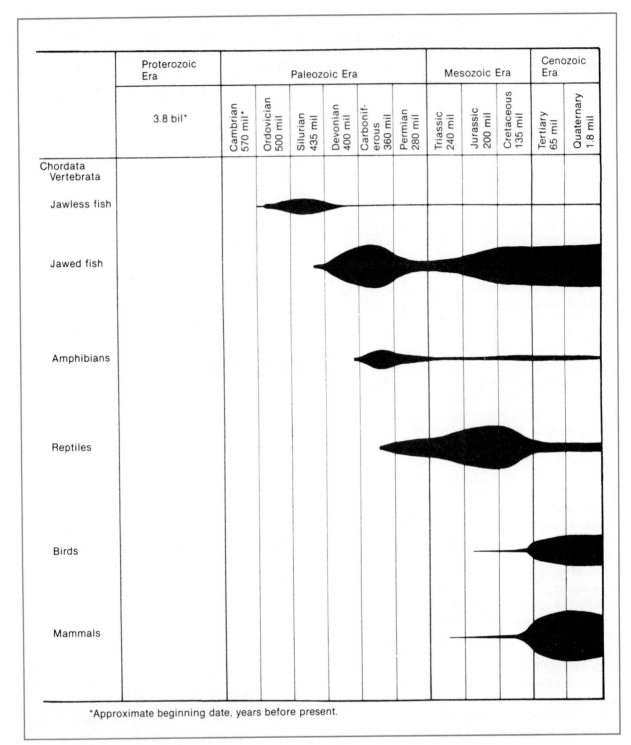

	Proterozoic Era	Paleozoic Era						Mesozoic Era			Cenozoic Era	
	3.8 bil*	Cambrian 570 mil*	Ordovician 500 mil	Silurian 435 mil	Devonian 400 mil	Carboniferous 360 mil	Permian 280 mil	Triassic 240 mil	Jurassic 200 mil	Cretaceous 135 mil	Tertiary 65 mil	Quaternary 1.8 mil
Chordata Vertebrata												
Jawless fish												
Jawed fish												
Amphibians												
Reptiles												
Birds												
Mammals												

*Approximate beginning date, years before present.

Figure 29.3. Fossil record of vertabrates

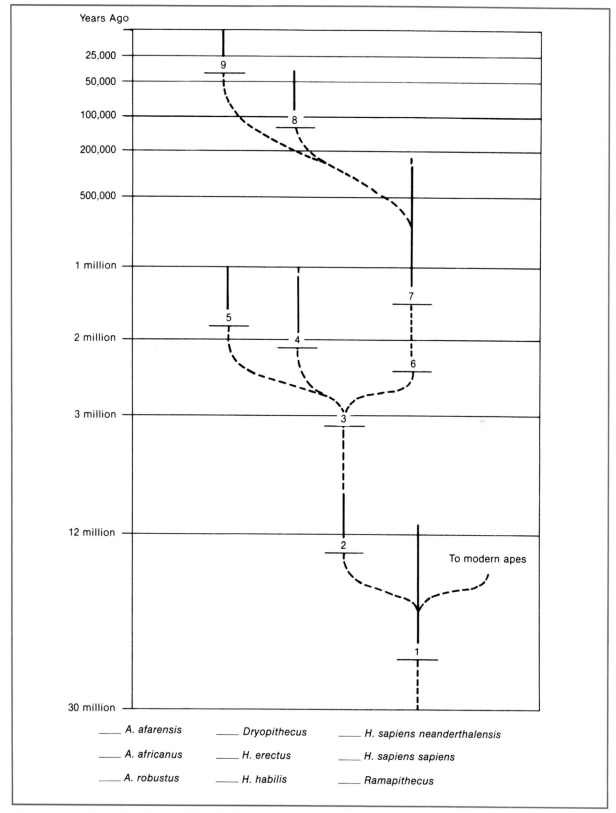

Figure 29.4. Possible hominid evolutionary relationships

Figure 29.4 shows one possible pattern of human evolution, and a discussion of the major fossils in this pattern follows.

Dryopithecus

This primitive ape lived 10 million to 25 million years ago in Africa, Asia, and Europe. The fossils suggest that *Dryopithecus* was primarily a tree dweller and that it also used grassland plants as food. The evolutionary line leading to modern apes is believed to originate with this form. *Dryopithecus* may be the ancestor of *Ramapithecus*.

Ramapithecus

Meager fossils of this form have been dated at 10 million to 14 million years ago. The jaws and teeth exhibit both ape and human characteristics. The **canines** are smaller than those of apes, but they are larger than those of humans. The **incisors** are relatively small. Wear on the low-crowned **molars** suggests sequential eruption, a human characteristic. The dental arch is unlike the narrow U-shaped dental arch of apes and could be a primitive form leading toward the parabolic dental arch of hominids.

Some authorities believe *Ramapithecus* is ancestral to the hominid line and place it near the point of ape and hominid separation. A few believe that it is just a progressive Miocene ape. Most agree, however, that it possesses hominid-like characteristics and apparently lived in a savannah (grassland with scattered trees) habitat.

Australopithecus

The earliest definitive hominid fossil is *Australopithecus afarensis* (named Lucy by her discoverers), which is 3–3.7 million years old. *A. afarensis* has been suggested as the ancestor of the later-appearing *Australopithecus* species and the earliest form of the genus *Homo*. In this view, neither *A. africanus* (2 million years ago) nor *A. robustus* (1.8 million years ago) is ancestral to the genus *Homo*. Some authorities believe *A. afarensis* is only an early form of *A. africanus* and that both *A. robustus* and *Homo* evolved from *A. africanus*. Another view is that *Homo* evolved much earlier so that neither *A. africanus* nor *A. afarensis* could be ancestral to *Homo*.

Australopithecus was 4–4.5 ft tall, erect, and bipedal and had a brain a bit larger than a chimpanzee. Some evidence suggests that *Australopithecus* used tools and hunted other animals for food. Plants are thought to have formed a major part of the diet, however.

Homo habilis

This is the first human. It lived 2–2.5 million years ago, and it coexisted with *A. africanus* and *A. robustus*. *A. africanus* may be ancestral to *H. habilis*, which in turn is thought to be ancestral to *Homo erectus*. Some authorities believe that *H. habilis* is only an advanced form of *A. africanus*. The skeletal features of *H. habilis*, especially the teeth and cranial size, are more human-like than are those of *Australopithecus*.

Homo erectus

Homo erectus lived from 300,000 to 1.5 million years ago. Numerous fossils of this species have been found in Africa, Asia, and Europe. *H. erectus* was taller (5 ft), larger brained, and more efficiently bipedal than was *Australopithecus*. Evidence indicates construction and use of stone tools, use of fire, and a hunter and gatherer lifestyle. The skull is characterized by (1) heavy brow ridges, (2) sloping forehead, (3) long, low-crowned cranium, (4) no chin, and (5) protruding face.

Homo sapiens neanderthalensis

This human lived from 40,000 to 125,000 years ago. Neanderthals were stocky with heavy muscles and about 5 ft tall. Skulls exhibit (1) heavy brow ridges, (2) large protruding faces, (3) no chin, and (4) a sloping, low-crowned, but large, cranium. They constructed stone, bone, and stick tools; used fire, wore clothing; were skilled hunters; and were able to cope with the frigid climate of the glacial periods. They also buried flowers with their dead. They coexisted briefly with modern humans before their extinction. The cause for their extinction is not understood.

Homo sapiens sapiens

Modern humans appeared about 40,000 years ago. The fossil record shows that they ranged throughout western and central Europe, although they may have originated elsewhere.

TABLE 29.1
Cranial Capacity (cc) of Hominids

Australopithecus afarensis	500
Australopithecus africanus	450–600
Australopithecus robustus	450–600
Homo habilis	650
Homo erectus	750–1,200
Homo sapiens neanderthalensis	1,100–1,300
Homo sapiens sapiens	1,400–1,700

The early modern humans represented by **Cro-Magnon man** were (1) excellent tool and weapon makers, using stone, sticks, bone, and antlers as raw materials; (2) skilled hunters; and (3) fine artists, as evidenced by cave paintings and sculptures. The radiation of these early modern humans ultimately produced the various races observed today.

Assignment 2

1. Examine Table 29.1.
2. Label Figure 29.4.
3. ***Complete item 2 on the laboratory report.***

SKULL ANALYSIS

Figure 29.5 indicates some of the major differences between gorilla and human skulls. Although this comparison is between two modern species, it does suggest the type of changes that should be present in intermediate forms when a common ancestry is assumed.

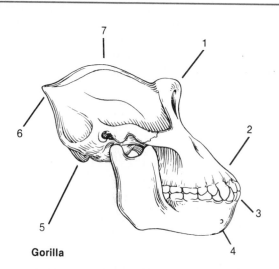

Gorilla

1. Heavy brow ridges; low crowned, sloping cranium
2. Muzzlelike face
3. Large canines
4. No chin
5. Posterior skull attachment
6. Occipital crest for attachment of heavy neck muscles.
7. Sagittal crest for attachment of heavy jaw muscles.

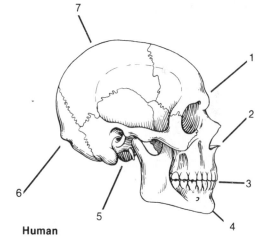

Human

1. Almost no brow ridges; high forehead and high-crowned cranium
2. Flattened face
3. Small canines
4. Well-developed chin
5. Central skull attachment
6. No occipital crest, neck muscles not as heavy and attached lower
7. No sagittal crest, lighter jaw muscles attached lower

Figure 29.5. Comparison of gorilla and human skulls

An understanding of hominid evolution is based on the study of fossil bones, teeth, and artifacts, such as tools, associated with the specimens. Analysis of the fossils involves detailed measurements and careful comparisons of the fossils. In this section, you will analyze certain skeletal materials and fossil replicas by making measurements and calculating indexes as well as conduct qualitative comparisons of the available specimens.

Indexes are useful for comparative purposes because they overcome the problems caused by differences in the size of the specimens. An index is a ratio calculated by dividing one measurement by another, and it indicates the proportional relationship of the two measurements. Calipers and meter sticks are to be used in making the measurements, and all measurements are to be in millimeters. Figures 29.6 and 29.7 show how to make the measurements. The indexes are calculated by dividing one measurement by another (carried to two decimal places) and multiplying by 100.

Handle the bones and fossil replicas carefully since they are easily damaged. Avoid making marks or scratches on the specimens. The measurements are best made by working in groups of two to four students.

Materials

Per student group
Calipers
Meter stick

Per lab
Skulls of modern humans
Skull replicas of:
    Australopithecus africanus
    Homo erectus
    Homo sapiens neanderthalensis
    Homo sapiens sapiens (Cro-Magnon)

Cranial (Cephalic) Index

When this index is determined for living forms, it is called a **cephalic index.** The term **cranial index** is used in reference to nonliving specimens. An index of less than 75 indicates a longheaded condition. An index of 80 or more identifies a roundheaded individual.

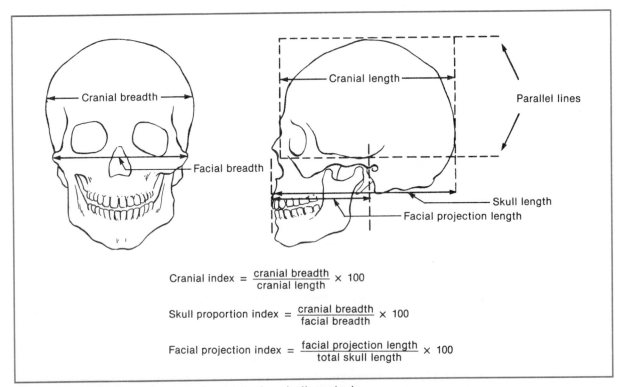

$$\text{Cranial index} = \frac{\text{cranial breadth}}{\text{cranial length}} \times 100$$

$$\text{Skull proportion index} = \frac{\text{cranial breadth}}{\text{facial breadth}} \times 100$$

$$\text{Facial projection index} = \frac{\text{facial projection length}}{\text{total skull length}} \times 100$$

Figure 29.6. Measurements to be made for skull analysis

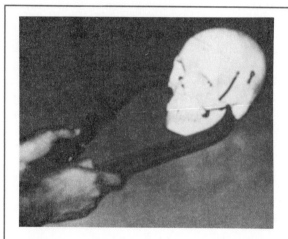
Calipers are used to establish the distance between two points.

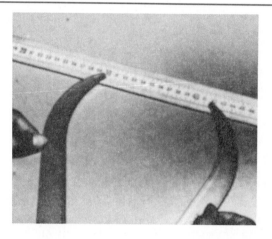
The distance between the two points is determined by placing the caliper tips on the meterstick and measuring the distance between their inside edges.

Figure 29.7. Procedures of making measurements for skeletal analysis

$$\text{Cranial index} = \frac{\text{cranial breadth}}{\text{cranial length}} \times 100$$

Cranial breadth: maximum width of the cranium

Cranial length: maximum distance between the posterior surface and the small prominence (glabella) between the brow ridges

Skull Proportion Index

The skull is composed of the face and the cranium, and the skull proportion index identifies the proportional relationship between these two components. The greater the value of the index, the larger the cranium is in relation to the face.

$$\begin{matrix} \text{Skull} \\ \text{proportion} \\ \text{index} \end{matrix} = \frac{\text{cranial breadth}}{\text{facial breadth}} \times 100$$

Cranial breadth: maximum breadth of the cranium

Facial breadth: maximum distance between the lateral surfaces of the cheekbones (zygomatic arches)

Facial Projection Index

A projecting, muzzlelike face is a primitive condition among primates. This index identifies the degree of facial projection in a specimen. The greater the value of the index, the greater the degree of facial projection.

$$\begin{matrix} \text{Facial} \\ \text{projection} \\ \text{index} \end{matrix} = \frac{\text{facial projection length}}{\text{total skull length}} \times 100$$

Facial projection length: distance between the anterior margins of the auditory canal and upper jaw (maxilla)

Total skull length: maximum distance between the posterior surface of the cranium and the anterior margin of the upper jaw (maxilla)

Assignment 3

1. Determine the cephalic index for each member of your laboratory group. Exchange data with all groups in the class and list the indexes for each person in Table 29.2. Determine the range and average index value for the class.
2. ***Complete items 3a–3d on the laboratory report.***
3. Determine the (1) cranial index, (2) skull proportion index, and (3) facial projection index

TABLE 29.2
Cephalic Indexes of Class Members

1. _____	6. _____	11. _____	16. _____	21. _____	26. _____
2. _____	7. _____	12. _____	17. _____	22. _____	27. _____
3. _____	8. _____	13. _____	18. _____	23. _____	28. _____
4. _____	9. _____	14. _____	19. _____	24. _____	29. _____
5. _____	10. _____	15. _____	20. _____	25. _____	30. _____

Range _____   Average _____

for each of the skulls available. ***Record your data in item 3e on the laboratory report.***

4. Compare the available skulls as to the following characteristics and rate each on a 1–5 scale. ***Record your responses in item 3e on the laboratory report.***

Skull and vertebra attachment
    1 = most posterior
    5 = most central

Brow ridges
    1 = most pronounced
    5 = least pronounced

Length of canines
    1 = longest
    5 = no longer than incisors

Forehead
    1 = most sloping
    5 = best developed

Chin
    1 = no chin
    5 = best developed

5. ***Complete the laboratory report.***

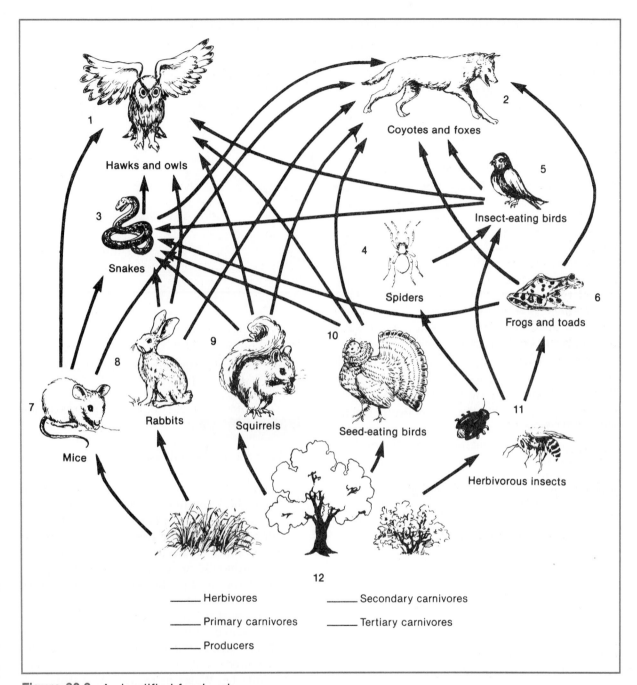

Figure 30.2. A simplified food web

posed to a variety of concentrations of pollutant in order to determine the concentration where a harmful effect is clear.

In this section, you will use *Daphnia magna* (Figure 30.4), a small freshwater crustacean, as an indicator organism to investigate the effect of pollution. *Daphnia* feed on microscopic algae

and protists, and, in turn, serve as a prime food source for small fish. They have a jerky swimming motion powered by rapid strokes of paired, feathery antennae, and while swimming they sweep tiny food organisms into their mouths with their thoracic legs.

Daphnia normally live in water with a pH of

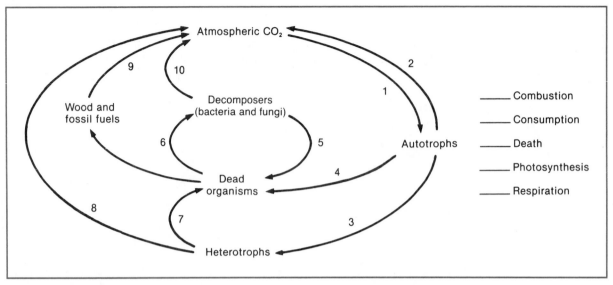

Figure 30.3. The carbon cycle

5.5–7.0, at summer temperatures of 15–25 °C, and with very little or no pesticide content. You will test the sensitivity of *Daphnia* to acid pollutions of pH 4 and pH 2, thermal pollutions of 30 °C and 35 °C, and very low concentrations of a common insecticide, Diazinon Plus (Ortho). This insecticide is normally applied at a concentration of about 0.25%. You will use only 5 or 10 drops of a 0.12% solution per 15 ml water in the experiment.

Work out a division of labor among members of your lab group for setting up the experiment.

Materials

Per student group
Beaker, 250 ml
Dropping bottle of Diazinon Plus in spring or
  pond water, 0.12%
Glass marking pen
Graduated cylinder, 25 ml or 50 ml
Hot plate
Medicine dropper modified to transfer *Daphnia*
pH test paper strips
Rubber stoppers to fit test tubes, 2
Test tubes, 4
Test-tube rack
Thermometer, Celsius

Per lab
Cultures of *Daphnia*
Spring or pond water

Spring or pond water adjusted with sulfuric acid to: pH 4 and pH 2

Assignment 3

1. Your instructor will assign your group to prepare either Tubes 1, 2, 4, and 6 or Tubes 1, 3, 5, and 7 that are described in Step 3 below. After the experiment, you are to exchange results with other groups.
2. Fill a 250-ml beaker two-thirds full with tap water. Place it on a hot plate and slowly heat it to either 30 or 35 °C as assigned by your instructor. Adjust the hot plate so the temperature remains constant.
3. Use a marking pen to number the test tubes as assigned. Add 15 ml of the solutions as follows. Use either spring or pond water, *not* tap water.
   Tube 1: Water at room temperature
   Tube 2: Water to be heated to 30 °C
   Tube 3: Water to heated to 35 °C
   Tube 4: Water adjusted to pH 4 with sulfuric acid
   Tube 5: Water adjusted to pH 2 with sulfuric acid
   Tube 6: Water plus 10 drops of the pesticide solution
   Tube 7: Water plus 5 drops of the pesticide solution

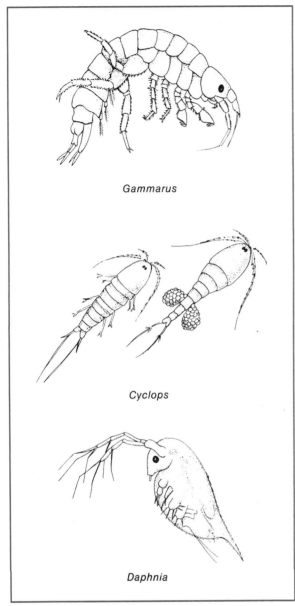

Gammarus

Cyclops

Daphnia

Figure 30.4. Crustaceans in the aquatic eco-systems

4. Insert a rubber stopper into Tubes 6 and 7 and invert the tubes a couple of times to mix the contents. Remove and rinse the stoppers.
5. Measure and record the water temperature of Tube 1. Determine the pH of water in Tube 1 by dipping a pH paper test strip into the water, quickly withdrawing it, and comparing the color of the wet strip with the color chart on the pH test strip dispenser. Record the pH.
6. Place as assigned either Tube 2 or Tube 3 in

the water bath at the assigned temperature. Wait 5 min for temperature equilibration.
7. Use a modified medicine dropper to transfer 5 *Daphnia* from a culture jar into each test tube, starting with Tube 1. Record the time of transfer for each tube.
8. Observe the behavior and survival of the *Daphnia* in the tubes after exposure times of 15, 30, and 60 min. At each observation, count the number of living and dead *Daphnia*. **Record your results and the totals for the class in item 3a, and complete item 3 on the laboratory report.**

BEHAVIORAL ECOLOGY

If you have looked carefully at organisms in an ecosystem, you have noticed that they tend to occupy specific **habitats,** particular places where they live. For example, some plants live in sunlit areas, others thrive in shade, and some prefer partial sunlight. Animals also exhibit habitat preferences, and some of their preferences result from responses to specific stimuli. One task of biologists is to determine an organism's habitat and how the organism is adapted to it.

In this section, you will investigate the responses to certain stimuli that may explain why sowbugs (wood lice) tend to be found under logs, rocks, and fallen leaves. You will test the null hypothesis that light and moisture have no effect on the orientation of sowbugs. The experiments will offer alternative locations for sowbugs to congregate: (1) light or shade, or (2) moisture or dryness.

Materials

Per student group
Sowbugs
Black construction paper
Dropping bottle of tap water
Laboratory lamp, preferably fluorescent
Masking tape
Petri dish, plastic, 14 cm in diameter
Filter paper, 14.5 cm in diameter

Assignment 4

1. Obtain a large Petri dish and fit a circle of filter paper into the bottom of it. This pro-

vides a dry habitat for testing the sowbugs' response to light.

2. Use black construction paper to make a rooflike light shield that will shade half of the Petri dish. The roof must be low enough to provide heavy shade, but high enough so you can see under it to observe the sowbugs. You may need to anchor the light shield to the lab table with masking tape.

3. Position the lamp so it shines on the Petri dish. If it has an incandescent light source, keep it at least 3 ft from the Petri dish to prevent heat from affecting your experiment.

4. *Complete items 4a and 4b on the laboratory report.*

5. Place 5 sowbugs in the Petri dish and replace the lid. Before you place it under the light shield, observe the movement of the sowbugs. Then slide it under the light shield so that half of the dish is heavily shaded to provide a choice for the sowbugs of light or shade. Observe the sowbugs and *record their positions (i.e., light or shade) at 3-min intervals for 15 min in item 4c, and complete item 4d on the laboratory report.*

6. Remove the light shield and turn off the laboratory lamp. Remove the lid of the Petri dish and add 5 drops of water to one spot at the edge of the filter paper. Replace the lid. The sowbugs now have a choice between moist and dry areas. Observe the sowbugs and *record their positions (i.e., dry or moist) at 3-min intervals for 15 min in item 4e, and complete the laboratory report.*

POPULATION GROWTH

After completion of the laboratory session, you should be able to:
1. Distinguish between population growth and population growth rate.
2. Compare theoretical and realized population growth curves.
3. Describe the role of the following in population growth:
   a. Biotic potential
   b. Environmental resistance
   c. Density-dependent factors
   d. Density-independent factors
   e. Environmental carrying capacity
4. Compare the growth of human and natural animal populations.
5. Compare the growth of human populations in major regions of the world.
6. Define all terms in bold print.

A **population** is a group of interbreeding members of the same species within a defined area. Populations possess several unique characteristics that are not found in individuals: **density, birth rate, death rate, age distribution, biotic potential, dispersion,** and **growth form.** One of the central concerns in ecology is the study of population growth and factors that control it.

Natural populations are maintained in a state of dynamic equilibrium with the environment by two opposing factors: biotic potential and environmental resistance. **Biotic potential** is the maximum reproductive capacity of a population that is theoretically possible in an unlimiting environment. It is never realized except for brief periods of time. **Environmental resistance** includes all **limiting factors** that prevent the biotic potential from being attained.

The limiting factors may be categorized as **density-dependent factors,** the effects of which increase as the population increases. Space, food, water, waste accumulation, and disease are examples of such factors. In contrast, **density-independent factors** exert the same effect regardless of the population size. Climatic factors and the kill of individual predators generally function in this manner.

Assignment 1

Complete item 1 on Laboratory Report 31 that begins on page 411.

GROWTH CURVES

There are two basic types of population growth patterns. **Theoretical population growth** is the growth that would occur in a population if the biotic potential of the species were realized, that is, if all limiting factors were eliminated. This type of growth does not occur in nature except for very short periods of time. **Realized population growth** is the growth of a population that actually occurs in nature. Let's consider both types of population growth in bacteria.

Theoretical Growth Curves

In the absence of limiting factors, bacteria exhibit tremendous theoretical population growth potential. Bacterial population size is measured as the number of bacteria in a milliliter (ml) of nutrient broth, a culture medium. Assume that bacterial cells divide at half-hour intervals. If the initial population density were 10,000 bacteria per milliliter of nutrient broth (10×10^3 bacteria/ml), a half hour later the population would be 20,000 bacteria per milliliter (20×10^3 bacteria/ml), and the population would have doubled. See Table 31.1. In another half hour, the population would double again to 40×10^3 bacteria/ml. In an unlimiting environment, this population would double every half hour. If the population size is calculated for each half-hour interval and plotted on a graph, a line joining the points on the graph yields a **theoretical growth curve.**

Realized Growth Curves

Natural populations exhibit realized growth that results from the opposing effects of biotic potential and environmental resistance. Table 31.2 indicates the growth of a bacterial population in a tube of nutrient broth—a limited environment which results in the ultimate death of the entire population. The population was sampled at half-hour intervals, and no additional nutrients were added.

TABLE 31.2
Realized Growth of a Bacterial Population

Time (hr)	Population Density (10^3 bacteria/ml)
0.0	10
0.5	20
1.0	40
1.5	80
2.0	150
2.5	290
3.0	450
3.5	520
4.0	520
4.5	515
5.0	260
5.5	80
6.0	0

Assignment 2

1. Determine and record the theoretical growth of the bacterial population at half-hour intervals and record the data in Table 31.1.
2. *Plot the theoretical growth in item 2a on the laboratory report.*
3. *Plot the realized growth curve in item 2a on the laboratory report.* Use a different symbol or color for this curve to distinguish it from the theoretical curve.
4. Compare the theoretical and realized population growth curves. The theoretical curve is a J-shaped curve. The realized growth curve starts off in the same manner, but it is soon changed by environmental resistance. As the limiting factors take effect, the curve becomes S shaped. In the artificial environment of a test tube, the bacterial population dies out. In natural populations where materials can cycle and energy is continuously available, the population size levels off where it is at equilibrium with the environment.

TABLE 31.1
Theoretical Growth of a Bacterial Population

Time (hr)	Population Density (10^3 bacteria/ml)
0.0	10
0.5	20
1.0	_____
1.5	_____
2.0	_____
2.5	_____
3.0	_____
3.5	_____
4.0	_____

5. Study Figure 31.1, which compares generalized theoretical and realized population growth curves. The theoretical growth curve exhibits exponential growth with the slope of the curve becoming ever steeper. In contrast, the effect of environmental resistance causes the realized growth curve to become S shaped.

   Several parts of the realized growth curve are recognized. Growth is slow, initially, in the **lag phase** due to the small number of organisms present. This is followed by the **exponential growth phase** when the periodic doubling of the population yields explosive growth. As environmental resistance takes effect, the growth gradually slows, leading to the **inflection point** that indicates the start of the **decreasing growth phase** as the curve turns to the right. Continued environmental resistance causes the curve to level off at the **carrying capacity** of the environment, where the population size is in balance with the environment.

   The carrying capacity is the maximum population that can be supported by the environment—a balance between biotic potential and environmental resistance. The population remains stable because births equal deaths.

6. ***Complete item 2 on the laboratory report.***

HUMAN POPULATION GROWTH

One of the major concerns of modern society is the rapid growth of the human population in recent decades. Table 31.3 shows the estimated size of the world population at various times in history and projects a rather conservative growth in the near future.

Growth Rate

Ultimately, a population is limited by a decrease in **birth rate,** an increase in **death rate,** or both. In natural nonhuman populations, an increase in death rate is the usual method. This allows the environment to select for those better adapted individuals who will contribute the greater share of the genes to the next generation. While this mode of selection occurred in preindustrialized human societies, it has been highly modified, but not stopped, by technology in modern societies.

Concern over the growth rate of the human population gained worldwide attention in the 1960s and persists today. The growth rate is the difference between the number of persons born (birth rate) and the number who die (death rate) per year. For humans, the growth rate usually is expressed per 1,000 persons. For example, if a

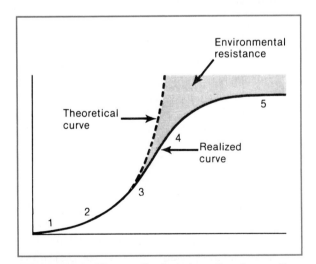

Figure 31.1. Theoretical and realized growth curves. 1 = lag phase, 2 = exponential growth phase, 3 = inflection point, 4 = decreasing growth phase, 5 = carrying capacity.

TABLE 31.3
Estimated World Human Population

Year	Population Size (millions)
8000 B.C.	5
4000	86
1 A.D.	133
1650	545
1750	728
1800	906
1850	1,130
1900	1,610
1950	2,515
1960	3,019
1970	3,698
1980	4,450
1990	5,292
2000	6,400*

* Conservative projection.

human population shows 25 births and 10 deaths per 1,000 persons in a year, the growth rate may be determined as follows:

$$\frac{25 - 10}{1,000} = \frac{15}{1,000} = \frac{1.5}{100} = 1.5\% \text{ growth rate}$$

The growth rate reached a peak of 2% in 1965 and declined to 1.7% in 1984, but this slight decline does not mean that the population growth is no longer a concern.

The growth rate is often expressed as the **doubling time,** that is, the time required for the population to double in size. In 1850, the population required 135 yr to double, but this was reduced to only 35 yr in 1965. All of the efforts to reduce population growth only extended the doubling time to 41 yr by 1984.

The doubling time of a population is determined by dividing 70 yr (a demographic constant) by the growth rate. For example, if the world population growth rate is 1.7%, the doubling time (d) would be determined as follows:

$$d = \frac{70 \text{ yr}}{\text{growth rate}} = \frac{70 \text{ yr}}{1.7} = 41 \text{ yr}$$

This means that the entire world population will double in only 41 years. If the present standard of living is to be maintained, the available resources must also double in 41 years. Is this likely? Are the earth's resources infinite?

Obviously, the standard of living varies throughout the world. Developed (industrialized) countries have a relatively high standard of living, and developing countries have a relatively low standard of living. The standard of living is usually expressed as the resources per person and is determined by dividing the gross national product (GNP) by the population size. The contrast in the standard of living in developed and developing countries is evident in the GNP per capita for the United States and India in 1988.

United States = $19,840
India = $329

Assignment 3

1. Use the data in Table 31.3 to *plot the human population growth curve in item 3a on the laboratory report.*
2. *Complete items 3b–3i on the laboratory report.*
3. Calculate the growth rate and doubling time for the human population in the countries or regions shown in Table 31.4.
4. *Complete the laboratory report.*

TABLE 31.4
1990 Human Population Data

Country or Region	Population Size (10^6)	Birth Rate (per 10^3)	Death Rate (per 10^3)	Growth Rate (% inc./yr)	Doubling Time (yr)
World	5,292	27	10	1.7	41.2
Africa	642	45	15	_____	_____
Asia	3,113	28	9	_____	_____
Europe*	498	13	11	_____	_____
Latin America	448	29	7	_____	_____
North America	276	15	9	_____	_____
Oceania	26	19	8	_____	_____
C.I.S. (formerly the USSR)	289	18	11	_____	_____

* Excludes the Commonwealth of Independent States (formerly the USSR).
Source: 1990 Demographic Yearbook. Published by the United Nations, 1992.

PART VI

LABORATORY REPORTS

LABORATORY REPORT 1

Orientation

1. Laboratory Procedures

a. Place an X by those activities that are to be completed before coming to the laboratory.

_____ Perform the experiments.	_X_	Color-code diagrams.
_X_ Learn the meaning and spelling of new terms.	_X_	Tear lab report out of lab manual.
_X_ Label diagrams.	_X_	Answer lab report questions on background information.
_X_ Understand the objectives.		

b. Place an X by those events in the laboratory that should be brought to the attention of your instructor.

_X_ Problems with equipment.	_X_	Minor injury.
_X_ Spillage of a liquid.	_X_	Directions not understood.
_X_ Breakage of glassware.	_X_	Problems with supplies.

2. Biological Terms

a. Use Appendix A to determine the literal meaning of these terms.

Biology _study of life_
Cardiac _pertaining to the heart_
Hypodermic _pertaining to under or below the skin_
Erythrocyte _red cell_
Dermatitis _skin inflamation_

b. Use Appendix A to construct terms with these literal meanings.

Pertaining to within a cell _intracytal endocytal_ _entocytal -ac, -alis, -an, -ar, -ary_
Study of the heart _cardiology_
Nerve disease _neurosis_ _*neuropathy_
Cancerous tumor _carcinoma_
Pertaining to a single cell _haplocytal? - haplocytac?_

3. Measurements: Length

a. Indicate the name and value of these metric symbols.

Symbol	Name of Unit	Value of Unit
ml	_millileter_	_1/1000 th of a liter (0.001 liter)_
cm	_centimeter_	_1/100 th of a meter (0.01 meter)_
mm	_millimeter_	_1/1000 th of a meter (0.001 meter)_
kg	_kilogram_	_1,000 grams_

b. Measure the diameter of a penny in millimeters. _18_ mm
Convert your measurement to centimeters and meters. _1.8_ cm; _0.018_ m

c. Measure the length of your little finger.   _____57_____ mm;   _____5.7_____ cm

d. Complete the chart showing the number of students with each finger length.

Length in mm	No. of Students	Length in mm	No. of Students	Length in mm	No. of Students	Length in mm	No. of Students
51	_____	59	I	67	_____	75	_____
52	_____	60	III	68	_____	76	_____
53	I	61	_____	69	_____	77	_____
54	_____	62	_____	70	_____	78	_____
55	II	63	I	71	_____	79	_____
56	II	64	_____	72	_____	80	_____
57	III	65	IIII	73	_____	81	_____
58	_____	66	_____	74	_____	82	_____

e. Using the data from the chart in item 3d, indicate:
The range of little finger lengths.   _____53_____ mm to _____65_____ mm
The average of little finger lengths.   (59.6)   _____60_____ mm
(The average is calculated by dividing the sum of the finger lengths by the number of fingers
measured.)
$$18\overline{)1073.000}\quad\frac{59.611\ldots}{}$$

f. A graph presents data in a visual way. Plot the data from the chart in item 3d on the graph
below. Draw a heavy vertical line from each dot to the horizontal axis, producing a bar graph.
Another way of showing data in a graph is by drawing a curve that fits the dots. Try it.

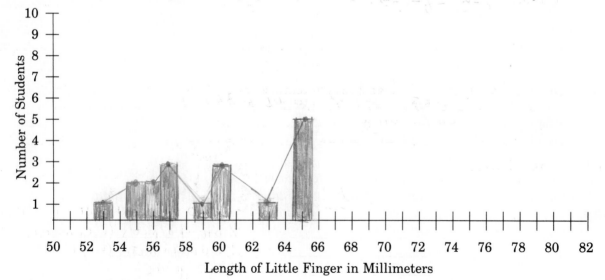

Length of Little Finger in Millimeters

g. Referring to the graph, what is the most common finger length?   _____65_____ mm
   How many students have little finger lengths within 1 mm of the
   average finger length?   _____4_____
   These students compose what percentage of the class?   _____22%_____%
   (Percentage is calculated by dividing the number of students in this category by the number of
   students in the class and multiplying by 100.)

4. Measurements: Mass

a. Record the mass of the wood blocks.   #7 above   #9 above
   #1 __47.6__ g; #2 __50.8__ g; #3 _____ g

b. A physician directs a patient to take 1 g of vitamin C each day. How many 250-mg tablets must
   be taken each day?   __4 – 250mg tablets__

c. It is recommended that adults eat 0.8 g (8 dg) of protein per kilogram of body mass each day. What
   should be the minimum daily protein intake of a man with a weight of 185 pounds? (Hint: First
   convert 185 pounds to kilograms.)   __67__ g

   185 lbs ÷ 2.2 = 84.09 kg
   84.09 kg × 0.8g = 67.272

5. Measurements: Volume

a. Record the dimensions of the wood blocks in centimeters.
   #1 L _4.9_ ; W _4.1_ ; D _3.7_   #2 L _5_ ; W _4.2_ ; D _3.5_   #3 L ___ ; W ___ ; D ___

b. Calculate the volume of the three wood blocks in cubic centimeters.
   #1 __74.33__ cm³   #2 __73.5__ cm³   #3 _____ cm³

   84 ml water

   47.6/74.33   50.8/73.5

c. Calculate the density of each wood block.
   #1 _.64_   #2 _.69_   #3 _____

d. What is the volume of the test tube? _____23_____ ml

e. How many drops are in 1 ml of water? _____22_____

f. Measure the mass of the 50-ml graduated cylinder _92.8 w/ring_ _89.3 w/out ring_ g

Measure the mass of graduated cylinder + 30 ml of water. _122.4 "_ _118.6_ g

Calculate the mass of the water, only. _29.6 w/_ _29.3 w/o_ g

6. Measurements: Temperature

a. Using a Celsius thermometer, measure the temperature of the following and convert the temperature to °F.

Air in the room. _____23_____ °C; _____73.4_____ °F

Cold tap water. _____20_____ °C; _____68_____ °F

b. Normal body temperature is 37 °C (98.6 °F). If a patient's temperature is 38 °C, what is the temperature in °F? __68.4_____ °F

7. Scientific Method

a. Why is a control group important in an experiment? _It is a good way to show_ _validation of the results — if you can consistantly show_ _that mice taking vitamin C have lower cancer rates — then the vit. C_ _can be proven to reduce risk of cancer._

b. State two new key questions raised by the results and conclusion of the hypothetical example of the scientific method.

✱ 1. _If 50 mg of vitamin C reduced risk of getting cancer by_ _30%, would a higher daily dose (100 mg) reduce risk even further?_

2. _What are the side effects of a daily dose of 50 mg of_ _vitamin C?_

c. Using one of these key questions, state a new hypothesis. _Increasing daily vitamin_ _C intake from 50 mg to 100 mg will decrease the risk of_ _cancer even more._

State a new prediction. _If increasing daily dose of vitamin C to 100 mg_ _reduces cancer risk more than 50 mg, then the mice receiving_ _100 mg of vitamin C will have fewer instances of cancer than those_ _taking 50 mg and still fewer yet then those not taking any._

d. Describe your controlled experiment. _____ _Same as before except a third group of mice will be added_ _and they will be given 100 mg of vitamin C daily — all_ _other variables are controlled._

e. What results would support your null hypothesis? _If the mice @ 50 mg and_ _@ 100 mg have the same or higher cancer rates._ _(100 mg mice have_

LABORATORY REPORT 2

The Microscope

1. The Compound Microscope

a. List the labels for Figure 2.1

1. _Revolving nosepiece_
2. _Objective_
3. _Stage_
4. _Iris diaphragm lever_
5. _Condenser control knob_
6. _Light switch_
7. _Ocular_
8. _Body tube_
9. _Arm_
10. _Stage clamp_
11. _Condenser_
12. _Fine-focusing knob_
13. _Coarse-focusing knob_
14. _Light source_
15. _Base_

b. Write the term that matches each meaning.

1. Used as handle to carry microscope — _arm_
2. Lenses attached to the nosepiece — _objective lenses_
3. Concentrates light on the object — _condenser_
4. Lens you look through — _ocular lens_
5. Platform on which slides are placed — _stage_
6. Rotates to change objectives — _revolving nosepiece_
7. The shortest objective — _scanning objective_
8. The longest objective — _high-power objective (some have oil immersion)_
9. Control knob used for fine focusing — _fine-focusing knob_
10. Control knob used for rough focusing — _coarse-focusing knob_
11. Controls amount of light entering condenser — _iris diaphragm control lever_

c. List the power of the ocular and objectives on *your* microscope, and calculate the total magnification for the combinations noted.

Magnification		Total
Ocular	**Objective**	**Magnification**
10X	4X scanning	40X
	10X low power	100X
	40X high power	400X
	100X oil immersion	1000X

1000 microns in 1 millimeter
_1,000 μm = 1 mm_

Unit	
25 μm (milimicrons)	
10 μm	
2.5 μm	
1 mm (micron)	

```
    2
   75
   25
  375
  150
 1875
```

d. Write the term that matches each meaning.

1. Tissue used to clean lenses — _lens paper_
2. Objective with the least working distance — _oil immersion_
3. Slide with an attached cover glass — _prepared slide (permanent)_
4. Objective with the largest field — _scanning objective (4X)_

2. Initial Observations

a. Indicate the direction the image moves when the slide is moved:

To the left _____to the right_____ Away from you _____toward you_____

b. Indicate the steps to be used in focusing on an object with the high-power objective. _____ using course adjkts (& looking from side) move stage as high up as you can without touching it to the slide, then look through ocular & fine _c_

start w/ these stage farthest away condenser full up

c. Describe the best way to relocate an object that is "lost" while viewing with the high-power objective. _____re-center object in slide directly over condenser_____ (back the stage down if necessary) (make sure_____

d. Draw the letter *i* as it appears when observed with each of the following:

Unaided Eye	**4× Objective**	**10× Objective**	**40× Objective**

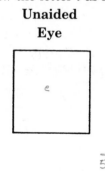
$\frac{23}{25}$
$\overline{575}$

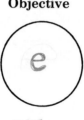
$\frac{59}{\times 10}$
$\overline{590}$
575 µm

590 µm

$\frac{230}{2.5}$
$\overline{575}$
575 µm

3. Depth of Field

Based on your observations, draw a side view of a large spine from a fly wing at these total magnifications.

40×

$\frac{25}{2}$

1,825 µm = 1.825 mm

$\frac{182}{\times 10}$
$\overline{1820}$

100×

$\times \frac{10}{5}$

50 µm = 0.050 mm

4. Diameter of Field

a. Indicate the estimated diameter of field at each *total magnification* for your microscope.

Magnification	Diameter of Field
_40_ ×	_4.5_ mm
_100_ ×	_1.8_ mm
_400_ ×	_0.45_ mm

b. Indicate the estimated lengths of objects that extend across:

1. Two-thirds of the field at 40× _____3.0_____ mm _____3000_____ µm
2. 25% of the field at 100× _____0.45_____ mm _____450_____ µm
3. Half of the field at 400× _____0.225_____ mm _____225_____ µm
4. 80% of the field at 40× _____3.6_____ mm _____3600_____ µm

c. Determine and record these measurements:

The length of the letter *i* including the dot _____ mm _____ µm

The diameter of the dot _____ mm _____ µm

5. Applications

a. When viewing the crossed hairs and the top hair is in sharp focus, is the other hair visible or in sharp focus at the following total magnifications?

Magnification	Visible	Sharp Focus
40×	yes	yes
100×	yes	no
400×	yes, just @ edge	no

b. Estimate the diameter of the blond hair. _____ 250 _μm

c. Describe the shape of the blond hair. ___cylindrical_____

d. On the basis of your observations, are the following characteristics increased, decreased, or unchanged when magnification is increased?

Illumination _decreased_   Depth of field _increased_ ←

Working distance _decreased_   Diameter of field _decreased_

e. Draw five or six organisms observed in the pond water slides. Indicate the approximate size, color, and speed of motion of each. Make your drawings large enough to show details.

6. The Dissecting Microscope

a. When using the dissecting microscope, indicate the direction the image moves when the coin is moved

To the left _____   Toward you _____

b. Measure and record the diameter of field at each magnification of the dissecting microscope.

Magnification	Diameter of Field
20×	_____ mm
40×	_____ mm

7. Review

a. Contrast a prepared slide and a wet-mount slide.

Prepared slide _____

_____

Wet-mount slide _____

_____

b. How should prepared slides by handled? _____

_____

c. Describe how to determine the total magnification when using any objective of your microscope.

_____

_____

d. Describe how to determine the length of an object observed with your microscope. _____

_____

_____

e. How should light intensity be adjusted when viewing nearly transparent specimens? _____

_____

_____

LABORATORY REPORT 3

Student _____

Lab Instructor _____

The Cell

1. Prokaryotic Cells

Sketch the appearance of the prokaryotic cells of bacteria and cyanobacteria at 400× as observed from your slides.

Bacteria   Cyanobacteria

2. Eukaryotic Cell Structure

a. Write the term that matches each statement.
1. Sites of photosynthesis — *chloroplasts*
2. Filaments of DNA and protein — *chromosome*
3. Control center of the cell — *nucleus*
4. Sites of aerobic cellular respiration — *mitochondria*
5. Supports plant cells — *plastids?*
6. Separates the nucleus and cytoplasm — *nuclear envelope?*
7. Sites of protein synthesis — *rough ER*
8. Sacs of strong digestive enzymes — *lysosomes*
9. Packages materials in vesicles for export from the cell — *golgi apparatus*
10. Controls passage of materials into and out of the cell — *cell membrane*
11. Semiliquid substance in which cellular organelles are embedded — *cytoplasm*
12. Large fluid-filled organelle in mature plant cells — *central vacuole*
13. Channels for the movement of materials within the cell — *rough & smooth endoplasmic reticulum (ER)*
14. Assembles precursors of ribosomes within the nucleus — *nucleolus*
15. Nuclear components containing the genetic code controlling cellular processes — *Chromosomes (DNA)*

b. List the labels for Figures 3.2 and 3.3

Figure 3.2
1. Secretory vesicle
2. Golgi apparatus
3. Smooth endoplasmic (ER)
4. Chromatin granules
5. Nucleolus
6. Nuclear envelope
7. Nucleus
8. Lysosome
9. Vacuole
10. Cilia
11. Cell membrane
12. Cytoplasm
13. Microfilaments
14. Ribosome
15. Microtubules
16. Centrioles
17. Mitochondrion
18. Rough endoplasmic reticulum (ER)

307

Figure 3.3

1. _Rough ER_   7. _Vesicle_   13. _Ribosome_
2. _Nuclear envelope_   8. _Cell wall_   14. _Microfilaments_
3. _Chromatin granules_   9. _Cytoplasm_   15. _Microtubules_
4. _Nucleolus_   10. _Cell membrane_   16. _Chloroplast_
5. _Nucleus_   11. _Mitochondrion_   17. _Central Vacuole_
6. _Golgi apparatus_   12. _Smooth ER_

3. Onion Epidermal Cells

a. List the labels for Figure 3.4.

1. _Cell wall_   3. _Nucleus_   5. _Central Vacuole_
2. _Cytoplasm_   4. _Vacuolar membrane_   6. _Cell membrane_

b. Diagram three adjacent onion epidermal cells as observed on your slide and label the parts that you see.

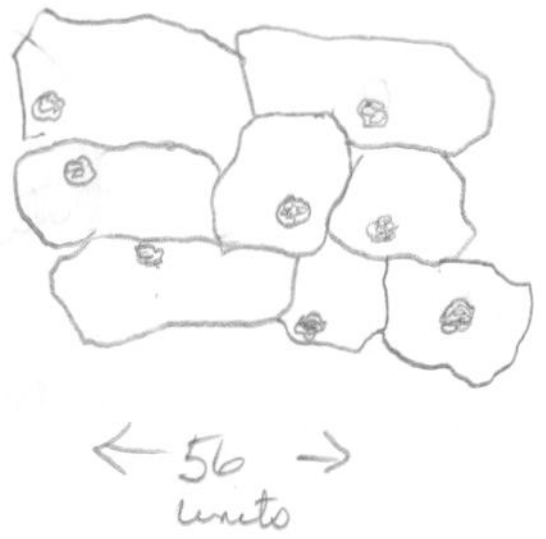

33 units
×10 low 100×
33 units _____
330 μ wide

56 units
×10
_____
560 μ long

← 56 → units

c. Determine and record the average length and width of three adjacent cells.

Width _____ 0.33 _____ mm   Length _____ 0.56 _____ mm

d. Examination of your slide will show that a few nuclei do not appear next to a cell wall. Explain that observation. _____

_____

4. _Elodea_ Leaf Cells   _get young leaf @ tip_   (genus)

a. What structure gives an _Elodea_ cell its shape? _cell wall_

b. Describe the shape of a cell. _elongated rectangle, some elongated octagon, thick_

c. What organelles are green? _Cell wall, cytoplasts_

d. What structure fills the greatest volume in a cell? _Vacuole_

e. Draw a spine cell at the edge of the leaf at 100×. Label the observed parts.

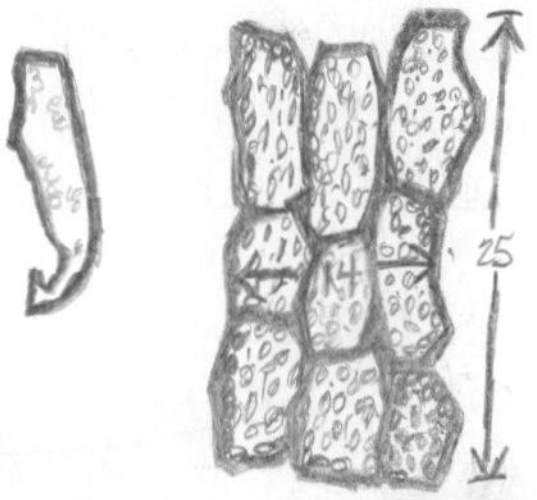

14 units
× 10 @ 100× low
_____
140 μ wide

25 units
×10 @ 100× low
_____
250 μ long

f. Are the cells in each layer of the leaf about the same size? _Yes_

g. Determine and record the average width and length of three cells in the upper cell layer.

Width _____ 0.14 _____ mm   Length _____ 0.25 _____ mm

h. List the labels for Figure 3.6.

1. _Chloroplast_   4. _Vacuole_   6. _Cell membrane_
2. _Cell wall_   5. _Vacuolar membrane_   7. _Cytoplasm_
3. _Nucleus_

5. Human Epithelial Cells

a. What structures observed in onion epidermal cells are *not* present in human epithelial cells.
   Cell wall, central vacuole, vacuolar membrane.

b. Indicate the description that best describes the human epithelial cells.
   Thin and platelike _✓___   Thick and boxlike _____

c. Are human epithelial cells thinner than *Elodea* cells? _*yes*_____

d. Diagram a few cells from your slides and label the parts observed.

$$20 \text{ units}$$
$$\times 2.5 \text{ high } (400\times)$$
$$\overline{100}$$
$$\underline{40}$$
$$50.0 \ \mu \text{ diameter}$$

e. List the labels for Figure 3.7.
   1. *Cell membrane*   2. *Nucleus*   3. *Cytoplasm*

6. Amoeba

a. Diagram the *Amoeba* on your slide and label the parts observed.

Not Available

b. Is the cell membrane rigid or flexible? _____

c. Is the *Amoeba* alive? __*yes*_____   Why do you think so? _*self-maintenance*_
   (digests food) & self-replication (asexual reproduction)

d. Would you use this same evidence to determine if cells in your body are alive? _____
   Explain. _____

7. Review

a. Name the two characteristics of living organisms that distinguish them from nonliving things.
   self-maintenance _self-replication_

b. Indicate the structural and functional unit of life. _a single cell_

c. Based on your study and observations, indicate the presence of the following structures in animal and plant cells by placing an X in the appropriate spaces.

Structure	Animal	Green Plant	Nongreen Plant
Cell membrane	✓	✓	✓
Cell wall		✓	✓
Central vacuole		✓	✓
Centrioles	✓		
Chloroplasts		✓	
Chromatin granules	✓	✓	✓
Cytoplasm	✓	✓	✓
Endoplasmic reticulum	✓	✓	✓
Lysosomes	✓		
Mitochondria	✓	✓	✓
Microtubules	✓	✓	✓
Nucleolus	✓	✓	✓
Nucleus	✓	✓	✓
Ribosomes	✓	✓	✓

LABORATORY REPORT 4

Chemistry of Cells

1. Introduction

a. Write in the space provided the term that matches the definition.

_____ A substance composed of two or more elements chemically combined.

_____ A substance that cannot be broken down into a simpler substance by chemical means.

_____ The smallest particle of an element that exhibits the characteristics of that element.

_____ The smallest particle of a compound that exhibits the characteristics of that compound.

b. Complete this sentence. At its most fundamental level, life consists of _____

_____

_____

2. pH

a. Write in the space provided the term that matches the definition.

_____ A substance releasing OH^- ions when placed in water.

_____ A substance releasing H^+ ions when placed in water.

_____ A measure of the relative concentrations of H^+ and OH^- ions in a solution.

_____ A substance that combines with or releases H^+ ions, keeping the pH of a solution rather constant.

b. For these pH values, *circle* the acids and *underline* the bases.

   1. 5   2. 1   3. 10   4. 8   5. 6.5   6. 12   7. 3   8. 7.5

Rank the numbers of the acids and bases in order of increasing strength.

   Acids _____ , _____ , _____ , _____ .   Bases _____ , _____ , _____ , _____ .

c. Measure and record the pH values for these solutions.

Household ammonia _____   Lemon juice _____   Vinegar _____

Detergent solution _____   Alka-Seltzer _____   Mouthwash _____

Unknown #1 _____   Unknown #2 _____   Unknown #3 _____

d. Indicate the number of drops of 1.0% HCl required to decrease the pH of distilled water and buffer solution from pH 7 to pH 6.

Distilled water _____   Buffer solution _____

Explain your results. _____

_____

_____

3. Biological Molecules

a. Prepare a set of standards for the chemical tests. Record your results in the chart below.

Standards

Test	Organic Compound Testing for	Tube No. and Substance Tested	Results
Iodine	Starch	1A. Starch	
		1B. Water	
Benedict's	Sugars	2A. Glucose	
		2B. Water	
Paper spot	Lipids	— Corn oil	
		— Water	
Sudan IV	Lipids	3A. Corn oil	
		3B. Water	
Biuret	Proteins	4A. Albumin	
		4B. Water	

b. Perform the chemical tests on the substances provided. Compare your results with the standards and record your results in the chart below.

Unknowns

Substance Tested	Iodine	Benedict's	Paper Spot	Sudan IV	Biuret	Organic Compounds Present
Onion						
Potato						
Apple						
Egg white						

Test Results (column group header over Iodine, Benedict's, Paper Spot, Sudan IV, Biuret)

* + = positive;   − = negative

c. How do you explain the negative results when some plant cells were tested for the presence of protein, since all cells contain proteins in cell membranes, chromosomes, ribosomes, and enzymes? _____

4. Enzyme Action

a. What are the chemical subunits composing an enzyme? _____

_____

b. What determines the specificity of an enzyme? _____

_____

c. What determines the three-dimensional shape of an enzyme? _____

_____

d. What is the role of enzymes in cells? _____

_____

e. Write the equation for the chemical reaction that catalase catalyzes.
   Circle the substrate and *underline* the products.

f. Record the thickness (mm) of the foam layer after 1 minute.

Tube	Thickness of Foam Layer	Tube	Thickness of Foam Layer
1. Water		4. Apple	
2. Liver		5. Onion	
3. Steak		6. Potato	

What is the function of Tube 1 containing water and peroxide? _____

State a conclusion from your results. _____

_____

_____

_____

Among those tested, do animal or plant tissues exhibit greater catalyase activity? _____

_____

How do you explain this? _____

_____

_____

_____

g. Record the thickness (mm) of the foam layer after 1 minute.

Tube	Temperature	Thickness of Foam Layer
1		
2		
3		
4		

State a conclusion from your results. _____

_____

_____

_____

h. Record the thickness (mm) of the foam layer after 1 minute.

Tube	pH	Thickness of Foam Layer
1	4	
2	6	
3	8	
4	10	

State a conclusion from your results. _____

_____

_____

i. Among temperatures tested, which totally inactivated catalase?

_____

_____

j. Among pH values tested, which totally inactivated catalase?

_____

_____

k. At the molecular level, what causes inactivation of catalase?

_____

_____

l. Does catalase seem to be active in a broad or narrow range of:

temperatures? _____

pH values? _____

m. How would you determine the optimum pH for liver catalase? _____

_____

_____

_____

_____

_____

_____

LABORATORY REPORT 5

temperature
pressure

Diffusion and Osmosis

1. Brownian Movement
   a. Is Brownian movement a living process? _*no*_
   b. Explain why the dye particles moved in water. *molecules of liquids & gases* *are in constant random motion*

2. Diffusion and Temperature
   a. What natural phenomenon enables diffusion to occur? *the constant random* *motion of molecules*

   b. Define diffusion. *the net movement of the same kind of molecules* *from an area of the higher concentration to an area of* *their lower concentration.*

   c. In a 10% glucose solution, what is the percentage of water? *90%*
   d. What is the independent variable in the experiment? *temperature*
   e. What is the dependent variable in the experiment? *rate of diffusion*
   f. Record the water temperature for each beaker. A _*77*_ °C   B _*12*_ °C
   g. After 15 min, compare the rate of diffusion in beakers A and B.
   @ 2:30 *hot* A _*completed diffused*_   *cd* B _*not completely*_
   h. Do the results support the null hypothesis? _*no*_
   i. What can you conclude about the relationship between temperature and the rate of diffusion?
   rate of diffusion increases when temperature increases *– because of faster movement*
   j. How would you test this new hypothesis? _____

3. Diffusion and Molecular Mass
   a. What is the independent variable in the experiment? _*mass*_
   b. What is the dependent variable? _*rate of diffusion*_
?   c. Record the diameter (mm) of the colored circles after 1 hr.
   5 cm   Potassium permanganate _*0.02*_ mm *3 mm* Methylene blue _*0.0125*_ mm
   d. Did the results of the experiment *faster* support the null hypothesis? _____
   e. What can you conclude about the relationship between molecular mass and the rate of diffusion? _*heavier takes more time to diffuse than a lighter*_ *molecule*

4. Diffusion and Molecular Size

   a. After 20 min, describe any color change in your setup. *inside sac – blue* *dark*

   b. Explain any color change. *because iodine in test tube diffused into sac*

   c. Did starch diffuse from the sac? *no* — How do you know? *starch* *test w/ iodine – no color change — no color outside sac*

   d. Did glucose diffuse from the sac? *yes* — How do you know? *benedictine* *solution – turned brown/red*

   e. Considering starch, (glucose,) iodine, and water, *circle* those substances whose molecules are able to pass through the sac.

5. Osmosis

   a. Define osmosis. *the diffusion of water through a semipermeable or selectively permeable membrane.*

   b. If 10% and 5% salt solutions are separated by a membrane permeable to both salt and water, indicate by an arrow the direction that:

      salt will diffuse. | 10% sol'n → 5% sol'n |

      water will diffuse. | 10% sol'n ← 5% sol'n |

   c. Consider 100 ml of 10% sucrose and 10 ml of 50% sucrose. Which has the: *10 ml* *5 ml*
     greater concentration of sucrose? *10 ml of 50% sucrose*
     greater quantity of sucrose? *100 ml of 10% sucrose*

   d. What is the independent variable? *(% of solution) sucrose – concentration*

   e. What is the dependent variable? *rate of osmosis*

   f. What is the purpose of the sac of water? *control – for comparison of weights* *(need to know how much the same sized sac of water weighs to determine mass of solutes...)*

   g. Record the mass of the sacs in the chart below.

	20% Sucrose	10% Sucrose	100% Water
After 1 hr	_____ g	_____ g	_____ g
Start	_____ g	_____ g	_____ g
Increase	_____ g	_____ g	_____ g
% increase*	_____	_____	_____

$$* \; \% \text{ increase} = \frac{\text{mass increase}}{\text{mass at start}}.$$

h. Do the results support the null hypothesis? _____

i. What can you conclude from the results? _____
_____
_____
_____

6. Osmosis and Living Cells

a. Considering hypotonic and hypertonic solutions, which solution has the:

greater concentration of water? *hypotonic*

greater concentration of solutes? *hypertonic*

b. Indicate with an arrow the direction that water moves between hypotonic and hypertonic solutions separated by a selectively permeable membrane.

| Hypotonic solution | ----⫶-> | Hypertonic solution |

c. Consider a 5% sucrose solution separated from a 10% sucrose solution by a semipermeable membrane.

Which solution is hypotonic? *5% sucrose*

Which solution is hypertonic? *10% sucrose*

d. Indicate the surrounding fluid that caused the celery sticks to be:

more flexible. *10% NaCl*   more crisp. *water*

Explain what produced this result. *water filled cells so the stalk was crisp & rigid ; salt leached cells of water so the membranes pull away from walls (shrink), leaving the stalk flacid.*

e. Draw the distribution of chloroplasts in an *Elodea* cell mounted in:

Water   **Salt Solution**

f. Explain what happened to *Elodea* cells mounted in salt solution. *same as celery — hypertonic salt solution caused water to leave the cells, they shrink & pull away from cell wall*

g. Explain the basis of preserving foods with salt or sugar. _____
_____
_____

LABORATORY REPORT 6

Photosynthesis

1. Introduction
   a. Write the term that matches the phrase.
      1. Organisms synthesizing organic nutrients _____
      2. Form of energy captured by photosynthesis _____
      3. Form of energy formed by photosynthesis _____
      4. End product of photosynthesis _____
      5. Storage form of carbohydrates in plants _____
      6. Source of O_2 formed by photosynthesis _____
      7. Source of oxygen atoms in glucose formed by photosynthesis _____
      8. Source of hydrogen atoms in glucose formed by photosynthesis _____
      9. Source of carbon atoms in glucose formed by photosynthesis _____
   b. List the key events in each reaction:
   Light reaction _____
   _____
   Dark reaction _____
   _____

2. Carbon Dioxide and Photosynthesis
   a. What is the independent variable? _____
   Indicate the presence or absence of the independent variable.
      Plant A _____   Plant B _____
   b. What is the dependent variable? _____
   c. Record your results: the presence or absence of starch.
      Plant A _____   Plant B _____
   d. Do the results support the null hypothesis? _____
   State a conclusion from your results. _____
   _____
   _____
   _____
   _____

3. Light and Photosynthesis
  a. What is the independent variable? _____
  b. What is the dependent variable? _____
  c. Trace or sketch your leaf, showing the position of the leaf shield.

Indicate your results by showing on the diagram of your leaf where starch is present and absent.
  d. Do the results support the null hypothesis? _____
  State a conclusion from your results. _____
  _____
  _____

4. Chlorophyll and Photosynthesis
  a. What is the independent variable? _____
  b. What is the dependent variable? _____
  c. Trace or sketch half a leaf, showing the distribution of chlorophyll.

Indicate your results by showing on the diagram of your leaf where starch is present and absent.
  d. Do the results support the null hypothesis? _____
  State a conclusion from your results. _____
  _____
  _____

  e. Indicate the presence or absence of sugar.
     Green leaf tissue _____   Nongreen leaf tissue_____
     Explain your results. _____
     _____
     _____
     _____

5. Chloroplast Pigments
   a. Attach your chromatogram and label the pigment lines of carotene, chlorophyll a, chlorophyll b, and xanthophyll.

   b. Indicate the chlorophyll type that was:
   Most soluble _____   Least soluble _____
   c. Explain why all leaf pigments are not visible in a healthy leaf.
   _____
   _____

   d. How do you explain the appearance of yellow pigments in many plant leaves in autumn?
   _____
   _____

   e. Examine Figure 6.5 and determine:
   1. The three parts of the spectrum responsible for most of photosynthesis.
   _____   _____   _____

   2. Colors of light absorbed primarily by chlorophyll a.
   _____   _____

   3. Colors of light absorbed primarily by chlorophyll b.
   _____   _____

   4. Explain why most leaves appear green. _____
   _____
   _____

6. Light Intensity and the Rate of Photosynthesis
   a. What is the independent variable? _____
   b. What is used to indicate the rate of photosynthesis? _____
   _____
   Why is this a good indicator of photosynthetic action? _____
   _____

c. Record your data in the chart below.

Light Intensity and the Rate of Photosynthesis

Distance from Light Source (cm)	Readings (ml)		ml O_2/3 min	ml O_2/min
	Start	Stop		
30	———	———	———	———
	———	———	———	———
	———	———	———	———
	Average		———	———
60	———	———	———	———
	———	———	———	———
	———	———	———	———
	Average		———	———
90	———	———	———	———
	———	———	———	———
	———	———	———	———
	Average		———	———
120	———	———	———	———
	———	———	———	———
	———	———	———	———
	Average		———	———

d. Do the results support the null hypothesis? _____ State a conclusion from your results.

_____

_____

_____

_____

e. Explain why your conclusion should be modified to "within the range of values tested"?

_____

_____

_____

_____

LABORATORY REPORT 7

Cellular Respiration

1. Introduction

a. Write the summary equation for the aerobic respiration of glucose. *Underline* the reactants and *circle* the products.

b. Indicate the products of anaerobic cellular respiration of glucose in:

Animals _____   Yeast _____

c. Indicate the number of ATP molecules produced per glucose molecule respired by:

Aerobic respiration _____   Anaerobic respiration _____

d. What molecule provides immediate energy for cellular work? _____

e. Some energy released by cellular respiration is captured in high-energy phosphate bonds. What happens to these high-energy phosphates? _____
_____
_____

What happens to released energy that is not captured? _____
_____
_____

f. What regulates cellular respiration so that energy is released gradually? _____
_____
_____

2. Respiration and Carbon Dioxide Production

a. Record the results of your experiments in the table below.

Carbon Dioxide Production in Animals and Germinating Seeds

Tube	Contents Plus Bromthymol Blue	Color of Bromthymol Blue After Test Interval	CO_2 Concentration	
			Increase	No Change
Exp. 1: Humans				
1	Exhaled air	_____	_____	_____
2	Atmospheric air	_____	_____	_____
Exp. 2: Germinating Peas and Crickets				
1	Germinating peas	_____	_____	_____
2	Live crickets	_____	_____	_____
3	Atmospheric air	_____	_____	_____

b. What is the purpose of tube 2 in Experiment 1 and tube 3 in Experiment 2? _____

c. Do the results support the null hypothesis? _____

d. State a conclusion from the results of Experiment 1. _____

e. State a conclusion from the results of Experiment 2. _____

3. Respiration and Heat Production

a. State your prediction. _____

b. State your null hypothesis. _____

c. Which bottle serves as the control? _____

d. Record the temperature in each vacuum bottle.

    Bottle 1 _____ °C   Bottle 2 _____ °C   Bottle 3 _____ °C

e. Do the results support your null hypothesis? _____

f. State a conclusion from your results. _____

4. Temperature and Respiration Rate: Peas and Crickets

a. Record the data collected for the rate of respiration in germinating pea seeds and crickets at 10 °C, room temperature, and 40 °C. *Add your room temperature to the chart.

Oxygen Consumption in Pea Seeds and Crickets

Organism	Temp. (°C)	ml O$_2$/3 min 5 Replicates 1	2	3	4	5	Average ml O$_2$/3 min	Total Mass (g)	Average ml O$_2$/hr/g
Peas	10								
	*								
	40								
Crickets	10								
	*								
	40								
Control	10								
	*								
	40								

b. Do the results for peas support the null hypothesis? _____
   State a conclusion from the results. _____
   _____

c. Do the results for crickets support the null hypothesis? _____
   State a conclusion from the results. _____
   _____

d. Which organism has the greater average rate of aerobic respiration per gram of body mass? _____

5. Temperature and Respiration Rate: Frog and Mouse

a. Record the data collected for the rate of respiration in a frog and a mouse at 10 °C, room temperature, and 40 °C. *Add your room temperature to the chart.

Oxygen Consumption in a Frog and a Mouse

Organism	Temp. (°C)	1	2	3	4	5	Average ml O_2/3 min	Mass (g)	Average ml O_2/hr/g
Frog	10								
	*								
	40								
Mouse	10								
	*								
	40								

Columns 1–5 are headed: *ml O_2/3 min — 5 Replicates*

b. What is the independent variable? _____
c. Do the results support the null hypothesis? _____
d. Plot the average respiration rate (ml O_2/hr/g) of each organism tested in the graph below. Select and record a different symbol for each organism so your graph may be easily read.

Symbols
Frog:
Mouse:

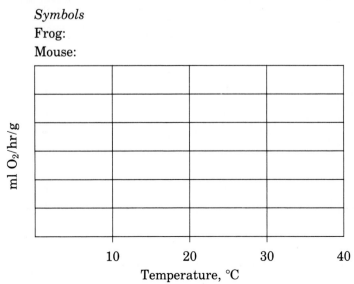

ml O_2/hr/g

Temperature, °C

10   20   30   40

e. State a conclusion about the effect of temperature on the rate of aerobic respiration in the frog.

_____

_____

_____

_____

_____

_____

_____

f. State a conclusion about the effect of temperature on the rate of aerobic respiration in the mouse. _____

_____

_____

_____

_____

_____

g. Explain the different responses to 10 °C by the frog and the mouse.

_____

_____

_____

_____

_____

_____

LABORATORY REPORT 8

Student _____

Lab Instructor _____

Cell Division

1. Introduction

Write the type of cell division that matches each phrase.

1. Occurs in prokaryote cells — *binary fission*
2. Occurs in haploid eukaryote cells — *mitotic (mitosis)*
3. Occurs in diploid eukaryote cells — *mitotic & meiotic both occur in diploid*
4. Forms cells with identical genetic composition — *mitotic*
5. Forms cells with half the chromosome number of the parent cell — *meiotic*

2. Mitotic Division

a. Write the term that matches each phrase.

1. Compose a replicated chromosome — *Interphase*
2. Division of the cytoplasm — *Cytokinesis*
3. 5–10% of the cell cycle — *Mitosis — M stage*
4. 90–95% of the cell cycle — *Interphase*
5. Stage of interphase where chromosomes replicate — *Synthesis - S stage*
6. Members of a chromosome pair — *sister chromatids*
7. Diploid chromosome number in humans — *46*
8. Haploid chromosome number in humans — *23*

b. Write the mitotic stage that matches each phrase.

1. Nuclear membrane and nucleolus disappear — *Prophase*
2. Spindle is formed — *Metaphase*
3. Separation of sister chromatids — *books Prophase ? teacher* — *Anaphase*
4. Daughter cells are formed — *Prophase (Cytokinesis)*
5. Rod-shaped chromosomes are first visible — *Prophase*
6. Chromosomes line up on equatorial plane — *Metaphase*
7. New nuclei are formed — *Telophase*
8. Daughter chromosomes migrate to opposite poles of the cell — *Anaphase*

c. Draw whitefish blastula cells in the stages noted below as they appeared on your slide. Label pertinent structures.

Interphase	Prophase	Metaphase

high power

Early Anaphase	Late Anaphase	Telophase

d. Draw onion root tip cells in the following stages as they appeared on your slide. Label pertinent structures.

Prophase *Metaphase* **Anaphase**   **Telophase**

e. Describe how mitotic cell division in plants differs from that in animal cells. ① *most plants don't have centrioles just spindle fibers* ② *cell wall prevents formation of cleavage furrow during cytokinesis – cell plates form* ③ *doesn't always into to*

f. In the onion root tip, do the daughter cells occupy the same column of cells as the parent cell? *yes* Is it the same for all cells in the root tip? *no*

g. Describe the process of mitotic cell division in your own words without using the names of the phases. *one cell (parent) divides into two identical cells (daughter cells) – each identical to parent (diploid 2N) each of which go on to become parent cells and so on.*

3. Meiotic Division

a. Considering meiosis in humans, indicate the:

1. Number of chromosomes in the parent cell — 46
2. Number of chromatids in daughter cells formed by first division — 46
3. Number of chromosomes in daughter cells formed by first division — 23
4. Number of chromosomes in daughter cells of second division — 23
5. Number of haploid cells formed by meiotic division of parent cell — 4

b. Diagram the arrangement of the chromosomes in mitosis and meiosis as they would appear in the phases noted below where the diploid (2n) chromosome number is 4.

Mitosis

Metaphase

(23 pair lined up @ equator)

Anaphase

separated & migrated to poles

Meiosis I

Metaphase

tetrads @ equator

Anaphase

separate pulled by spindle fibers

Meiosis II

Metaphase

dyads @ equator

Anaphase

c. Indicate the number of chromosomes in human cells formed by:
Mitotic cell division ___46___ Meiotic cell division ___23___

d. Summarize in your own words the process of meiosis without using the names of the phases.

one parent cell goes through 2 cell divisions to end up as 4 daughter cells. During meiosis I the chromosomes pair forming tetrads (4 chromatids each) and separate (2 cells 23 pairs chrom) then in meiosis II they split again to finally form 4 daughter cells w/ 23 chromosomes each (haploid N)

e. Meiotic division is the process leading to the formation of specific cells in animals and plants. Name these cells.
Animals ___gametes (sperm & egg)___ Plants ___meiospores___

f. Indicate the number of chromosomes in a human:
Egg cell ___23___ Sperm ___23___

g. (True/False)
The zygote formed by the union of haploid gametes receives:

A random number of chromosomes from each parent ___false   23 from each___

Equal numbers of chromosomes from each parent ___true___

One member of each chromosome pair from each parent ___true___

Both members of each chromosome pair from each parent ___false___

4. Summary

a. Summarize the primary role of mitotic cell division in:

Unicellular organisms ___reproduction___

Infant humans ___growth___

Adult humans ___growth & repair — as worn out or damaged cells die they are replaced by newly formed cells___

b. Summarize the role of meiotic cell division in:

Animals ___reproduction (sexual) — eggs & sperm___

Plants ___sexual reproduction — production of spores___

LABORATORY REPORT 9

Heredity

1. Fundamentals

Write the term that matches each phrase.

1. Traits passed from parents to progeny   _____
2. Part of DNA coding for a specific protein   _____
3. Contain homologous genes   _____
4. Alternate forms of a gene   _____
5. Alleles of a gene pair are identical   _____
6. Alleles of a gene pair are different   _____
7. Observable form of a trait   _____
8. Genetic composition determining a trait   _____
9. Allele expressed in heterozygote   _____
10. Allele not expressed in heterozygote   _____

2. Monohybrid Crosses with Dominance

a. In corn plants, plant height is controlled by one gene with two alleles: tall (T) and dwarf (t). How many alleles that control the trait for height are present in each cell nucleus of a corn plant? _____

How many alleles that control the trait for height are present in the nucleus of each gamete formed by a corn plant? _____

b. Indicate the genotypes of these corn plants.

homozygous tall _____   heterozygous tall _____   dwarf _____

c. Indicate the genotypes of possible gametes of these corn plants.

homozygous tall _____   heterozygous tall _____   dwarf _____

d. Determine the predicted phenotype ratio in progeny of a cross between two monohybrid tall corn plants.

     Parent phenotypes   _____   _____

     Parent genotypes   _____   _____

     Gametes   ____   ____   ____   ____

     Punnett square

     Genotype ratio   _____

     Phenotype ratio   _____

e. Examine the corn seedlings that are progeny of a cross of two monohybrid tall corn plants. Record the number of:

Tall plants _____   Dwarf plants _____

Determine the observed phenotype ratio of tall plants to dwarf plants by dividing the number of plants of each type by the number of dwarf plants. Round your answers to whole numbers.

_____ Tall ÷ _____ dwarf = _____   _____ Dwarf ÷ _____ dwarf = 1

Record the observed phenotype ratio: _____ Tall : 1 dwarf

Is the observed ratio different from the predicted ratio? _____

If so, explain why this may have happened. _____

_____

f. Is it possible to observe differences between homozygous and heterozygous tall corn plants?

_____

How can you determine their genotypes? _____

g. In a test cross, a tall corn plant is crossed with a _____ corn plant. Indicate the genotype of the tall corn plant when:

all the progeny are tall.   _____

half the progeny are tall.   _____

h. Indicate the possible genotypes of parent pea plants in crosses yielding the following ratios.

Phenotype Ratio	**Parent Genotypes**
1. All white flowers	_____ × _____
2. All purple flowers (3 possibilities)	_____ × _____
	_____ × _____
	_____ × _____

i. Mary has freckles, but her husband Dick does not. Mary's father has freckles but her mother does not. What is the probability that Mary and Dick's child will have freckles? _____

	Mary	**Dick**
Parent phenotypes	_____	_____
Parent genotypes	_____	_____
Gametes	____ ____	____ ____

Punnett square

Genotype ratio   _____

Phenotype ratio   _____

j. Newlyweds Bill and Sue are nonfreckled. Since each had one parent who had freckles, they wonder what the probability is of their children having freckles. What would you tell them?

_____

_____

_____

k. Judy and Tom are wondering what kind of hairline their soon-to-be-born baby will have. Judy has a straight hairline but Tom has a widow's peak. What is Judy's genotype? _____
What are the two possible genotypes for Tom? _____
What is the probability of their baby having a widow's peak if Tom is:
　　homozygous? _____　　　　　heterozygous? _____

l. The table below lists several human traits that are determined by a simple dominant/recessive mode of inheritance. Record your phenotypes and the frequency of the phenotypes among members of your class.

Human Dominant/Recessive Traits Determined by a Single Gene*

Trait	Phenotype	Your Phenotype	Number in Class with Phenotype
Handedness	Right handed*		
	Left handed		
Ear lobes	Free*		
	Attached		
Freckles	Freckled*		
	Nonfreckled		
Hair line	Widow's peak*		
	Straight		
Little finger	Bent*		
	Straight		
Tongue roll	Yes*		
	No		
Rh factor	Rh$^+$*		
	Rh$^-$		

*Indicates dominant traits.

m. For each trait, is the dominant phenotype always more abundant among class members?

_____

n. Assuming your class to be representative of the general population, what can you conclude about the relationship between dominant and recessive phenotypes and their frequencies in the population? _____

_____

3. Monohybrid Crosses with Codominance
a. What is the genotype of a pink-flowering snapdragon? _____

b. Indicate the predicted genotype and phenotype ratios in progeny from crossing two pink-flowering snapdragons.
　Genotype ratio _____
　Phenotype ratio _____

c. What type of cross would you use to produce progeny that are 100% pink-flowering?
_____

d. What is the probability that a child with sickle-cell anemia will be born to parents that are each heterozygous for sickle-cell? _____

4. **Multiple Alleles: ABO Blood Types**
   a. Indicate the expected genotype and phenotype ratios for these matings:
      1. $I^A I^B \times ii$
         Genotype ratio: _____
         Phenotype ratio: _____
      2. $I^A i \times I^B i$
         Genotype ratio: _____
         Phenotype ratio: _____
   b. (True/False)
      _____ A type O child may have two type A parents.
      _____ A type O child may have one type AB parent.
      _____ A type B child may have two type AB parents.
      _____ A type AB child may have one type O parent.
   c. Ann (type A, Rh$^+$ blood) is suing Joe (type AB, Rh$^-$ blood) for child support claiming that he is the father of her child (type B, Rh$^+$ blood). On the basis of blood type, it is possible that Joe is the father? _____ What ABO blood type(s) would Joe have to possess to *prove* that he is *not* the father? _____

5. **Dihybrid Crosses with Dominance**
   a. Determine the predicted phenotype ratio in children of dihybrid parents that are both right-handed and possess free earlobes.
      Parent phenotypes _____   _____
      Parent genotypes   _____   _____
      Gametes   ____ ____ ____ ____   ____ ____ ____ ____

      Punnett square

	____	____	____	____
____				
____				
____				
____				

      Phenotype ratio   _____ right-handed   _____ left-handed
                                free earlobes   free earlobes
                        _____ right-handed   _____ left-handed
                                attached earlobes   attached earlobes

b. Determine the predicted phenotype ratio in children of a woman who is dihybrid for right-handed and free earlobes and a man who is left-handed with attached earlobes.

Parent phenotypes _____   _____

Parent genotypes _____   _____

Gametes _____ _____ _____ _____   _____ _____ _____ _____

Punnett square _____ _____ _____ _____

Phenotype ratio _____ right-handed   _____ left-handed
free earlobes   free earlobes
_____ right-handed   _____ left-handed
attached earlobes   attached earlobes

6. Linked Genes

a. Indicate the types of gametes produced by an AaBb dihybrid when the:

genes are not linked. _____

genes are linked. _____

b. Determine the predicted genotype and phenotype ratios in children of a noncarrier woman with normal vision and a color-blind man.

Parent phenotypes _____   _____

Parent genotypes _____   _____

Gametes _____ _____   _____ _____

Punnett square _____ _____

Genotype ratio _____

Phenotype ratio _____

c. From which parent does a color-blind son inherit the gene for color blindness? _____

7. Pedigree Analysis

Examine the pedigrees shown below, and determine whether the inherited trait (■ or ●) is dominant, recessive, or sex-linked recessive; □ = male; ○ = female.

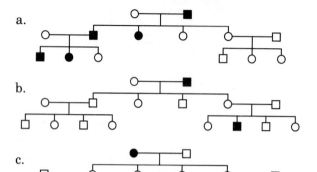

a. _____

b. _____

c. _____

8. Polygenic Inheritance

List several human traits that seem to be determined by polygenes. _____

_____

9. Chi-Square Analysis

a. In the example of chi-square determination in Table 9.6, what is the hypothesis being tested?

_____

_____

Would the hypothesis be supported by a chi square of 3.952? _____

b. Record your results and do a chi-square analysis.

Chi-Square Analysis of Progeny from a Monohybrid Cross

Phenotype	Actual Results	Expected Results	Deviation (d)	(d^2)	(d^2/e)
Purple kernels	_____	_____	_____	_____	___/___ = ___
White kernels	_____	_____	_____	_____	___/___ = ___

$$\Sigma(d^2/e) = \text{\_\_\_\_}$$

$$\chi^2 = \text{\_\_\_\_}$$

p falls between _____ and _____.

Is the predicted ratio supported by chi square? _____

Molecular and Chromosomal Genetics

1. DNA

a. Write the term that matches each phrase.
1. Molecule containing genetic information
2. Sugar in DNA nucleotides
3. Base that pairs with adenine
4. Base that pairs with cytosine
5. Chemical bonds joining complementary nitrogen bases
6. Number of nucleotide strands in DNA
7. Two molecules forming sides of the DNA "ladder"

DNA – deoxyribonucleic acid
deoxyribose (C₅)
thymine (T)
guanine (G)

hydrogen bonds
2
sugar & phosphate

b. In DNA replication, what determines the sequence of nucleotides in the new strands of nucleotides that are formed? *the complementary pairing of nitrogen bases*

c. After determining the pairing pattern of the nitrogenous bases in Figure 10.2, add the missing bases to this hypothetical strand of DNA.

A = adenine, C = cytosine, G = guanine, T = thymine

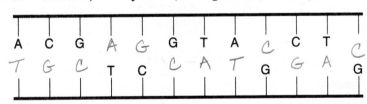

A	C	G	A	G	G	T	A	C	C	T	C
T	G	C	T	C	C	A	T	G	G	A	G

d. At the left is a hypothetical segment of DNA. After replication it forms two strands as shown at the right. Add the missing bases in the replicated strands.

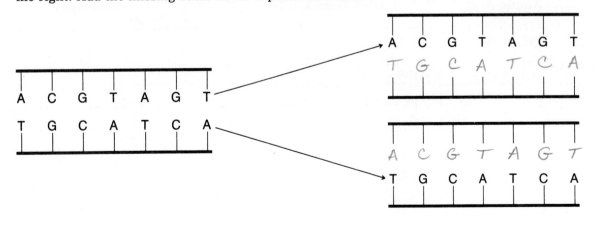

2. RNA Synthesis

a. Write the term that matches each meaning.
1. Sugar in an RNA nucleotide   *ribose*
2. Number of nucleotide strands in RNA   *single*
3. RNA base that pairs with adenine   *uracil*
4. Template for RNA synthesis   DNA

b. The figure below shows a portion of DNA molecule whose strands have separated to synthesize mRNA. The bases are shown for the strand that is *not* serving as the template. Add the bases of the DNA strand serving as the template and the bases of the newly formed RNA molecule.

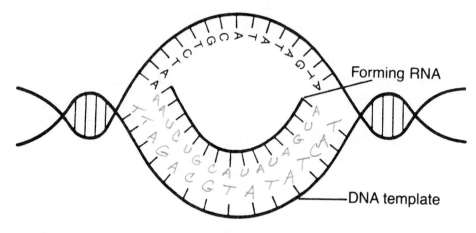

Forming RNA

DNA template

3. Protein Synthesis

a. Write the term that matches each meaning.
1. Carries the genetic code to ribosomes   mRNA
2. Formed of an mRNA base triplet   codon
3. Specifies a particular amino acid   base triplet
4. Sites of protein synthesis   ribosome
5. Carries amino acids to ribosome   tRNA
6. tRNA triplet that pairs with codon
7. Codon that starts polypeptide synthesis   AUG
8. Codons that stop polypeptide synthesis   UUA, UAG, UGA

b. Indicate the DNA base triplet and tRNA anticodons for the mRNA codons in Figure 10.4.

Amino Acid	DNA Base Triplet	Anticodon
Tryptophan	UGA, UGG	ACU, ACC
Lysine	AAA, AAG	UUU, UUC
Valine	GUU, GUC, GUA, GUG	CAA, CAG, CAU, CAC
Methionine	AUG	UAC

c. Using Table 10.1, indicate the possible sequences of mRNA codons that code for the polypeptides below.

methionine - **lysine** - **histidine** - **tryptophan** - **glutamine** - **tryptophan** - **(stop)**

| AUG | AAA | CAU | _____ | CAA | UGA | _____ |
| | AAG | CAC | | CAG | UGC | _____ |

d. Molecular biologists use probes of mRNA to locate specific gene loci on DNA segments. Use the codon sequences determined in item 3c as probes to locate the gene (complementary base sequence) in the segment of template (single-strand) DNA shown below. *Circle* the located gene and indicate the probe that was successful in locating it.

-A-G-T|T-C-T|T-A-C|C-C-T|G-A-A|C-G-G|C-A-T|T-C-A|

-G-A-C|A-T-T|C-C-G|C-T-A|G-A-C|C-A-T|T-A-C|T-T-C|

-G-T-A|A-C-C|G-T-T|A-C-C|A-T-C|G-T-A|T-C-A-

Successful probe _____

e. Consider the following hypothetical gene. Indicate the mRNA codons that it forms and the sequence of amino acids produced in the polypeptide chain.

-T-A-C|T-T-A|G-A-A|A-T-A|C-C-G|A-A-G|A-C-T-

mRNA codons

_____

Amino acid sequence

_____

f. For the following mutations, show the effect by indicating the mRNA codons and the amino acid sequence in the polypeptide.

1. Base substitution

-T-A-C|T-T-A|G-A-G|A-T-A|C-C-G|A-A-G|A-C-T-

mRNA codons

_____

Amino acid sequence

_____

2. Triplet addition

-T-A-C|T-T-A|G-G-T|G-A-A|A-T-A|C-C-G|A-A-G|A-C-T-

mRNA codons

_____

Amino acid sequence

_____

3. Base addition

        -T-A-C⏐T-T-A⏐G-A-A⏐A-T-C-A⏐C-C-G⏐A-A-G⏐A-C-T-

mRNA codons

_____

Amino acid sequence

_____

4. Base deletion (site indicated by arrow)

                           ↓

        -T-A-C⏐T-T-A⏐G-A-A⏐A-T⏐C-C-G⏐A-A-G⏐A-C-T-

mRNA codons

_____

Amino acid sequence

_____

g. What seems to determine the magnitude of a mutation's effect? _____

_____

_____

_____

h. Summarize how DNA controls cellular functions. _____

_____

_____

_____

_____

_____

4. **Chromosomal Abnormalities**
   a. Prepare karyotypes for patients A, B, and C on the following three pages.
   b. Indicate the sex and chromosomal abnormality, if any, for each patient whose karyotype you analyzed.

Patient	Sex	Abnormality
A	_____	_____
B	_____	_____
C	_____	_____

   c. Amniocentesis is used to obtain fetal cells from the amniotic fluid for chromosomal analysis. If you were a prospective parent of a fetus with a chromosomal abnormality, what factors would you consider in deciding whether to abort the fetus? _____

   _____

   _____

   _____

Human Karyotype Analysis Form
Patient A

A

B

1   2   3   4   5

C

6   7   8   9   10   11   12   X

D

E

13   14   15   16   17   18

F

G

19   20   21   22   Y

Sex of subject _____   Number of chromosomes _____

Chromosomal disorder _____

Human Karyotype Analysis Form
Patient B

A

B

-        -

| 1 | 2 | 3 | 4 | 5 |

C

- -

| 6 | 7 | 8 | 9 | 10 | 11 | 12 | X |

D

E

-        -

| 13 | 14 | 15 | 16 | 17 | 18 |

F

G

-        -

| 19 | 20 | 21 | 22 | Y |

Sex of subject _____   Number of chromosomes _____

Chromosomal disorder _____

Human Karyotype Analysis Form
Patient C

A

B

1	2	3	4	5

C

6	7	8	9	10	11	12	X

D

E

13	14	15	16	17	18

F

G

19	20	21	22	Y

Sex of subject _____   Number of chromosomes _____

Chromosomal disorder _____

Student _____

LABORATORY REPORT 11

Lab Instructor _____

Organization of the Human Body

1. Organs and Organ Systems

Match the organ systems to the phrases below.

a. Cardiovascular   e. Lymphatic   i. Reproductive, female
b. Digestive   f. Muscular   j. Respiratory
c. Endocrine   g. Nervous   k. Skeletal
d. Integumentary   h. Reproductive, male   l. Urinary

Organs

_C_ Pancreas, ovaries, testes
_B_ Stomach, liver, intestines
_K_ Bones, cartilages, ligaments
_E_ Spleen, lymph nodes, tonsils
_J_ Trachea, bronchi, lungs
_C_ Thyroid and adrenal glands
_H_ Testes, vas deferens, urethra
_G_ Sensory receptors, nerves

_G_ Brain and spinal cord
_D_ Epidermis and dermis
_A_ Heart, blood vessels
_L_ Kidneys, urinary bladder
_I_ Uterus, oviducts
_B_ Esophagus, liver
_L_ Urethra, ureters
_B_ Pancreas, salivary glands

Functions

_C_ Chemical coordination of body functions.
_A_ Transports materials throughout the body.
_G_ Coordination of body functions by neural impulses.
_D_ Protects body from ultraviolet radiation.
_F_ Contractions enable body movements.
_B_ Converts nonabsorbable food into absorbable nutrients.
_L_ Removes metabolic wastes and excess minerals from the body.
_J_ Exchange of gases between atmospheric air and blood.
_E_ Collects, cleans, and recycles extracellular fluid.
_K_ Supporting framework of the body.
_D_ Prevents excessive evaporative water loss from body surface.
_L_ Produces urine and removes it from the body.

2. Body Cavities

a. List the labels for Figure 11.1.

1. Dorsal Body
2. Cranial
3. Spinal
4. Diaphragm
5. Thoracic
6. Ventral Body

7. Abdominopelvic
8. Abdominal
9. Pelvic
10. Left Pleural
11. Right Pleural
12. Pericardial

b. Match the organ with the body cavity in which it is located.

1. Abdominal	3. Pelvic	5. Thoracic
2. Cranial	4. Spinal	6. None

_____ Brain   _____ Liver   _____ Bronchi

_____ Urinary bladder   _____ Heart   _____ Stomach

_____ Adrenal gland   _____ Pancreas   _____ Rectum

_____ Spinal cord   _____ Lungs   _____ Thyroid gland

_____ Testes   _____ Spleen   _____ Uterus

_____ Esophagus   _____ Kidney   _____ Trachea

_____ Duodenum   _____ Colon   _____ Ovaries

3. Epithelial Tissues

a. Draw the following epithelial tissues as they appear on your slides net 100×.

Simple Columnar   **Simple Cuboidal**

**Pseudostratified
Ciliated Columnar**   **Stratified Squamous**

b. Match the epithelial tissue with the phrase.

1. Simple squamous	3. Simple ciliated columnar	5. Pseudostratified ciliated columnar
2. Simple cuboidal	4. Simple nonciliated columnar	6. Stratified squamous

_____ Forms kidney tubules   _____ Lines digestive tract

_____ Lines upper respiratory passages   _____ Lines oviducts

_____ Contains goblet cells   _____ Lines blood vessels

_____ Epidermis of skin   _____ Lines body cavities

4. Connective Tissues

   a. Draw the following connective tissues as they appear on your slides at 100×.

Loose Fibrous   **Adipose**   **Dense Fibrous**

Conn. Tissue   **Tissue**   **Conn. Tissue**

Hyaline

Cartilage   **Fibrocartilage**   **Bone**

   b. Match the connective tissue with each phrase.

1. Loose fibrous	4. Hyaline cartilage	7. Bone
2. Adipose	5. Elastic cartilage	8. Blood
3. Dense fibrous	6. Fibrocartilage	

     _____ Forms ligaments and tendons   _____ Forms cartilage of the nose

     _____ Supports trachea and larynx   _____ Attaches skin to muscles

     _____ Fat storage tissue   _____ Supports external ear

     _____ Forms intervertebral discs   _____ Has a liquid matrix

     _____ Forms dermis of skin   _____ Covers ends of long bones

5. Muscle Tissue

   a. Draw the following muscle tissues as they appear on your slides. Indicate the magnification for each drawing.

Smooth Muscle   **Skeletal Muscle**   **Cardiac Muscle**

b. Match the muscle tissue with the phrase.

1. Skeletal   2. Cardiac   3. Smooth

_____ Multinucleated cells   _____ Striations present

_____ Striations absent   _____ Voluntary

_____ Involuntary   _____ Wall of heart

_____ Wall of stomach   _____ Attached to bones

6. Nerve Tissue

a. Draw one or two neuron cell bodies with their neuron processes as they appear on your slide. Indicate the magnification used.

b. Name the two types of cells composing nerve tissue.

_____   _____

Which cells conduct neural impulses? _____

Which cells are supportive cells? _____

c. Which neuron process conducts impulses:

toward the cell body? _____

away from the cell body? _____

Student _____

LABORATORY REPORT 13

Lab Instructor _____

Circulation of Blood

1. The Heart

a. List the labels for Figure 13.1.

1. Aorta
2. Superior vena cava
3. Pulmonary trunk
4. Pulmonary semilunar valve
5. Right atrium
6. Tricuspid atrioventricular valve
7. Right ventricle
8. Inferior vena cava
9. Pulmonary veins
10. Left atrium
11. Aortic Semilunar Valve
12. Bicuspid Atrioventricular Valve
13. Chordae Tendinae
14. Ventricular septum
15. Left ventricle
16. Papillary muscle

b. Write the term that matches the phrase.

1. Returns blood to right atrium — Superior vena cava

   inferior vena cava
2. Returns blood to left atrium — left & right pulmonary veins
3. Separates the ventricles — ventricular septum
4. Pumps blood into pulmonary trunk — Right ventricle
5. Pumps blood into aorta — Left ventricle
6. Prevents backflow of blood into the right atrium — Tricuspid atrioventricular valve
7. Prevents backflow of blood into the left atrium — Bicuspid atrioventricular valve
8. Prevents backflow of blood into the right ventricle — pulmonary semilunar valve
9. Prevents backflow of blood into the left ventricle — aortic semilunar valve
10. Fibers restraining AV valve cusps — Chordae tendinae
11. Heart muscle — papillary muscles
12. Valves whose closure produces the first heart sound — AV valves
13. Valves whose closure produces the second heart sound — semilunar valves

c. Describe how blood pressure changes in the ventricles open and close the heart valves.

In ventricular systole, the ventricles contract & the atria relax, the sudden increase in blood pressure w/in the ventricles closes the AV valves & pumps blood thru the semilunar valves into the arteries.

d. The heart cycle is the period from one ventricular contraction to the next ventricular contraction. Based on the heart sounds, does the heart relax between contractions? *yes*
Explain. *there is a slight pause after the two sounds — to*
allow blood time to fill up chambers again &
valves to open/close

2. **Dissection of a Sheep Heart** *cows*
   a. What arteries carry blood to supply the heart with nutrients?
   Superior and Inferior vena cava
   b. Which has a thicker wall, the aorta or anterior vena cava?

   c. Do atria or ventricles compose most of the heart? *ventricles*
   d. What vessels open into the left atrium? *pulminary veins*
   e. What vessels open into the right atrium? *superior & inferior vena cava*
   f. *Circle* the terms that describe the chordae tendineae.
      elastic   nonelastic   thick   thin   flexible   stiff
   g. Describe the arrangement and appearance of a semilunar valve. _____

   h. Which ventricle has the thicker wall? *left ventricle*
   Explain the basis of this condition. *the right ventricle pumps blood*
   only to the lungs, the left ventricle has to pump blood
   w/ enough force to reach all other parts of the body.

3. **Pattern of Circulation**
   a. Write the name of the type of blood vessels described.
      1. Carry blood toward the heart   *veins*
      2. Carry blood away from the heart   *arteries*
      3. Smallest arteries   *arterioles*
      4. Smallest veins   *venules*
      5. Smallest blood vessels   *capillaries*
   b. Trace the pathway of blood from the heart to the lungs and back to the heart. Right ventricle
      ⟶ *pulminary trunk*   ⟶ *pulminary arteries*
      ⟶ lungs ⟶ *pulminary veins*   ⟶ left atrium.
   c. Trace the pathway of blood from the heart to the intestine and back to the heart. Left ventricle
      ⟶ *descending aorta*   ⟶ *mesenteric arteries*
      ⟶ intestine ⟶ *hepatic portal vein*   ⟶ liver ⟶
      hepatic vein   ⟶ *inferior vena cava*   ⟶ right atrium.
   d. Describe the flow of blood through a capillary in a frog's foot.

e. Describe the control of blood flow through a capillary. _flow of blood into capillaries is controlled by a precapillary sphincter muscle that is governed by impulses from the autonomic nervous system & local chemical stimuli_ increase in local CO_2 opens sphincter & increases flow decrease " " reduces capillary blood flow.

f. By what process are materials exchanged between capillary blood and body cells? _by tissue (interstitial) fluid_

g. Draw from your slides cross sections of these vessels showing the relative diameter of the vessel interior and thickness of the wall.

Normal Vein	Normal Artery	Atherosclerotic Artery

h. Explain the hazards of atherosclerosis. _it is an accumulation of soft masses of fatty materials, particularly cholesterol, beneath the inner linings of arteries. Such deposits are called plaque, which protrudes into vessel as it develops, interfering w/ blood flow. Can cause clot to form — thrombus which can lead to thromboembolism._

i. What arteries carry deoxygenated blood? _pulmonary & (superior/inferior) vena cavae_
What veins carry oxygenated blood? _pulmonary & aorta_

j. List the labels for Figure 13.6.
1. _Superior Vena Cava_   4. _Right Atrium_   7. _Aorta_
2. _Pulmonary Arteries_   5. _Right Ventricle_   8. _Left Atrium_
3. _Pulmonary Veins_   6. _Inferior Vena Cava_   9. _Left Ventricle_

4. Blood Pressure

a. Explain the meaning of systolic and diastolic blood pressure.
Systolic _ventricular contraction — usually 120 ± 10 mm Hg_

Diastolic _ventricular relaxation usually 80 ± 10 mm Hg_

b. Explain the cause of the pulse. _the alternating increase & decrease of arterial blood pressure causes corresponding pulsating expansion & contraction of the elastic arterial walls_

should be 40-50 the between two numbers

c. Record your pulse rate (beats/min).

At rest _____63_____ /min⁹³   Immediately after exercise _____107_____ /min

How long did it take for your pulse rate to return to its resting rate? _____

_____

d. Record the average and range of the resting pulse rates, by gender, among your class members.

Males: Average ___62.2___ /min   Range ___52___ /min to ___75___ /min

Females: Average ___75.6___ /min   Range ___63___ /min to ___89___ /min

Total: Average ___68.9___ /min   Range ___57.5___ /min to ___82___ /min

Does there seem to be a gender difference in resting pulse rates? ___yes___

Explain. _____

_____

_____

_____

What do you think causes an increase in heart rate during exercise? _____

_____

How would you test your hypothesis? _____

_____

_____

_____

_____

e. Record your blood pressure:

At rest   _____121_____ mm Hg systolic   _____80_____ mm Hg diastolic

After exercise   _____140_____ mm Hg systolic   _____130_____ mm Hg diastolic

3 min later   _____114_____ mm Hg systolic   _____70_____ mm Hg diastolic

What do you think causes an increase in blood pressure during exercise? _____

_____

f. Explain how the sphygmomanometer works in the determination of blood pressure. _____

_____

_____

_____

_____

_____

LABORATORY REPORT 14

Blood

1. Blood Cells
a. Use colored pencils to draw 1 or 2 blood cells of these types as they appear on your slide.

Erythrocytes	Lymphocytes	Monocytes

Neutrophils	Eosinophils	Basophils

b. Draw sickled erythrocytes as they appear on your slide.

c. Write the term that matches the phrase.
   1. Basic function of blood
   2. Liquid portion of blood
   3. Blood cells lacking a nucleus
   4. Blood cells possessing a nucleus
   5. Tiny cell fragments in blood
   6. Blood cells containing hemoglobin
   7. Most abundant blood cells
   8. Most abundant leukocyte
   9. Small WBCs with little cytoplasm
   10. Phagocytic leukocytes

   11. Transport oxygen and carbon dioxide
   12. Produce antibodies
   13. Start the clotting process

(transport materials)
carries materials to & from body cells
plasma (92% water 8% proteins)
RBCs (erythrocytes)
WBCs
Thrombocytes
RBCs
Erythrocytes RBCs
Neutrophils
Lymphocytes
Neutrophils &
Monocytes
RBC's (loosely comb. w/hemoglobin)
B-lymphocytes
Thrombocytes — platelets

2. Differential White Cell Count

a. Record your tabulation of 100 WBCs and their percentages in normal blood.

Leukocyte	Tabulation	Percentage
Neutrophils		
Eosinophils		
Basophils		
Lymphocytes		
Monocytes		

(handwritten in right margin: 60, 2-, 0.5-, 20-, 3-8)

b. Record your tabulation of 100 WBCs and their percentages in blood of a patient with infectious mononucleosis.

Leukocyte	Tabulation	Percentage
Neutrophils		
Eosinophils		
Basophils		
Lymphocytes		
Monocytes		

c. Indicate the WBCs with higher than normal percentages during these conditions.
   1. Antigen-antibody reactions   *Lymphocytes & Basophils*
   2. Parasitic worm infestation   *Eosinophils*
   3. Acute bacterial infection   *Neutrophils*
   4. Chronic infection   *Monocytes*
   5. Viral infection   *Lymphocytes*

ABO BLOOD SIMUTYPE KIT
Student Instructions

Name: _C. Quillen_

PROCEDURE

NOTE. All blood and blood sera in this kit are simulated and contain no blood or blood products. These materials cannot be used to type actual human blood. If human anti-sera were available, however, these same instructions could be used to type actual human blood.

Each student will type one sample of blood. You will need one Hema-tag®, one microscope slide, one blue and one yellow toothpick. Take your turn to follow the steps outlined below. After using the slide, wash it with tap water and return it for 'others to use.

1. Perform test at room temperature. Lay a microscope slide over the anti-A and anti-B circles on a Hema-tag®. Place 1 drop of anti-A and anti-B in their respective circles on the microscope slide.

2. Add 1 drop of cells being tested to each drop of antiserum.

3. Mix the simulated cells and antisera with the appropriate toothpick within their respective circles, scraping the toothpick firmly against the microscope slide in the process. Continue

stirring for two minutes or until agglutination is noted.

4. Read for gross agglutination and record test results. Note that in this simulation precipitations are recorded as agglutinations.

POSITIVE: Cells are agglutinated.
NEGATIVE: Cells are not agglutinated.

RESULTS

Sample	Type
1. Jane Doe	A
2. John Doe, Sr.	O
3. John Doe, Jr.	AB
4. Wiley Smith	B

QUESTIONS

Jane is John Doe, Jr.'s mother. John Doe, Sr. claims that he is not John, Jr.'s father and suspects that Wiley Smith is the father.

Could John, Sr. be John, Jr.'s father?   *no*
Explain.
   O parent cannot have AB child (can have A, B, or O)

Could Wiley Smith be John Jr.'s father? *yes*
Explain.
   mother A, Smith B, child AB

3. **Blood Clotting Time**

   a. Record your blood clotting time. _____ min

   b. Is a short or long clotting time best? ~~short~~ neither _____
      Explain. ~~long~~ ~~leads~~ ~~prevent~~ excessive loss of blood
      _____ short- can lead _____

   c. People with atherosclerosis of the coronary arteries are often given medications to increase the
      clotting time. Explain the reason for such a practice. _____
      _____
      _____

4. **Volume of Packed Red Cells**

   a. What is anemia? condition in which oxygen-carrying capacity of blood is decreased

   b. Record your volume of packed red cells _____ %
      Is this value within the normal range for your gender? _____

   c. Is it possible to be anemic and have a normal VPRC? yes _____
      Explain. you could have a decreased concentration of RBC's _____
      _____

5. **Blood Typing**

   a. Indicate your ABO blood group. _____ A _____   Rh type ____ + ____

   b. Indicate compatibility (C) and incompatibility (I) of possible blood transfusions shown in the
      chart below.

Blood Type and Antigen of Donor	*Blood Type (and Antibodies) of Recipient*			
	O *(a,b)*	*A* *(b)*	*B* *(a)*	*AB* *(none)*
O	C	C	C	C
A	I	C	I	C
B	I	I	C	C
AB	I	I	I	C

   c. Which ABO blood type may receive blood from the other three types in emergencies? AB+
      This is the **universal recipient.**

   d. Which ABO blood type may donate blood to the other three types in emergencies? O-
      This is the **universal donor.**

   e. Considering *both* ABO and Rh blood types, indicate:
      Universal donor ___ O- ___   Universal recipient ___ AB+ ___

   f. Infants suffering from erythroblastosis fetalis may require a massive blood transfusion. What
      blood type or types should be given to a baby with a blood type of A,Rh+? A+, O+, A-, O-
      Explain your answer. baby already has RH factor +, so giving
      A or O in either RH is ok

LABORATORY REPORT 15

Gas Exchange

1. Respiratory System

a. List the labels for Figure 15.1.

1. _____   7. _____
2. _____   8. _____
3. _____   9. _____
4. _____   10. _____
5. _____   11. _____
6. _____

b. List in sequence the air passages that carry air into the lungs.

1. __Nostrils_____   6. _____
2. _____   7. _____
3. _____   8. _____
4. __Glottis_____   9. _____
5. _____   10. __Alveoli_____

c. Write the terms that match the phrases.

1. Site of gas exchange   _____
2. Contains vocal cords   _____
3. Warms and filters inhaled air   _____
4. Separates nasal and oral cavities   _____
5. Opening into larynx   _____
6. Air passageways entering lungs   _____
7. The windpipe   _____
8. The throat   _____
9. Closes glottis when swallowing   _____
10. Tiny air sacs in lungs   _____

d. Draw the following from your slide of trachea, x.s.

Cartilaginous Ring (40×)

Pseudostratified Ciliated Columnar Epithelium (400×)

e. Describe how the ciliated epithelium lining respiratory passages remove particles from inhaled air. _____

_____

_____

f. What is the function of the cartilaginous rings in trachea and bronchi? _____

_____

g. Are the cartilaginous walls of the larynx hard or soft? _____

h. Describe any movement of the larynx during swallowing. _____

_____

_____

i. Draw the microscope appearance of lung tissue at 40×.

      Normal Lung Tissue   **Emphysematous Lung Tissue**

j. By what process are oxygen and carbon dioxide exchanged between air and capillary blood in the lungs? _____

k. What is the value of a large respiratory surface area for gas exchange? _____

_____

_____

l. How is the respiratory surface area reduced in an emphysematous lung? _____

_____

_____

2. **Breathing Mechanics**

  a. Comparing the breathing mechanics model with the body, indicate the body parts simulated by the:

    bell jar _____   rubber sheet _____

    glass tubing _____   balloons _____

  b. When the rubber sheet of the breathing mechanics model is pulled down:

    the volume of the bell jar is _____ and the air

    pressure in the bell jar is _____ , which causes air

    to flow into the lungs.

  c. When the rubber sheet of the breathing mechanics model is pushed up:

    the volume of the bell jar is _____ and the air

    pressure in the bell jar is _____ , which causes air

    to flow out of the lungs.

d. During inspiration, atmospheric air pressure is _____ than air pressure in the lungs.

During expiration, atmospheric air pressure is _____ than air pressure in the lungs.

3. Lung Volumes

a. Calculate your tidal volume (TV).

Spirometer reading after 5 normal breathing cycles. _____

Spirometer reading divided by 5 equals TV. _____ ml

b. Calculate your vital capacity (VC).

Spirometer reading for each replicate: 1 _____ ml; 2 _____ ml; 3 _____ ml

Sum of the replicates _____ ml   Sum divided by 3 equals _____ ml

Is your vital capacity within the normal range? _____

c. Calculate your expiratory reserve volume (ERV).

Spirometer reading minus 1,000 ml for each replicate:

   1 _____ ml − 1,000 ml = _____ ml

   2 _____ ml − 1,000 ml = _____ ml

   3 _____ ml − 1,000 ml = _____ ml

Sum of the replicates _____ ml   Sum divided by 3 equals _____ ml

d. Calculate your inspiratory reserve volume (IRV) using the data from the above tests.

VC − (TV + ERV) = IRV

   VC _____ ml − (TV _____ ml + ERV _____ ml) = IRV _____ ml

e. Calculate your respiratory minute volume (RMV).

   TV _____ ml × breathing cycles/minute = RMV _____ ml/min.

f. Indicate by a check which of the following that you think will change during heavy breathing following exercise.

_____ Tidal volume   _____ Expiratory reserve volume

_____ Respiratory minute volume   _____ Inspiratory reserve volume

_____ Breathing cycles/minute   _____ Vital capacity

g. Test your hypotheses by running in place for 3 min and measuring those items that you have checked. Were your hypotheses confirmed? _____

Explain. _____

_____

_____

h. Emphysema patients have difficulty expelling air from their lungs because their residual volume is increased. What effect would this have on their:

Vital capacity? _____

Expiratory reserve volume? _____

Concentration of oxygen in the lungs? _____

Effectiveness of gas exchange? _____

LABORATORY REPORT 16

Digestion

1. The Digestive System

a. Write the term that matches the phrase.

1. Produces bile _____

2. Digestive fluid formed by stomach mucosa _____

3. End product of protein digestion _____

4. Where carbohydrate digestion begins _____

5. Site of decay of nondigestible materials _____

6. Absorption of nutrients is completed _____

7. Wavelike contractions of digestive tract _____

8. Separate oral and nasal cavities _____

9. Carries food from pharynx to stomach _____

10. Where digestion of food is completed _____

11. Break food into smaller pieces _____

12. Pushes food into pharynx in swallowing _____

13. Reabsorbs water from nondigestible material _____

b. List the labels for Figure 16.1.

1. _____ 4. _____ 7. _____

2. _____ 5. _____ 8. _____

3. _____ 6. _____

c. List the labels for Figure 16.2.

1. _____ 4. _____ 7. _____

2. _____ 5. _____ 8. _____

3. _____ 6. _____

d. List the labels for Figure 16.3.

1. _____ 7. _____ 13. _____

2. _____ 8. _____ 14. _____

3. _____ 9. _____ 15. _____

4. _____ 10. _____ 16. _____

5. _____ 11. _____ 17. _____

6. _____ 12. _____ 18. _____

2. Histology of the Small Intestine

a. List the four layers of the intestine from inside out.

1. _____   3. _____

2. _____   4. _____

b. In what way do villi affect the surface area of the intestinal mucosa? _____

_____

Why is this advantageous? _____

_____

c. What is the function of the smooth muscle tissues in the intestinal wall? _____

_____

_____

d. Draw the following as they appear on your slide.

2 or 3 Villi (40×)   **Columnar Epithelium (400×)**

3. Digestion of Starch

a. Record your results by placing an "S" for a positive test for starch or an "M" for a positive test for maltose in the appropriate squares.

	Tube Numbers and pH																	
	pH 5						pH 8						pH 11					
Temp.	1	1B	2	2B	3	3B	4	4B	5	5B	6	6B	7	7B	8	8B	9	9B
5 °C																		
37 °C																		
70 °C																		

b. List the control tubes expected to give starch tests that are:

positive _____ negative _____

What should be done if these results are not attained? _____

_____

c. Indicate the optimum pH and temperature, among those tested, for amylase activity.

pH _____ Temp. _____ °C

d. How do some pH values and temperatures inactivate amylase? _____

_____

e. Pancreatic amylase is a component of pancreatic juice that is released into the small intestine. What can you hypothesize about the pH of the digestive fluids in the small intestine? _____

_____

LABORATORY REPORT 17

Neural Control

1. Neurons

a. Write the term that matches the phrase.
1. Nerve cell _____
2. Neuron carrying impulses to CNS _____
3. Neuron carrying impulses from CNS _____
4. Composed of cranial and spinal nerves _____
5. Composed of brain and spinal cord _____
6. Cells providing support for nerve cells _____
7. Part of neuron containing nucleus _____
8. Process carrying impulses from cell body _____
9. Process carrying impulses toward cell body _____
10. Neuron with all parts located in the CNS _____
11. Junction of axon tip and adjacent neuron _____
12. Secretes neurotransmitter _____
13. Receives neurotransmitter _____
14. Forms myelin sheath in PNS neurons _____
15. Forms myelin sheath in CNS neurons _____
16. Necessary for regrowth of neuron processes _____
17. Spaces between Schwann cells _____
18. A stimulatory neurotransmitter _____
19. An inhibitory neurotransmitter _____
20. Cells located between neurons in the CNS _____

b. Explain the mechanism of one-way transmission of impulses at a synapse. _____
_____
_____

c. Draw a giant multipolar neuron and a nerve, x.s., as observed on the prepared slides. Label pertinent parts.

Neuron Cell Body   **Nerve, x.s.**

2. The Brain

a. Write the term that matches the phrase.

    1. Fissure between cerebral hemispheres   _____

    2. Cerebral lobe containing visual center   _____

    3. Outer portion of cerebrum   _____

    4. Small ridges on surface of cerebrum   _____

    5. Cerebral lobe containing hearing center   _____

    6. Protective membranes covering CNS   _____

    7. Remnant of a third eye   _____

    8. Endocrine gland attached to hypothalamus   _____

b. List the labels for Figure 17.4.

    1. _____   4. _____   6. _____

    2. _____   5. _____   7. _____

    3. _____

c. List the labels for Figure 17.7.

    1. _____   5. _____   9. _____

    2. _____   6. _____   10. _____

    3. _____   7. _____   11. _____

    4. _____   8. _____   12. _____

d. Matching

    1. Cerebrum   2. Cerebellum   3. Thalamus   4. Hypothalamus   5. Medulla oblongata

    6. Corpus callosum   7. Ventricles

    _____ Uncritical awareness of sensations   _____ Muscular coordination

    _____ Contains cerebrospinal fluid   _____ Controls water balance

    _____ Critical interpretation of sensations   _____ Largest part of mammalian brain

    _____ Regulates heart rate   _____ Controls body temperature

    _____ Connects cerebral hemispheres   _____ Intelligence, will

    _____ Breathing control center

e. Is the relative size of the following larger in humans or sheep?

    Cerebrum _____   Cerebellum _____

f. Matching

    1. What matter   2. Gray matter

    _____ Exterior of cerebrum   _____ Corpus callosum

    _____ Myelinated fibers   _____ Nonmyelinated fibers

    _____ Neuron cell bodies   _____ Exterior of the pons

3. Spinal Cord

a. List the labels for Figure 17.8.

    1. _____   5. _____   8. _____

    2. _____   6. _____   9. _____

    3. _____   7. _____   10. _____

    4. _____

b. Write the term that matches the phrase.
   1. Neuron carrying impulses to effector   _____
   2. Neuron receiving impulses from receptor   _____
   3. Neuron in dorsal root   _____
   4. Neuron in ventral root   _____
   5. Location of sensory neuron cell body   _____
   6. Connects sensory and motor neurons   _____
c. Describe the effect of an injury that cuts the:
   Dorsal root of a spinal nerve _____

   _____

   Ventral root of a spinal nerve _____

   _____

   Spinal cord _____

   _____

d. Diagram from your slide (1) a cat spinal cord, x.s., labeling gray and white matter, ventral
   fissure, and central canal, and (2) a motor neuron cell body and associated processes, labeling the
   nucleus and neurofibrils at the base of the processes.

   Cat Spinal Cord, x.s.   **Motor Neuron Cell Body**

4. Human Reflexes

a. Rate the strength of your patellar response as strong (3), moderate (2), weak (1), or none (0):
   Left leg: without diversion _____   with diversion _____
   Right leg: without diversion _____   with diversion _____
   Explain any differences in response. _____

   _____

   _____

b. If you touch a hot stove, you will reflexively jerk back your hand a split second before you feel the
   pain. Explain the delayed pain response. _____

   _____

   _____

c. Describe the effect on the size of the pupil when:
   The subject is looking into a darkened area. _____
   The subject is looking into a bright light. _____

d. When a penlight is shined into the left eye, what response is seen:
In the left eye? _____
In the right eye? _____
Explain your observations. _____
_____
_____

5. Reaction Time

a. Record the 5 replicates of your reaction time in milliseconds (msec).

_____   _____   _____   _____   _____

Calculate your average reaction time. _____

b. Record the average reaction time (msec) for each class member by gender.

Males			**Females**		
1. _____	7. _____	13. _____	1. _____	7. _____	13. _____
2. _____	8. _____	14. _____	2. _____	8. _____	14. _____
3. _____	9. _____	15. _____	3. _____	9. _____	15. _____
4. _____	10. _____	16. _____	4. _____	10. _____	16. _____
5. _____	11. _____	17. _____	5. _____	11. _____	17. _____
6. _____	12. _____	18. _____	6. _____	12. _____	18. _____

c. For each gender and the total class, indicate the range of values and calculate the average of the average reaction times.

Males: Range _____ msec to _____ msec   Average _____ msec
Females: Range _____ msec to _____ msec   Average _____ msec
Class: Range _____ msec to _____ msec   Average _____ msec

d. Do the results support the null hypothesis? _____
Does there seem to be a significant gender difference in reaction times? _____
Explain. _____
_____
_____

e. State a conclusion from the results. _____
_____
_____

f. In what types of activities is a faster reaction time advantageous?
_____

g. Indicate your reaction times for 5 replicates after practicing the test for 20 times.

_____   _____   _____   _____   _____

Calculate your average reaction time. _____
Did your reaction time decrease with practice? _____

LABORATORY REPORT 18

Sensory Perception

1. The Eye

a. List the labels for Figure 18.1.

1. _____	6. _____	11. _____
2. _____	7. _____	12. _____
3. _____	8. _____	13. _____
4. _____	9. _____	14. _____
5. _____	10. _____	15. _____

b. Write the term that matches each meaning.

1. Outer white fibrous layer _____

2. Fills space in front of lens _____

3. Controls amount of light entering eye _____

4. Transparent window in front of eye _____

5. Layer with photoreceptors _____

6. Controls shape of lens _____

7. Opening in center of iris _____

8. Fills space behind lens _____

9. Site of sharp, direct vision _____

10. Receptors for color vision _____

11. Carries impulses from eye to brain _____

12. Receptors for black and white vision _____

13. Membrane covering front of eye _____

14. Focuses light on retina _____

15. Junction of retina and optic nerve _____

c. *Circle* the terms that describe the sclera.

soft   firm   weak   strong   tough   easy to cut   rigid   flexible

d. *Circle* the terms that describe the vitreous humor.

watery   jellylike   transparent   murky

e. When looking through the lens across the room, what is different about the image? _____

_____

Does the lens magnify or reduce the print? _____

f. Is the retina thick or thin? _____

Is the retina firmly attached to the choroid layer? _____

What holds the retina in place in an intact eye? _____

2. Visual Tests

a. Record the distance from your eye to Figure 18.2 at which the blind spot is detected.

Left eye _____   Right eye _____

Explain any difference in the distances. _____

_____

b. Record the near-point distance for your eyes.

Left eye: without glasses _____   with glasses _____

Right eye: without glasses _____   with glasses _____

Explain any differences in the distances. _____

_____

How do your near-point values compare with your age? See Table 18.1.

_____

c. Is astigmatism present?

Left eye: without glasses _____   with glasses _____

Right eye: without glasses _____   with glasses _____

d. Record the acuity for each eye.

Left eye: without glasses __20/_____   with glasses __20/_____

Right eye: without glasses __20/_____   with glasses __20/_____

e. Record your responses to the color-blindness test in the table.

Responses to the Ishihara Color-Blindness Test Plates

Plate No.	Subject	Responses*		
		Normal	*Red-Green Color Blind*	*Totally Color Blind*
1	_____	12	12	12
2	_____	8	3	0
3	_____	5	2	0
4	_____	29	70	0
5	_____	74	21	0
6	_____	7	0	0
7	_____	45	0	0
8	_____	2	0	0
9	_____	0	2	0
10	_____	16	0	0
11	_____	Traceable	0	0
			Red Cones Absent	*Green Cones Absent*	
12	_____	35	5	3	0
13	_____	96	6	9	0
14	_____	Can trace two lines	Purple	Red	0

* 0 means the subject sees no pattern in the test plate.

Are you color blind? _____   If so, describe your type of color blindness. _____

_____

3. The Ear

a. List the labels for Figure 18.5.

1. _____   6. _____   11. _____
2. _____   7. _____   12. _____
3. _____   8. _____   13. _____
4. _____   9. _____
5. _____   10. _____

b. Write the term that matches each meaning.

1. Contains sound receptors _____
2. Bone in which inner ear is embedded _____
3. Contains receptors for static equilibrium _____
4. Connects middle ear to pharynx _____
5. Interprets impulses as sound _____
6. Contain receptors for dynamic equilibrium _____
7. Conduct vibrations across middle ear _____
8. Membrane separating outer and middle ear _____

4. Ear Physiology

a. Record the average distance for the watch-tick test.
   Left ear _____   Right ear _____   Class average _____

b. State your conclusion from the Rinne test.
   Left ear _____
   Right ear _____

c. Indicate your degree of wavering in the static equilibrium test. Use the terms *slight*, *moderate*, and *great* to indicate variations.
   Standing on both feet with arms at side
   Eyes open _____   Eyes closed _____
   Standing on both feet with arms extended
   Eyes open _____   Eyes closed _____

d. How important are visual clues in maintaining balance? _____

5. Skin Receptors

a. Indicate the minimal distances that provided a two-point sensation.
   Forearm _____   Back of neck _____
   Palm of hand _____   Fingertip _____
   What is the value of this distribution? _____
   _____
   _____

b. Indicate the adaptation time when using
   A single coin _____   Several coins _____
   What is the value of adaptation to stimuli? _____
   _____
   _____

c. What is the physiological basis for a boy looking for his hat only to discover that he is wearing it? _____
   _____

d. Indicate the smallest temperature differential that you could detect by immersing your finger in water. _____ °C

Indicate the range and average of the class.   Range _____ °C   Average _____ °C

e. Was the sensation stronger or weaker when your entire hand was immersed in the water? _____ Explain the sensation. _____ _____ _____

f. Did the sensation change with time when your hands were immersed in

Ice water? _____ Warm water? _____

Explain. _____ _____ _____

g. Describe the sensation when both hands were placed in tap water after immersion in ice and warm water, respectively. _____ _____

How do you account for this? _____ _____ _____

h. What is the physiological basis for a girl believing that the temperature of water in a swimming pool is "fine" after testing it with her foot but complaining that it is "freezing cold" after jumping in? _____ _____

6. **Taste**

a. For each of the tastes tested, do the locations of your taste receptors agree with those in Figure 18.7?

   Sweet _____   Sour _____   Salt _____

   Explain any differences. _____ _____ _____

b. Are taste receptors sensitive to dry sugar granules? _____

   Explain the role of saliva in taste sensations. _____ _____

c. Indicate the role of the following in sensory perception.

   Receptors _____ _____

   Brain _____ _____

   Nerves _____ _____

LABORATORY REPORT 19

Support and Movement

1. The Skeleton

a. List the labels for Figure 19.1.

1. _____   9. _____   16. _____
2. _____   10. _____   17. _____
3. _____   11. _____   18. _____
4. _____   12. _____   19. _____
5. _____   13. _____   20. _____
6. _____   14. _____   21. _____
7. _____   15. _____   22. _____
8. _____

b. Matching

1. Clavicle 2. Humerus 3. Femur 4. Coxal bones 5. Scapula 6. Ulna and radius
7. Tibia and fibula 8. Mandible 9. Skull 10. Sacrum 11. Sternum 12. Vertebrae
13. Tarsals 14. Carpals 15. Phalanges

_____ Ankle bones   _____ Form backbone
_____ Bones of lower arm   _____ Bone of upper arm
_____ Finger bones   _____ Bones of lower leg
_____ Shoulder blade   _____ Collarbone
_____ Wrist bones   _____ Breastbone
_____ Thighbone   _____ Lower jawbone
_____ Five fused vertebrae   _____ Hipbones

c. Write the terms that match the phrases.

1. Vertebrae of the neck   _____
2. Small bone in the lower leg   _____
3. Number of thoracic vertebrae   _____
4. Part of skull enclosing brain   _____
5. Forms the kneecap   _____
6. Fibrocartilage between vertebrae   _____
7. Movable bone of the skull   _____
8. Bones encasing spinal cord   _____
9. Bones protecting heart and lungs

d. Write the name of the bone(s) that articulates with the:

1. Scapula and sternum   _____
2. Coxal bone and tibia   _____
3. Scapula and ulna   _____
4. Sternum and thoracic vertebrae   _____
5. Carpals and phalanges   _____

371

e. Matching

1. Immovable 2. Slightly movable 3. Hinge 4. Gliding 5. Pivot 6. Ball and Socket

_____ Vertebrae 1 and 2   _____ Humerus/ulna

_____ Humerus/scapula   _____ Finger joints

_____ Joints between vertebrae   _____ Joints of cranial bones

_____ Femur/coxal bone   _____ Wrist and ankle

_____ Femur/tibia   _____ Lower jaw/skull

2. Sexual Differences of the Pelvis

a. Using Table 19.1, record your observations of a male and female pelvis and of the "unknown" pelvis.

Characteristic	Male	Female	Unknown
General structure			
Acetabula			
Pubic angle			
Sacrum			
Coccyx			
Pelvic brim			
Ischial spines			
Pelvic ratio			

b. What is the gender of the "unknown" pelvis? _____

c. What is the advantage of the adaptations of the female pelvis? _____

_____

3. Fetal Skeleton

a. What portions of the:

Long bones are formed of bone? _____

Cranial bones are formed of bone? _____

b. Carefully using the tape measure, determine the part of the fetal skeleton that has the largest circumference. _____

c. What is the advantage of the incomplete ossification of the cranial bones prior to birth? _____

_____

_____

4. Macroscopic Bone Structure

a. List the labels for Figure 19.3.

1. _____   4. _____   7. _____

2. _____   5. _____   8. _____

3. _____   6. _____

b. Write the term that matches the phrase.

1. Location of yellow marrow _____

2. Fills spaces in cancellous bone _____

3. Type of bone forming the diaphysis _____

4. Type of bone forming the epiphyses _____

5. Fibrous membrane covering bone _____

6. Forms blood cells _____

7. Cartilage between diaphysis and epiphyses _____

8. Protects articular surface of the bone _____

9. Site of growth in length _____

c. Distinguish between the epiphyseal disc and epiphyseal line. _____
_____
_____

d. What advantage is provided by the articular cartilage? _____
_____

e. Why is adequate dietary calcium necessary for the development of strong bones? _____
_____

f. Describe the appearance, feel, and function of the articular cartilage.

1. Appearance _____

2. Feel _____

3. Function _____

g. Circle the following terms that describe the ligaments.

pliable   stiff   weak   strong   elastic   nonelastic

5. Skeletal Muscles

a. Indicate the type of lever described.

1. CF lies between F and R _____

2. F lies between CF and R _____

3. R lies between CF and F _____

b. Which type of lever has the greatest mechanical advantage and can move the largest resistance with the least contraction force? _____

Explain. _____
_____
_____

c. Indicate the action of the antagonists of muscles that produce:

Flexion _____   Adduction _____

d. The stationary end of a contracting muscle is the: _____

The moving end of a contracting muscle is the: _____

e. Indicate the directional movement and lever class for each illustration in Figure 19.6.

Illustration	**Movement**	**Lever Class**
(a)		
(b)		
(c)		
(d)		
(e)		
(f)		
(g)		

f. Does the body seem to be adapted for great strength or speed of movement? _____
Explain. _____

_____

_____

6. Ultrastructure and Contraction

a. Draw the appearance of striations of the fibers before and after adding ATP, K^+, and Mg^{++}.

Before   **After**

b. Record the length of the fibers before and after adding ATP, K^+, and Mg^{++}.

Before _____ mm   After _____ mm

c. Write the term that matches the phrase.

1. Thick myofilament with cross-bridges   _____

2. Thin myofilaments attached to Z lines   _____

3. Form boundaries of sarcomere   _____

4. Myofilaments within the I band   _____

LABORATORY REPORT 20

Excretion

1. **Nitrogenous Wastes**
   a. List three forms of nitrogenous wastes excreted by animals.
   1. _____ 2. _____ 3. _____
   b. Which compound is most toxic? _____

2. **Urinary System**
   a. List the labels for Figures 20.1 and 20.2.

Figure 20.1	**Figure 20.2**	
1. _____	1. _____	8. _____
2. _____	2. _____	9. _____
3. _____	3. _____	10. _____
4. _____	4. _____	11. _____
5. _____	5. _____	12. _____
6. _____	6. _____	13. _____
7. _____	7. _____	14. _____
8. _____		

   b. Describe the function of each:
   Kidney _____
   Urethra _____
   Urinary bladder _____
   Ureter _____
   c. Diagram the structural relationship between the Bowman's capsule, glomerulus, afferent arteriole, and efferent arteriole.

   d. List in sequence the parts of a nephron.

      1. _____   3. _____

      2. _____   4. _____

3. Urine Formation

   a. Explain the adaptive value of increased blood pressure in the glomerulus. _____

   _____

   _____

   b. Describe what happens in each:

   Glomerular filtration _____

   _____

   Tubular reabsorption _____

   _____

   Tubular secretion _____

   _____

   c. Using the data in Table 20.1, what percentage of the blood passing through the kidneys becomes filtrate? _____ filtrate becomes urine? _____

   What happens to the rest of the filtrate? _____

   d. Why is a low volume of urine production important in humans and other terrestrial animals but unimportant in aquatic animals? _____

   _____

   e. Considering Table 20.2, explain how the high concentration of urea in urine (column 4) is attained. _____

   _____

   f. Does the kidney simply maintain the concentration of urea at a safe level or remove all of it from the blood? _____

   Explain. _____

   g. Comparing filtrate and urine, how many times is urea concentrated in the urine (columns 3 and 4)? _____

   How is that accomplished? _____

   h. Comparing the concentration of mineral salts in columns 1 and 5, very little is removed from the blood. What happens to most of the salts in the filtrate? _____

   i. Explain the protein concentrations in each column:

   Column 2 _____

   Column 5 _____

   j. Does glucose easily pass from the glomerulus into Bowman's capsule? _____

   How do you know this? _____

   What happens to the glucose in the filtrate? _____

   Explain the slightly lower concentration of glucose in column 5 as compared to column 1.

   _____

   _____

4. Urinalysis

a. Record the results of your analysis of the simulated urine samples in the table below. Use an X to indicate the presence of a component and record the color and specific gravity.

Characteristic	Tubes						
	1	*2*	*3*	*4*	*5*	*6*	*Your Urine*
Glucose							
Bilirubin							
Ketone							
Specific gravity							
Blood							
pH							
Protein							
Nitrite							
Color							

b. Indicate for each urine sample the health condition of the "patient" based on your urinalysis. Select the health conditions from the list that follows.

Tube 1 _____

Tube 2 _____

Tube 3 _____

Tube 4 _____

Tube 5 _____

Tube 6 _____

Your urine _____

Diabetes insipidis

This disorder is caused by a deficiency of antidiuretic hormone, which results in the inability of kidney tubules to reabsorb water. Symptoms are constant thirst, weight loss, weakness, and production of 4–10 liters of urine each day. The urine is very dilute (a low specific gravity) and is nearly colorless.

Diabetes mellitus

This disorder is caused by a deficiency of insulin, which results in a decreased ability for glucose to enter cells. Therefore, glucose accumulates in the blood and fats are used excessively in cellular respiration, producing ketones as a waste product. Symptoms include weakness, fatigue, weight loss, and excessive blood glucose levels. In uncontrolled diabetes, a urine sample contains excess glucose, ketones, and a low pH.

Hepatitis

Hepatitis is usually caused by a viral infection. In type A hepatitis, symptoms may include fatigue, fever, generalized aching, abdominal pain, and jaundice (yellowish tinge to the whites of the eyes due to excessive blood levels of bilirugin). Urine is dark amber in color due to excessive bilirubin.

Glomerulonephritis

In this kidney disease, excessive permeability of Bowman's capsule allows proteins and red blood cells to enter the filtrate. They are present in the urine since they cannot be reabsorbed.

Hemolytic anemia

A number of conditions cause the destruction of red blood cells, releasing hemoglobin into the blood plasma, for example, malaria and incompatible blood transfusions. The urine of such patients contains hemoglobin and may be red-brown or smoky in color.

Normal

See Table 20.3 for the composition of normal urine.

Strenuous exercise and a high-protein diet

Athletes in excellent health may produce urine with an acid pH, presence of a small amount of protein, and an elevated specific gravity due to the presence of the proteins.

Urinary tract infection

Bacterial urinary tract infections may involve the urethra (urethritis), urinary bladder (cystitis), ureters, or kidney. Production of alkaline urine promotes the growth of infectious bacteria. Urethritis often causes a burning pain during urination. Symptoms may include a low-grade fever and discomfort of the affected region. Urine will show a positive test for nitrite because bacteria in the urine convert nitrate, a normal component, into nitrite. Severe infections may also cause the urine to contain blood and pus (leukocytes), producing a cloudy urine.

c. Explain the relationship between the volume or urine, its specific gravity, and its color. _____

_____

_____

_____

_____

_____

_____

LABORATORY REPORT 21

Reproduction

1. Reproductive Systems

a. List the labels for Figure 21.1.

1. _____   6. _____   11. _____
2. _____   7. _____   12. _____
3. _____   8. _____   13. _____
4. _____   9. _____   14. _____
5. _____   10. _____   15. _____

b. Trace the path of sperm from the seminiferous tubules to the urethra. _____
_____
_____

c. List the labels for Figure 21.2.

1. _____   5. _____   9. _____
2. _____   6. _____   10. _____
3. _____   7. _____   11. _____
4. _____   8. _____

d. Contrast the function of the urethra in males and females.

Males _____

Females _____

e. Spermatozoa of all animals require a fluid medium in which to swim to the egg. How is this provided in humans? _____
_____
_____

f. Matching

1. Testes   2. Ovaries   3. Oviduct   4. Vas deferens   5. Penis   6. Prostate gland
7. Prepuce   8. Vagina   9. Uterus   10. Ejaculatory duct

_____ Secretion that activates sperm   _____ Male copulatory organ
_____ Produces eggs   _____ Carries eggs to uterus
_____ Site of embryo development   _____ Female copulatory organ
_____ Carries sperm to urethra   _____ Removed in circumcision
_____ Produces spermatozoa   _____ Lined with ciliated cells

g. Indicate the (1) site of production and (2) function of these hormones.

Testosterone   1. _____
                2. _____
Estrogen   1. _____
                2. _____
Progesterone   1. _____
                2. _____

2. Gametogenesis

a. Matching

1. Diploid 2. Haploid 3. Formed by first meiotic division

4. Formed by second meitoic division

_____ Spermatogonium   _____ Oogonium

_____ Spermatid   _____ Secondary oocyte

_____ First polar body   _____ Spermatozoa

_____ Secondary spermatocyte   _____ Primary oocyte

b. Draw the appearance of the following from your slides.

    Graafian Follicle   **Corpus Luteum**   **Sperm**

3. Birth Control

a. Contraceptives are birth control methods that prevent the union of sperm and a secondary oocyte. Which of the methods in Table 21.1 are *not* contraceptives? _____

_____

b. Which contraceptives provide a barrier to the entrance of sperm into the uterus? _____

_____

c. Which contraceptives are chemicals that kill sperm? _____

_____

d. Which contraceptive uses hormones to prevent ovulation? _____

_____

e. Which birth control method prevents the implantation of an early embryo? _____

_____

f. Explain why the rhythm method is not very effective. _____

_____

_____

g. Does using a condom guarantee that you will not be infected with human immunodeficiency virus (HIV) or pathogens of other sexually transmitted diseases? _____

Explain. _____

_____

_____

h. Induced abortion is the removal of a developing embryo or fetus from the uterus prior to term. Is it better to prevent unwanted pregnancies by abstinence or birth control or to rely on induced abortion? _____

Explain. _____

_____

_____

LABORATORY REPORT 22

Fertilization and Development

1. Introduction

Write the term that matches each phrase.

1. Penetration of egg by a sperm _____
2. Mitotic division of fertilized egg _____
3. Fusion of egg and sperm nuclei _____
4. Solid ball of cells (embryonic stage) _____
5. Hollow ball of cells (embryonic stage) _____
6. Cell formed by fertilization _____
7. Embryonic stage formed by gastrulation _____
8. Color of secretion containing sperm _____
9. Color of secretion containing eggs _____

2. Activation and Early Embryology

a. Write the term that matches each phrase.

1. Chemicals that attract sperm to eggs _____
2. Hemisphere of egg containing most yolk _____
3. Hemisphere of egg penetrated by sperm _____
4. Extension of egg engulfing sperm head _____
5. Prevents penetration by additional sperm _____
6. Time until most eggs are activated _____

b. From your slide, draw a few nonactivated and activated eggs. Show the distribution of sperm around the activated eggs. Label pertinent parts.

Nonactivated Eggs   **Activated Eggs**

c. Explain the distribution of the sperm. _____
_____

d. List the labels for Figure 22.5.

1. _____   4. _____   7. _____
2. _____   5. _____   8. _____
3. _____   6. _____   9. _____

e. Record the time required to reach each stage.

Two-cell stage _____   Four-cell stage _____   Eight-cell stage _____

f. Indicate the most abundant stage after each time period.

12 hr _____   24 hr _____   48 hr _____

3. Chordate Development

a. Indicate the plane (polar or equatorial) of the cleavage divisions.

1. First division   _____

2. Second division   _____

3. Third division   _____

b. Are the planes the same as in the sea urchin? _____

c. Why are the cells formed in the vegetal hemisphere larger than those of the animal hemisphere?

_____

_____

d. In which hemisphere does gastrulation occur? _____

Is this the same site as in the sea urchin? _____

e. Indicate the embryonic tissue that forms these structures.

1. Embryonic gut   _____

2. Mesodermal pouches   _____

3. Notochord   _____

4. Neural tube   _____

5. Lining of the coelom   _____

f. Matching

1. Allantois   2. Amnion   3. Chorion   4. Yolk sac

In Reptiles

_____ Envelops embryo   _____ Envelops yolk

_____ Embryonic urinary bladder   _____ Outermost membrane

_____ Gas exchange organ   _____ Provides "private pond" for

In Humans

_____ Envelops embryo   _____ Part of placenta

_____ Outermost membrane   _____ Part of umbilical cord

_____ Brings embryonic blood   _____ Provides "private pond" for
       vessels to placenta   embryo

g. Contrast the function of the allantois in reptiles and humans.

Reptiles _____

_____

Humans _____

_____

h. Contrast the function of the yolk sac in reptiles and humans.

Reptiles _____

_____

Humans _____

_____

4. Human Development

a. By what process are materials exchanged between embryonic and maternal bloods? _____

_____

b. Describe the function of the placenta and umbilical cord in humans.

Placenta _____

_____

Umbilical cord _____

_____

c. List the ages and distinctive visible developmental features of the fetuses observed.

Age	Characteristics
1	
2	
3	
4	
5	
6	
7	
8	
9	

d. At what age is an embryo implanted in the uterus? _____

e. At what age are germ layers formed? _____

f. At what age does the fetus resemble a human? _____

g. At 8 weeks, what percentage of the fetus' length consists of the head? _____

_____

h. At 8 weeks, are arms, legs, fingers, and toes evident? _____

i. What trend occurs in the head-body proportions during development? _____

_____

_____

_____

j. What is the duration of pregnancy in humans? _____

k. Physicians usually advise pregnant women to avoid all drugs during pregnancy. What is the rationale for such avoidance? _____

_____

_____

_____

l. Why does a fetus assume a "fetal position" in the uterus? _____

_____

_____

_____

m. Why is it important for a newborn baby to start breathing before the umbilical cord is cut?

_____

_____

_____

LABORATORY REPORT 23

Monerans, Protists, and Fungi

1. Monera

   a. Draw a few bacterial cells from your slide showing the three types of shapes in bacterial cells.

   Bacillus　　　　　　　　　**Coccus**　　　　　　　　　**Spirillum**

   b. Can you see the flagella of the motile bacteria? _____

   Can you see the cellular contents? _____

   c. What color is the chromoplasm in:

   Gloeocapsa? _____ 　　　 *Oscillatoria?* _____

   d. Why are these organisms considered colonial? _____

   _____

   _____

   e. Are cells of these cyanobacteria larger than bacterial cells? _____

   f. Do some strands of *Oscillatoria* exhibit movement? _____

   g. Draw these cyanobacteria as they appear on your slides.

   Gloeocapsa　　　　　　　　　　　　　　***Oscillatoria***

2. Animal-like Protists

   a. Draw the appearance of *Trichonympha* from your slide. Label the nucleus and flagella.

b. Are few or many *Trichonympha* present in the termite gut? _____

c. *Trichonympha* and a termite live in a mutualistically beneficial relationship. What does *Trichonympha* gain from the relationship? _____
_____

What does the termite gain from the relationship? _____
_____

d. Draw the appearance of *Pelomyxa* from your slide. Label a food vacuole, a contractile vacuole, a nucleus, and the plasma membrane.

e. Can *Paramecium* move backward as well as forward? _____
Is the anterior end pointed or rounded? _____
Does *Paramecium* rotate when swimming? _____

f. Do the cilia beat in unison or in coordinated groups? _____
How many food vacuoles are present in your specimen? _____
How many contractile vacuoles are present in your specimen? _____

3. Plantlike Protists

a. Is the flagellum of *Euglena* at the anterior or posterior end? _____
What gives the color to *Euglena*? _____
Is *Euglena* flexible or rigid? _____

b. Describe the distribution of *Euglena*. _____
_____

Explain the distribution. _____
_____

c. List the plantlike characteristics of dinoflagellates. _____
_____

d. Draw 1–2 of these organisms as observed on the prepared slides.

Dinoflagellates   **Diatoms**

e. How many species of diatoms are visible in the slide of diatomaceous earth? _____

4. Funguslike Protists
   a. Which stage of slime molds is the feeding stage? _____
   b. What stimulates formation of the reproductive stage? _____
   _____

5. Pond Water
   Draw 2–3 monerans and 4–5 protists observed in pond water. Indicate the magnification used and the division or phylum of each.

3. Fungi
   a. Write the term that matches the phrase.
      1. Filaments composing a fungus   _____
      2. Dormant reproductive cells   _____
      3. Nonreproductive body of a fungus   _____
      4. Reproductive structure containing
         spore-forming hyphae   _____
   b. Contrast saprotrophic and parasitic modes of nutrition.
      Saprotrophic _____
      _____
      Parasitic _____
      _____
   c. Describe how fungi obtain nutrients from organic matter. _____
      _____
      _____

d. Where is the youngest portion of a *Rhizopus* colony located? _____

_____

Indicate the color of: mature sporangia _____

immature sporangia _____ zygospores _____

e. Indicate whether the following are haploid (n) or diploid (2n) in *Rhizopus*.

Hyphae _____ Mitospores _____ Sporangia _____ Zygospore _____

f. What is the advantage of sexual reproduction? _____

_____

g. Draw a few budding yeast cells from your slide. Label a nucleus and a bud.

h. In *Penicillium,* are the youngest conidiospores at the center or near the edge of the colony?

_____

i. Are conidiospores formed with a sporangium? _____

Explain. _____

_____

j. Are conidiospores formed by mitotic or meiotic division? _____

k. Indicate the type of nuclei (n; n + n; 2n) in these cells of *Coprinus*.

 1. Cells of hyphae forming a mushroom mycelium   _____

 2. Cells of hyphae forming a mushroom fruiting body   _____

 3. Basidia after nuclear fusion   _____

 4. Basidiospores   _____

l. Are basidiospores formed by mitotic or meiotic division? _____

m. Draw a few basidia with basidiospores from you slide.

7. "Unknowns"

Write the numbers of the "unknown" specimens in the correct spaces.

_____ Bacteria   _____ Euglenoids

_____ Cyanobacteria   _____ Dinoflagellates

_____ Flagellated protozoans _____ Diatoms

_____ Amoeboid protozoans _____ Slime molds

_____ Ciliated protozoans   _____ Fungi

LABORATORY REPORT 24

Plants

1. Algae

a. Write the term that matches the phrase.
   1. Type of nutrition in algae _____
   2. Material forming cell wall _____
   3. Nutrient storage in green algae _____
   4. Organelles containing pigments _____
   5. Chlorophylls in green algae _____
   6. Chlorophylls in brown algae _____
   7. Chlorophylls in red algae _____

b. Is *Chlamydomonas* larger than bacterial cells? _____

c. Explain the distribution of *Chlamydomonas* in the partially shaded Petri dish. _____
_____
_____

d. What is the shape of a *Volvox* colony? _____

e. Describe the movement of *Volvox*. _____
_____
_____

   What enables this movement? _____

f. What is the shape of the chloroplast in a *Spirogyra* cell? _____

g. How many cells wide is each *Spirogyra* filament? _____
   How are new cells added to the filament? _____

h. In brown algae, what is the function of a:
   Holdfast _____
   Stipe _____
   Blade _____
   Air bladder _____

i. Considering the habitat of each, explain the robust body of brown algae in contrast to the delicate body of red algae. _____
_____
_____

j. In paired *Spirogyra* filaments, are the zygospores located randomly in either filament or do all zygospores occur in one filament? _____
_____

   Explain this distribution. _____
_____
_____

k. In alternation of generations:

Name the two adults in the life cycle. _____

What process forms gametes? _____ spores? _____

Which adult forms gametes? _____ spores? _____

2. Moss Plants

a. Write the term that matches the phrase.

1. Dominant generation   _____
2. Main photosynthetic organs   _____
3. Spore-forming generation   _____
4. Process-forming gametes   _____
5. Process-forming spores   _____
6. Organ-forming spores   _____
7. Organ-forming sperm   _____
8. Organ-forming egg   _____
9. Produced by spore germination _____
10. Anchors gametophyte to soil _____

b. Explain why the process of sperm transport restricts mosses to moist areas. _____

_____

_____

c. Draw the appearance of these structures from your slides.

Antheridia	**Archegonium**	**Sporangium**
with	**with**	**with**
Sperm	**Egg**	**Spores**

3. Ferns

a. Write the term that matches the phrase.

1. Type of nutrition in sporophytes   _____
2. Type of nutrition in gametophytes   _____
3. Dominant generation in life cycle   _____
4. Process-forming gametes   _____
5. Process-forming spores   _____
6. Anchors gametophyte to soil   _____

b. What is formed by the fusion of sperm and egg? _____

What does this cell develop into? _____

c. How many eggs are formed in an archegonium? _____

d. How are ferns better adapted to terrestrial life than mosses? _____
_____

e. In what way are ferns no better adapted to terrestrial life than mosses? _____
_____

f. What happens when sporangia dry out? _____
_____

g. Draw a sporangium from your slide.

4. Gymnosperms

a. Write the term that matches the phrase.
1. Dominant generation   *sporophyte*
2. Cones forming pollen grains   *male*
3. Cones forming female gametophytes   *megaspores*
4. Cones containing seeds   *female*
5. Develops into embryo sporophyte   *zygote*
6. Process-forming microspores   *meiosis*
7. Process-forming gametes   *mitosis*

b. Draw the following from your slides.

Pollen	**Female Gametophyte with Archegonium**	**Seed Section with Embryo Sporophyte**

c. How are pollen grains and seeds adapted for dispersal by wind?
Pollen _____
Seeds _____

d. Explain the advantage of pollination in the life cycle. _____
_____
_____

e. Explain the advantage of seeds in the life cycle. _____
_____
_____

5. Flowering Plants

a. Write the term that matches the phrase.
1. Cells that become pollen grains _____ *microspore*
2. Major pollinating agents _____ *petals*
3. Develops from functional megaspore _____ *female gametophyte*
4. Develops from pollen grain _____ *microspore*
5. Process-forming megaspores _____
6. Process-forming gametes _____ *gametophyte*
7. Composed of a mature, ripened ovary _____ *seed*
8. Dominant generation _____ *sporophyte*

b. Explain the role of flowers in the life cycle. _____ *attract insects & leads to pollenation, leading to fertilization of eggs by sperm node*

c. List the labels for Figure 24.8.
1. *piotal*   5. _____   9. _____
2. *stigma*   6. _____   10. _____
3. *style*   7. _____   11. _____
4. _____   8. _____

d. Are lily pollen grains adapted for wind transport? _____

e. How many compartments compose the ovary? _____
What is located within each compartment? _____

f. List the name and indicate the type of each numbered fruit.

    a. Dry dehiscent   b. Dry indehiscent   c. Fleshy

	Name	Type		Name	Type
1.			4.		
2.			5.		
3.			6.		

g. What part of a pea flower develops into a pea pod? _____
What part of a pea flower develops into a pea seed? _____

h. Explain why a corn kernel is a fruit and not a seed. _____

i. Where is starch concentrated in the:
Bean seed? _____   Corn fruit? _____

j. Explain the role of fruits in the life cycle. _____

6. "Unknowns"

Match the numbers of the "unknown" plants with the correct group.

_____ Algae   _____ Ferns   _____ Angiosperms
_____ Mosses   _____ Gymnosperms

Student _____

Lab Instructor _____

Structure of Flowering Plants

1. External Structure

a. List the labels for Figure 25.2.

1. _____ 4. _____ 7. _____
2. _____ 5. _____ 8. _____
3. _____ 6. _____ 9. _____

b. Write the term that matches the phrase.

1. Site of leaf attachment to a stem   _____
2. Type of leaf venation in dicots   _____
3. A leaf stalk   _____
4. Type of leaf venation in monocots   _____
5. Arrangement of vascular bundles in:
   dicots   _____
   monocots   _____
6. Number of cotyledons in seeds of:
   dicots   _____
   monocots   _____
7. Arrangement of flower parts in:
   dicots   _____
   monocots   _____
8. Type of root systems in:
   dicots   _____
   monocots   _____

c. Compare the *Coleus* and corn seedlings.

Structure	Coleus	Corn
Root type		
Leaf venation		
Arrangement of vascular bundles		

d. Identify as a dicot or monocot.   *Coleus* _____   Corn _____

   e. Identify the "unknowns" as dicots or monocots.

      1. _____   3. _____   5. _____

      2. _____   4. _____   6. _____

2. Roots

   a. Write the term that matches the phrase.

      1. Root-tip region forming new cells   _____

      2. Root-tip region of greatest growth   _____

      3. Root-tip region of root hairs   _____

      4. Tissue transporting water and minerals   _____

      5. Tissue transporting organic nutrients   _____

      6. Tissue forming branch roots   _____

      7. Tissue of water-impermeable cells   _____

   b. List the three functions of roots. _____

_____

   c. Draw a young root of a germinated seed with root hairs at 40× and an epidermal cell with its root hair at 100×. Label the oldest and youngest root hairs.

           Young Root   **Root Hair**

   d. Are the cells of an *Allium* root tip arranged in an orderly pattern or in a nonordered mass? Explain. _____

_____

   e. Draw the arrangement of roots in tap root and fibrous root systems.

        Tap Root System   **Fibrous Root System**

   f. What are adventitious roots? _____

_____

   g. Which type of root system is best for:

     preventing erosion of surface soil? _____

     reaching deep water sources? _____

3. Stems

a. Describe the function of stems. _____
_____
_____

b. Describe the arrangement of vascular bundles in a cross section of a *Zea* stem. _____
Are they more abundant near the periphery of the stem? _____
Explain any mechanical advantage in their arrangement? _____
_____
_____

c. In a *Medicago* vascular bundle, is xylen closer to the epidermis or pith? _____
Which vascular tissue has larger cells? _____
has thicker walls? _____ provides greater support? _____

d. In a young *Quercus* stem, what tissue produces new (secondary) xylem and phloem? _____
Is phloem formed interior or exterior to this tissue? _____
What is the outermost tissue in the young stem? _____

e. In an old *Quercus* stem, is the newest xylem near the pith or vascular cambium? _____
Is the oldest phloem next to the cortex or vascular cambium? _____
List the tissues forming the bark. _____
What tissue forms wood in a tree stem? _____

f. How can you distinguish spring wood and summer wood in an annual ring? _____
_____
Does spring or summer wood provide the greatest growth? _____

g. What forms the cork cells of the bark? _____
What is the function of cork cells? _____

i. Explain why removing a strip of bark completely around a tree stem causes the three to die.
_____
_____

j. What is the age of the demonstration section of tree stem? _____

4. Leaves

a. Write the term that matches the phrase.
1. The broad, flat part of a leaf   _____
2. A leaf stalk   _____
3. Venation in monocots   _____
4. Basic venation in dicots   _____

b. Indicate the type of venation of the numbered "unknown" leaves.
1. _____   4. _____
2. _____   5. _____
3. _____   6. _____

   c. Write the term that matches the phrase.
      1. Photosynthetic tissue of a leaf   _____
      2. Tiny openings in the epidermis   _____
      3. Waterproof coating of epidermis   _____
      4. Compose a leaf vein   _____
                                      _____
                                        _____
      5. Control size of stomata   _____

   d. Explain the advantage of the palisade mesophyll cells being closely packed together. _____
_____
_____

   e. Explain the advantage of the many spaces between spongy mesophyll cells. _____
_____
_____

   f. Draw a stoma showing the arrangement of the guard cells and adjacent epidermal cells. Label the stoma, guard cells, and chloroplasts.

   g. Explain how stomata are opened and closed by the guard cells. _____
_____
_____

   h. What advantage is there in having stomata only in the lower epidermis? _____
_____
_____

   i. What advantage is there in having stomata open during the day? _____
_____
_____

   j. Draw a portion of a *Zea* (corn) leaf, x.s., from your slide. Label the epidermis, stomata, mesophyll, and vein.

LABORATORY REPORT 26

Simple Animals

1. Sponges
   a. Which type(s) of sponge skeletons are soft and pliable? _____
   b. How does a sponge obtain food? _____
   _____
   c. What creates the water current flowing through a sponge? _____
   _____

2. Coelenterates
   a. Body form exhibited by *Hydra*. _____
   Body form exhibited by jellyfish. _____
   b. In *Obelia,* indicate the body form that reproduces:
   sexually _____   asexually _____
   c. When in a feeding position and waiting for prey, is *Hydra* extended or contracted? _____
   _____ inverted or upright? _____
   d. In *Hydra,* which tissue layer contains stinging cells? _____
   e. Describe the feeding behavior of *Hydra.* _____
   _____
   _____
   _____
   _____

3. Flatworms
   a. Which end of *Dugesia* is more sensitive to touch? _____
   Is movement of *Dugesia* fast or slow? _____ constant or jerky? _____
   b. Describe the two-step digestive process in coelenterates and flatworms.
   1. _____
   2. _____
   c. What is the advantage of a highly branched gastrovascular cavity? _____
   _____
   d. Does *Dugesia* prefer shade or lighted areas? _____
   e. Draw the appearance of these tapeworm structures from your slides.

   Scolex   **Gravid Proglottid**   **Encysted Larvae**

f. How can human infestation by beef tapeworms be prevented? _____
_____

g. Are humans the intermediate or final host of the beef tapeworm? _____

h. Give the name and infestation site in humans for each fluke observed.

1. _____
_____

2. _____
_____

3. _____
_____

4. Roundworms

a. Is *Turbatrix* visible to the naked eye? _____

Does *Turbatrix* move slowly or rapidly? _____

b. How can human infestation with *Ascaris* be prevented? _____
_____

c. What is the flat, ribbonlike structure within *Ascaris?* _____

d. *Ascaris* has a tube-within-a-tube body plan. What forms the "outer tube?" _____

the "inner tube?" _____

To what system do most of the internal organs of *Ascaris* belong? _____
_____

e. How are humans infected with *Trichinella?* _____
_____

How can infestation be prevented? _____
_____

Are *Trichinella* larvae visible to the naked eye? _____

5. Review

a. Match the animal group with the characteristic.

1. Sponges   2. Coelenterates   3. Flatworms   4. Roundworms

_____ Tissue level   _____ Lack contractile fibrils

_____ Bilateral symmetry   _____ Radial symmetry

_____ Organ level   _____ Organ system level

_____ Filter feeders   _____ Tube-within-a-tube body plan

_____ Gastrovascular cavity   _____ Asymmetrical

_____ All adults immobile   _____ Some adults immobile

_____ False coelom   _____ Saclike body plan

_____ Cellular-tissue level   _____ Some are human parasites

b. Place the numbers of the "unknown" animals in the correct spaces below.

_____ Sponge   _____ Flatworm

_____ Coelenterate   _____ Roundworm

Student _____

LABORATORY REPORT 27

Lab Instructor _____

Mollusks, Annelids, and Arthropods

1. Mollusks

a. List the four major characteristics of mollusks.

_____   _____

_____   _____

b. Write the term that matches the phrase describing a clam.

1. Covers visceral mass; secretes shell _____
2. Brings water into mantle cavity _____
3. Organs of gas exchange _____
4. Organs collecting food from water _____
5. Used to bury clam in mud or sand _____

c. Matching

1. Monoplacophorans 2. Chitons 3. Snails and slugs 4. Tooth shells
5. Clams and oysters 6. Octopi and squids

_____ Image-forming eyes   _____ Conical shell open at each end

_____ Radula present   _____ Shell of eight plates

_____ Remnants of segmentation _____ Shell of two valves

_____ All predaceous   _____ Filter feeders

_____ Digging foot   _____ Crawl on flattened foot

_____ Marine only   _____ Foot modified into tentacles

d. Place the numbers of the "unknown" mollusks in the correct spaces.

_____ Monoplacophorans   _____ Snails and slugs

_____ Chitons   _____ Clams, oysters, and mussels

_____ Tooth shells   _____ Octopi and squids

2. Annelids

a. What is the primary distinguishing characteristic of annelids? _____

_____

b. Write the term that matches the phrase regarding earthworms.

1. Name of body segments _____
2. Number of setae per segment _____
3. Secretes the egg case _____
4. Type of circulatory system _____
5. Large white structures near hearts _____
6. Remove metabolic wastes _____
7. Grinds food into tiny pieces _____
8. Number of hearts _____

c. Matching

1. Oligochaetes   2. Polychaetes   3. Leeches

_____ Head with simple eyes   _____ Blood-sucking parasites

_____ Few setae   _____ Parapodia and many setae

_____ Hermaphroditic   _____ Superficial rings on somites

_____ Suckers for attachment   _____ Marine only

d. Place the numbers of the "unknown" annelids in the correct spaces.

_____ Oligochaetes   _____ Polychaetes   _____ Leeches

3. Arthropods

a. List the major characteristics of arthropods. _____

_____

_____

b. Write the term that matches the phrase regarding a crayfish.

1. Number of legs   _____

2. Body divisions   _____

3. Number of antennae   _____

4. Type of eyes   _____

5. Type of circulatory system   _____

6. Protects the gills   _____

7. Number of appendages per body segment   _____

8. Large yellowish gland around stomach   _____

c. Write the term that matches the phrase regarding a grasshopper.

1. Number of legs   _____

2. Body part legs are attached to   _____

3. Number of tympani   _____

4. Number of spiracles per abdominal segment _____

5. Number of antennae   _____

d. Matching

1. Arachnids   2. Crustaceans   3. Centipedes   4. Millipedes   5. Insects

_____ Three pairs of legs   _____ Compound eyes only

_____ Two pairs of antennae   _____ Four pairs of legs

_____ Cephalothorax and abdomen   _____ Head, thorax, and abdomen

_____ Elongate, dorsally convex body _____ Body segments fused in pairs

_____ Compound and simple eyes   _____ No antennae

_____ Elongate, flattened body   _____ Five pairs of legs

_____ Wings attached to thorax   _____ Simple eyes only

e. Place the numbers of the "unknown" arthropods in the correct spaces.

_____ Arachnids   _____ Centipedes   _____ Insects

_____ Crustaceans   _____ Millipedes

Echinoderms and Chordates

1. Echinoderms
a. Write the term that matches the phrase.
1. Type of symmetry in adult _____
2. Body surface where mouth is located _____
3. Formed from the embryonic blastopore _____
4. Major distinctive characteristic _____
5. Type of skeleton _____

b. Matching
1. Brittle stars 2. Sea cucumbers 3. Sea lilies 4. Sea stars 5. Sea urchins

_____ Star-shaped with broad-based arms

_____ Star-shaped with narrow-based arms

_____ Attached by stalk from aboral surface

_____ Arms used for locomotion

_____ Elongate forms lacking arms and spines

_____ Globose forms with movable spines; mouth with five teeth

_____ Ambulacral grooves with tube feet

_____ Ambulacral grooves without tube feet

c. Write the numbers of the "unknown" echinoderms in the correct spaces.

_____ Brittle stars   _____ Sea cucumbers   _____ Sea lilies

_____ Sea stars   _____ Sea urchins

2. Chordates
a. List the four distinguishing characteristics of chordates. _____

_____

_____

Are they all present in: larval tunicates? _____

adult tunicates? _____   adult amphioxus? _____

b. What is the function of gills in tunicates and amphioxus? _____

_____

c. What is the function of gills in fish? _____

d. What trend is evident from lamprey to perch in the number of:

gill slits? _____   fins? _____

e. If a shark stops swimming, it will sink, but a perch will not.

Explain why this occurs. _____

_____

_____

f. Do most fish utilize internal or external fertilization? _____

internal or external development? _____

g. Draw a cross section of lamprey and shark vertebrae. Label vertebra, nerve cord, and notochord, if present.

Lamprey   **Shark**

h. Do amphibians utilize internal or external fertilization? _____
   internal or external development? _____
i. How are reptiles better adapted to terrestrial life than amphibians? _____
   _____

j. Embryos of reptiles, birds, and mammals develop pharyngeal slits that never become functional in the adult. How do you explain this? _____
   _____

k. What are the advantages of homeothermy? _____
   _____

   Indicate any disadvantages. _____
   _____

l. What are the advantages of internal embryonic development in mammals? _____
   _____

m. Matching
   1. Jawless fish (modern) 2. Cartilaginous fish 3. Bony fish 4. Amphibians
   5. Reptiles 6. Birds 7. Mammals

   _____ Diaphragm and hair   _____ Feathers, horny beak
   _____ Notochord in adult   _____ Cartilaginous endoskeleton
   _____ Leathery shelled eggs   _____ Intrauterine development
   _____ No scales, jaws, or operculum   _____ Lungs, paired limbs, no scales
   _____ Epidermal scales   _____ Dermal scales
   _____ Operculum, swim bladder   _____ Subterminal mouth, paired fins
   _____ Poikilothermic   _____ Homeothermic
   _____ Two-chambered heart   _____ Four-chambered heart
   _____ Seven pairs of gill slits   _____ Bony endoskeleton
   _____ Milk-secreting glands   _____ Claws or nails
   _____ Hard-shelled eggs   _____ Lightweight dermal scales
   _____ Rib cage   _____ Three-chambered heart

n. Write the numbers of the "unknown" vertebrates in the correct spaces.
   _____ Jawless fish   _____ Amphibians   _____ Birds
   _____ Cartilaginous fish   _____ Reptiles   _____ Mammals
   _____ Bony fish

Student ______________________

LABORATORY REPORT 29

Lab Instructor ______________________

Evolution

1. The Fossil Record

a. What groups of organisms were present during the Proterozoic era? ______________________

____

What groups were most abundant? ______________________

What groups are found in the oldest fossils? ______________________

b. Indicate the period of first appearance for these groups:

Chordates ______________________ Angiosperms ______________________

Psilopsids ______________________ Arthropods ______________________

Reptiles ______________________ Bryophytes ______________________

Gymnosperms ______________________ Ferns ______________________

Jawed fish ______________________ Amphibians ______________________

Birds ______________________ Mammals ______________________

c. Examine the fossils available and complete the chart below.

Fossils

Number	*Name*	*Period of First Appearance*	*Period of Greatest Abundance*	*Period of Extinction*
1				
2				
3				
4				
5				
6				
7				
8				
9				
10				

   d. Indicate the period when these groups invaded the land:
   Vascular plants _____ Arthropods _____
   Vertebrates _____
   e. What groups show a decrease in diversity in the Permo-Triassic Crisis? _____
   _____

   f. Indicate the vertebrate group that was:
   First to appear _____ Last to appear _____
   g. What vertebrate group declined in the Cretaceous during the Laramide Revolution? _____
   _____

2. **Human Evolution**
   a. List the labels for Figure 29.4.

   1. _____   6. _____
   2. _____   7. _____
   3. _____   8. _____
   4. _____   9. _____
   5. _____

   b. Write the term that matches each phrase.
   1. Earliest hominid   _____
   2. First human   _____
   3. Continent where humans evolved   _____
   4. First modern man   _____
   5. First human living on three continents   _____

3. **Skull Analysis**
   a. Plot the cephalic indexes of members of the class.

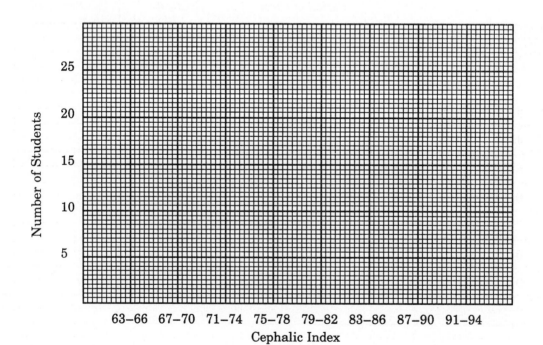

b. Indicate the following the data from your class.

Average cephalic index _____ Range of indexes _____

c. Considering the data in item 3b, what difficulties are encountered in trying to make firm conclusions about a single fossil specimen? _____

_____

_____

d. Why are indexes better than simple measurements for comparing fossil specimens? _____

_____

_____

e. Record the results of your skull analyses in the table below.

Skull Comparison

Characteristics	Chimpanzee	A. africanus	H. erectus	H. sapiens neanderthalensis	H. sapiens sapiens (Cro-Magnon)	H. sapiens sapiens (modern)
Cranial breadth						
Cranial length						
Cranial index						
Facial breadth						
Skull proportion index						
Facial projection length						
Skull length						
Facial projection index						
Skull and vertebra attachment						
Brow ridges						
Canine length						
Forehead						
Chin						

f. Which of the fossil hominid skulls has a cranial (cephalic) index that falls within the range of indexes for the class? _____

_____

_____

g. Have there been any major skull changes in humans during the last 40,000 years? _____

Explain your response. _____

_____

_____

h. Summarize the human evolutionary trends based on your analysis.

Cranium size _____

_____

_____

Face _____

_____

_____

Brow ridges _____

_____

_____

Skull attachment _____

_____

_____

LABORATORY REPORT 30

Ecological Aspects

1. Energy Flow

a. Write the term that matches each meaning.
   1. All populations in an area _____
   2. All populations and the nonliving environment in
      an area _____
   3. Members of a species in an area _____
   4. Ultimate source of energy _____
   5. Animals feeding on plants _____
   6. Animals feeding on animals _____
   7. Convert sunlight into chemical energy of organic
      molecules _____
   8. Obtain nutrient energy from dead organisms and
      organic wastes _____

b. If only 10% of the energy in one trophic level is passed to the next trophic level, what happens to the remaining 90%? _____

c. What percentage of the energy captured by photosynthesis is available to top carnivores? _____ %

d. What is the ultimate source of energy? _____

e. List the organisms (by number) in Figure 30.2 according to their roles in energy flow. Note: A carnivore may occupy more than one role.
   Producers _____ Herbivores _____
   Primary carnivores _____ Secondary carnivores _____
   Tertiary carnivores _____

f. If humans were added to the food web in Figure 30.2, what role(s) would they fulfill? _____

g. What would happen to the populations in the food web:
   If pesticides significantly reduced the populations of herbivorous insects? _____

   If trapping significantly reduced the populations of coyotes and foxes? _____

h. Is it possible to significantly reduce any population in the food web without affecting the other members of the community? _____
Explain. _____
_____
_____

i. An important ecological role filled by certain organisms is not shown in Figure 30.2. What is the role? _____

2. **Cycling of Materials**
   a. Match the steps in the carbon cycle with these processes.

   Combustion _____   Consumption _____

   Death _____   Photosynthesis _____

   Respiration _____

   b. Ignoring fossil fuels, what is the nonliving reservoir of carbon?
   _____

   What group of organisms can use carbon dioxide for their carbon needs?
   _____

   c. In what form must humans obtain carbon? _____

3. **Environmental Pollution**
   a. Record your results and the total for the class.

		Tubes						
		1	*2*	*3*	*4*	*5*	*6*	*7*
15 min	Number alive							
	Number dead							
30 min	Number alive							
	Number dead							
60 min	Number alive							
	Number dead							

   b. What characteristics are desirable in an indicator organism? _____
   _____
   _____

   c. Indicate the temperature and pH of the water in Tube 1.
   Temperature _____ °C   pH _____

d. What were the first behaviors that indicated some *Daphnia* were adversely affected by the pollutants? _____

_____

e. What is the function of Tube 1 in the experiment? _____
What effect would the death of *Daphnia* in Tube 1 have on the outcome of the experiment?

_____

f. State a conclusion from the results regarding the resistance of *Daphnia* to:
Thermal pollution _____

_____

_____

Acid pollution _____

_____

_____

Pesticide pollution _____

_____

_____

g. Does *Daphnia* seem to be a good indicator organism? _____
Explain. _____

_____

_____

h. What is the role of *Daphnia* in an aquatic food web? _____

i. If only the *Daphnia* population of a lake is reduced by pollution, what would you predict to be the effect of *Daphnia* reduction on populations of:
Microscopic algae? _____
Small fish? _____
Larger fish? _____

j. Is it possible that *Daphnia* that survived exposure to these pollutants for 1 hr may:
Not survive exposure for hours, days, or weeks? _____
Not survive exposure to two or more pollutants at the same time? _____

4. **Behavioral Ecology**
   a. Describe the normal movement of sowbugs. _____

   _____

   Does their speed of movement vary? _____
   Explain. _____

   _____

   Do they start and stop frequently? _____
   b. What observations would support the null hypothesis? _____

   _____

   _____

c. Record your observations.

Elapsed Time	Number of Sowbugs	
	In Shade	In Light
3 min		
6 min		
9 min		
12 min		
15 min		
Ave./3 min		

d. Do your results support the null hypothesis? _____
   State a conclusion from your results. _____
   _____
   _____

e. Record your observations.

Elapsed Time	Number of Sowbugs	
	In Moist Area	In Dry Area
3 min		
6 min		
9 min		
12 min		
15 min		
Ave./3 min		

f. Do the results support the null hypothesis? _____
   State a conclusion from your results. _____
   _____
   _____

LABORATORY REPORT 31

Population Growth

1. Introduction

Write the term that matches each meaning.

1. Sum of all limiting factors _____
2. Maximum theoretical reproductive capacity _____
3. Limiting factors whose effect is unchanged by population size _____
4. Limiting factors whose effect changes proportionately with population size _____

2. Growth Curves

a. Plot the theoretical and realized growth curves below.

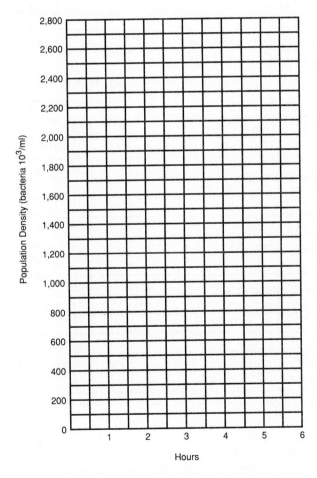

Theoretical and realized growth curves of bacterial populations

b. Indicate the time interval when the growth rate was greatest.
   Theoretical growth _____ Realized growth _____

c. In the realized growth curve:
   When did cell deaths most nearly equal cell "births"? _____
   Why did the realized bacterial population die out? _____
   _____
   What limiting factors determined the realized population growth pattern? _____
   _____
   Were density-dependent or density-independent factors involved? _____
   _____

d. In item 2a, show the shape the realized growth curve would have taken if additional nutrient broth was added at the fifth hour. Use a red pencil to draw the new curve.

e. Write the term that matches each meaning regarding Figure 31.1.
   1. Growth curve with a J shape   _____
   2. Growth curve with an S shape   _____
   3. Phase of explosive growth   _____
   4. Point where limiting factors take effect   _____
   5. Phase of equilibrium between biotic potential
      and environmental resistance   _____
   6. Where birth equals death   _____
   7. Where resources per capita are least   _____

3. Human Population Growth

a. Plot the human population growth curve below using data in Table 31.3.

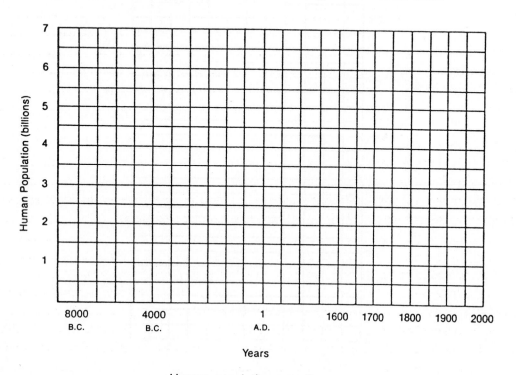

Human population growth curve

b. Does the human population growth curve resemble the theoretical growth curve or the realized growth curve? _____

c. In what phase of the realized growth curve does the human population curve appear to be? _____

d. What events in human history have enabled the exponential growth of the human population? _____ _____

e. Are human populations subject to the interaction of biotic potential and environmental resistance like natural populations? _____

f. List any limiting factors controlling the size of natural populations that do not affect human populations. _____

g. Humans are subject to certain limiting factors that are not observed in other animal populations. Name two of these. _____ _____

h. If nothing is done to check the growth of the human population, will deaths ultimately equal births? _____ What will be the shape of the population growth curve? _____

i. Ultimately, there are only two ways to decrease population growth. Name them. _____ _____

Which is preferable? _____

j. Record the growth rate and doubling time as calculated for Table 31.4

	Growth Rate (%/yr)	Doubling Time (yr)
Africa	_____	_____
Asia	_____	_____
Europe	_____	_____
Latin America	_____	_____
North America	_____	_____
Oceania	_____	_____
CIS	_____	_____

k. What region has the:
Fastest population growth? _____
Slowest population growth? _____

l. The rapid growth rate in developing countries has resulted from a decreased death rate without a proportionate decrease in birth rate. What has brought this about? _____ _____

m. What relationship exists between population size and resources per person? _____

n. Are the earth's resources infinite? _____ Does it seem likely that resources can increase to keep pace with the projected population growth? _____

o. The standard of living is based on available resources per person. If a country's population is increasing at a rate of 2.0% per year, what rate of increase in productivity is required to maintain the same standard of living? _____

p. At the carrying capacity, deaths equal births and resources are just adequate to maintain the existing population size. Would it be advantageous to curtail the human population (zero growth) before it levels off at the carrying capacity due to natural causes? _____ Explain your response. _____

_____

_____

_____

q. Are developing countries likely to improve their standard of living without decreasing their population growth? _____ Explain your answer. _____

_____

_____

_____

r. Famine has been a serious problem in Africa in recent years. What has caused the famine?

_____

_____

Food was provided by developed countries to help the people faced with starvation. Was the donated food a real or artificial increase in the carrying capacity of the environment?

_____

What is the long-range solution to this problem? _____

_____

_____

_____

_____

_____

_____

s. If you were a policymaker in the United States government, what policies would you propose to stimulate a decrease in population growth:
In developing countries receiving foreign aid? _____

_____

_____

_____

_____

In the United States? _____

_____

_____

_____

_____

Appendix A

COMMON PREFIXES, SUFFIXES, AND ROOT WORDS

Prefixes of Quantity

amphi-, diplo-	both, double, two
bi-, di-	two
centi-	one-hundredth
equi-, iso-	equal
haplo-	single, simple
hemi-, semi-	one-half
hex-	six
holo-	whole
quadri-	four
milli-	one-thousandth
mono-, uni-	one
multi-, poly-	many
oligo-	few
omni-	all
pento-	five
tri-	three

Prefixes of Direction or Position

ab-, de-,ef-, ex-	away, away from
acro-, apici-	top, highest
ad-, af-	to, toward
antero-, proto-	front
anti-, contra-	against, opposite
archi-, primi-, proto-	first
circum-, peri-	around
dia-, trans-	across
deutero-	second
ecto-, exo-, extra-	outside
endo-, ento-, intra-	inner, within
epi-	upon, over
hypo-, infra-, sub-	under, below
hyper-, supra-	above, over
inter-, meta-	between
medi-, meso-	middle
pre-, pro-	before, in front of
post-, postero-	behind
retro-	backward
ultra-	beyond

Miscellaneous Prefixes

a-, an-, e-	without, lack of
chloro-	green
con-	together, with
contra-	against
dis-	apart, away
dys-	difficult, painful
erythro-	red
eu-	true, good
hetero-	different, other
homo-, homeo-	same, similar
leuco-, leuko-	white

meta-	change, after
micro-	small
neo-	new
necro-	dead, corpse
pseudo-	false
re-	again
sym-, syn-	together, with
tachy-	quick, rapid

Miscellaneous Suffixes

-ac, -al, -alis, -an, -ar, -ary	pertaining to
-asis, -asia, -esis	condition of
-blast	bud, sprout
-cide	killer
-clast	break down, broken
-elle, -il	little, small
-emia	condition of blood
-fer	to bear
-ia, -ism	condition of
-ic, -ical, -ine, -ous	pertaining to
-id	member of group
-itis	inflammation
-lysis	loosening, split apart
-oid	like, similar to
-logy	study of
-oma	tumor
-osis	a condition, disease
-pathy	disease
-phore, -phora	bearer
-sect, -tome, -tomy	cut
-some, -soma	body
-stat	stationary, placed
-tropic	change, influence
-vor	to eat

Miscellaneous Root Words and Combining Vowels

andro	man, male
arthr, -i, -o	joint
aut, -o	self
bio	life
blast, -i, -o	bud, sprout
brachi, -o	arm
branch, -i	gill
bronch, -i	windpipe
carcin, -o	cancer
cardi, -a, -o	heart

carn, -i, -o	flesh
cephal, -i, -o	head
chole	bile
chondr, -i, -o	cartilage
chrom, -at, -ato, -o	color
coel, -o	hollow
crani, -o	skull
cuti	skin
cyst, -i, -o	bag, sac, bladder
cyt, -e, -o	cell
derm, -a, -ato	skin
entero	intestine
gastr, -i, -o	stomach
gen	to produce
gyn, -o, gyneco	female
hem, -e, -ato	blood
hist, -io, -o	tissue
hydro	water
lip, -o	fat
mere	segment, body section
morph, -i, -o	shape, form
myo	muscle
neph, -i, -o	kidney
neur, -i, -o	nerve
oculo	eye
odont	tooth
oo, ovi	egg
oss, -eo, osteo	bone
oto	ear
path, -i, -o	disease
phag, -o	to eat
phyll	leaf
phyte	plant
plasm	formative substance
pneumo, -n	lung
pod, -ia	foot
proct, -o	anus
soma, -to	body
stasis, stat, -i, -o	stationary, standing still
sperm, -a, -ato	seed
stoma, -e, -to	mouth, opening
therm, -o	heat
troph	food, nourish
ur, -ia	urine
uro, uran	tail
viscer	internal organ
vita	life
zoo, zoa	animal

Appendix B

COMMON METRIC UNITS AND TEMPERATURE CONVERSIONS

Common Metric Units

Category	Symbol	Unit	Value	English Equivalent
Length	km	kilometer	1,000 m	0. 62 mi
	m	meter*	1 m	39.37 in.
	dm	decimeter	0.1 m	3.94 in.
	cm	centimeter	0.01 m	0.39 in.
	mm	millimeter	0.001 m	0.04 in.
	μm	micrometer	0.000001 m	0.00004 in.
Mass	kg	kilogram	1,000 g	2.2 lb
	g	gram*	1 g	0.04 oz
	dg	decigram	0.1 g	0.004 oz
	cg	centigram	0.01 g	0.0004 oz
	mg	milligram	0.001 g	
	μg	microgram	0.000001 g	
Volume	l	liter*	1 l	1.06 qt
	ml	milliliter	0.001 l	0.03 oz
	μl	microliter	0.000001 l	

* Denotes the base unit.

TEMPERATURE CONVERSIONS

Fahrenheit to Celsius

$$°C = \frac{5}{9}(°F - 32)$$

Celsius to Fahrenheit

$$°F = \left(\frac{9}{5}°C\right) + 32$$

417